THE
MATHEMATICS
OF THE
ELEMENTARY
SCHOOL

THE MATHEMATICS OF THE ELEMENTARY SCHOOL

EDWARD G. BEGLE

School of Education
Stanford University

McGRAW-HILL BOOK COMPANY

NEW YORK   ST. LOUIS   SAN FRANCISCO   AUCKLAND   DÜSSELDORF
JOHANNESBURG   KUALA LUMPUR   LONDON   MEXICO   MONTREAL   NEW DELHI
PANAMA   PARIS   SÃO PAULO   SINGAPORE   SYDNEY   TOKYO   TORONTO

This book was set in Helvetica Light by Progressive Typographers.
The editors were A. Anthony Arthur and Shelly Levine Langman;
the designer was J. E. O'Connor;
the production supervisor was Sam Ratkewitch.
The drawings were done by Vantage Art, Inc.
The printer was The Murray Printing Company;
the binder, Rand McNally & Company.

THE MATHEMATICS OF THE ELEMENTARY SCHOOL

2 3 4 5 6 7 8 9 0 M U R M 7 9 8 7 6 5

Library of Congress Cataloging in Publication Data

Begle, Edward Griffith, date
    The mathematics of the elementary school.

    1. Mathematics—1961-   I. Title.
QA39.2.B44   513′.133   74-13226
ISBN 0-07-004325-6

To Elsie,
Cornelia, Sarah, James,
Emily, Elsie, Edward, and Douglas

CONTENTS

PREFACE

This text is devoted to those topics in mathematics that are important to elementary school teachers, either because the topic is already a standard part of the elementary school curriculum or because there is a good chance that it will be added to that curriculum within the next few years.

Many of these topics were in the curriculum when you were in elementary school and so you are already familiar with them. However, you must remember that an idea that seems simple to you now may not seem nearly so simple to a child meeting it for the first time. Consequently, these mathematical ideas are developed in this text from their very beginnings. It is important that you, as a prospective teacher, review these topics carefully to make sure that the ideas and their interrelationships are crystal clear to you.

A few topics may be new to you, such as flowcharting and probability. However, only the beginning ideas of each topic are taken up here and we know by experience that they are quite feasible for the upper elementary grades.

While this text is intended for elementary school teachers, no explicit attention is paid to techniques for teaching these ideas to children. However, this text has been designed to be in conformity with the basic principle which underlies good teaching of mathematics at any level.

This is the principle of meaningful learning. Long experience as well as careful research have shown that students learn mathematical skills more easily, retain them longer, and apply them more effectively to practical problems if they are led to understand them before they are asked to practice them. Consequently, the major emphasis in this text is to make sure that the mathematical ideas underlying the skills are clear and understandable.

The best way of making a mathematical idea clear is to show how it originated. Most of the mathematical ideas which are appropriate for the elementary school are abstracted directly from particular aspects of the world around us. We have tried to make clear these connections between the real world and the world of mathematical ideas and to present these ideas in the same spirit in which they should be taught to young children.

Since the stress is on ideas, it is necessary to read this text with care. But passive reading is not enough. Many of the ideas relate to physical activities on real objects. When studying these ideas, you should act them out. You will, incidentally, learn another pedagogical fact that will be useful. You will find that very simple, inexpensive props such as marbles, bottle tops, strips of cardboard, etc., are enough to clarify these ideas and that expensive apparatus is not needed.

In order to help you concentrate on the ideas in the text, a number of questions are interspersed throughout each chapter. Each is identified by a ■ in front of it. When you come to one of these, try to answer it. If you cannot, review the immediately preceding paragraphs and then try the question again. If you still cannot answer it, make a note to ask your instructor about it.

Acknowledgment   Much of the material in this text has been borrowed, often with little or no change, from various publications of the School Mathematics Study Group. Permission to do so is gratefully acknowledged. It is specifically noted that no endorsement by SMSG is thereby implied.

Edward G. Begle

THE
MATHEMATICS
OF THE
ELEMENTARY
SCHOOL

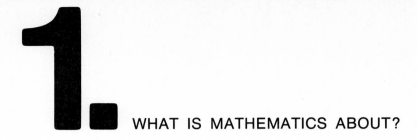

WHAT IS MATHEMATICS ABOUT?

Mathematics consists of a number of special ways of looking at the world we live in. Each of these special ways of looking is designed to help us focus on certain aspects of the world, and to help us ignore other aspects, in order to make it easier to solve the problem we are faced with or to answer the question we are being asked at the moment.

SETS In our everyday lives many of the questions which we ask or are asked begin with the words "how many." The ways of looking at the world which help us to answer these questions lead to the part of arithmetic that deals with whole numbers. We will take a careful look at this part of arithmetic in the next few chapters, but first we need to take a more careful look at those parts of the world around us that the "how many" questions refer to.

Any "how many" question is asked about some specific collection of things, as is illustrated in the following:

How many eggs are there in the refrigerator?
How many more payments are left on the car?
How many shopping days are there until Christmas?

Notice that these questions are not about any particular egg in the refrigerator, or any particular payment on the car, or any particular shopping day, but rather about *the collection* of all eggs in the refrigerator, *the collection* of all payments left on the car, and *the collection* of all shopping days until Christmas.

■ **Question 1** Does it make sense to ask how many sips there are in a glass of water?

The English language is rich in words used to refer to special collections. For example:

a flock of chickens
a herd of sheep
a span of horses
a shelf of books
a kit of tools

In mathematics, it is convenient to use a single word to refer to any kind of collection and this word is "set." As we shall see, all mathematical ideas involve sets of one kind or another to some extent. Consequently, it is well to start with a careful look at various aspects of the concept of a set. First, there is a certain amount of terminology to be learned. To begin with, we say that each object in a particular set is an *element* or a *member* of that set. Thus, for example, if we are thinking about the set consisting of the seven days of the week, Monday is an element (or a member) of this set.

If we wish to specify a particular set, one standard way of doing it is to write down the names of the individual members of the set and to enclose these names in braces. Thus, for example,

{Alaska, California, Hawaii, Oregon, Washington}

specifies the set of U.S. states which touch the Pacific Ocean. Sometimes, however, we use shortcuts. For example,

{the states of the United States}

is shorter to write than the list of the 50 names of all of the states, but it still stands for the same set. Another kind of shortcut is illustrated by

{1, 2, 3, . . . , 100}

This is a shortcut way of specifying the set of the first 100 whole numbers. We have to be careful when using such shortcuts to be sure that what we write gives a clear-cut way of deciding whether any particular object is in the set or not. For example, is this a set: {the seven handsomest presidents of the United States}? The answer is no, since it is not uncommon for two people to disagree as to whether or not a particular person is handsome.

■ **Question 2** Does this specify a set: {all children in New York City as of 11:59 P.M., December 24, 1973}?

Two other comments need to be made about this way of specifying sets. The first is that the order in which the members of the set are listed makes no difference. Thus, {D, E, R} and {R, E, D} specify the same set. Also, if one of the members of the set is listed more than once, we have a problem. Whenever we are required, as we are in a book like this or in a textbook for elementary school students, to use symbols or illustrations instead of real things, there is inevitably some ambiguity. This is because the context does not always make it clear whether we are talking about the symbols themselves or about the real things which they symbolize.

Thus, for example, a crossword puzzle fan would think of "deer" as a four-letter word. Someone else, however, might observe that only three different letters of the English alphabet are represented in the word "deer." The first person is thinking of the symbols themselves while the second person is thinking about the letters which they represent.

The same problem can arise when we use pictures instead of symbols. Consider the two sets in Figure 1.1. Are we to understand that a single specific cat happens to be a member of both sets? Or are there twin cats, one in each set?

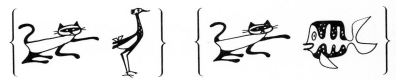

Figure 1.1

The only way an author can overcome this inherent ambiguity is to be quite explicit as to whether the symbols themselves or the things to which they refer are to have priority. In this text we give priority to the things to which the symbols refer. In particular, if the same symbol (or picture) appears in the specification of a set, we consider that one thing is being considered which has been named twice. Thus the set

{George Washington, first President of the U.S.A.}

consists of just one element, even though he had been specified twice. Similarly the set

{d, e, e, r}

contains just three members, the fourth, fifth, and eighteenth letters of the English alphabet.

■ **Question 3** How many members are there in the set {member, member, set, member, set}?

In each of the sets we have specified so far the members were all quite similar to each other. This is not necessary. The collection of all the objects in my wife's purse is a perfectly good example of a set.

Quite often we are interested in more than one set at a time. For example, in a geography class we might be interested in the connection between the set mentioned above, of U.S. states which touch the Pacific Ocean, and the set of U.S. states which border on Mexico:

{Arizona, California, New Mexico, Texas}

It would be inconvenient to have to write the full specification of each of these two sets each time we want to mention the set. To get around this, we give each of the sets a name. The custom is to use capital letters of the English alphabet for these names. Thus we might decide to use A as the name of the first of these sets and B as the name of the second. Then we would write:

$A = \{$Alaska, California, Hawaii, Oregon, Washington$\}$

$B = \{$Arizona, California, New Mexico, Texas$\}$

Another observation: We note that each of the sets we have specified so far contains several different members. This need not be the case. A set can consist of just one element. For example, the set consisting of all the U.S. states which both touch the Pacific Ocean and border on Mexico has only one element: California. The set of all U.S. states which touch both the Pacific Ocean and the Atlantic Ocean is empty. The usual notation for this empty set is $\{\ \ \}$. The brackets indicate that a set is being specified, but since nothing appears between the brackets, the set has no members.

■ **Question 4** How many members are in the set {the states which touch both Canada and Mexico}?

There are many more ideas about the set concept that we will have to consider, but we will postpone discussions of these ideas until we are ready to use them.

COMPARING SETS We said above that sets are what the "how many" questions are asked about. As a first step in learning how to find the answers to such questions, we need to consider a certain way of comparing two sets.

Suppose that we have two specific sets of objects[1] A and B, for example $A = \{\triangle, \bigcirc, \square\}$ and $B = \{\times, +\}$. We can think of pairing the individual members of A with those of B. To carry out this operation, we choose one member, in any way we wish, from the first set and at the same time one member from the second set. We put these two objects aside. Next we repeat the process, choosing one of the remaining members of the first set and one of the remaining members of the second. We put these aside, and then continue. We keep going until we run out of members of one of the sets (or perhaps both at the same time). For example, we could start a one-to-one pairing of the members of A and B by choosing the $\bigcirc$ from A and the $\times$ from B. We put them aside and for the second step we choose the $\triangle$ from A and the $+$ from B. Now we are finished since we have used all the members from B even though the $\square$ is left in A.

Another way of picturing this is to connect the members of the first set with the members of the second set that they are paired with. The example above can be pictured as in Figure 1.2.

When we pair the elements of two sets, there are only three possible outcomes. In the first place, we might run out of members of the two sets at the same time. In this case, we say that the sets *match*. For

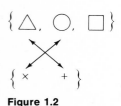

Figure 1.2

[1] For typographical convenience, we will often use various geometric shapes as the elements of the illustrative sets in this and later chapters.

example, these two sets match: $\{\triangle, \bigcirc\}$ and $\{\times, +\}$, but these two do not: $\{\triangle, \bigcirc, \square\}$ and $\{\times, +\}$.

■ **Question 5** Show two different ways of pairing the elements of $A = \{$animal, vegetable$\}$ and $B = \{1, 2\}$.

In the second place, we might use up all the elements of the second set before running out of members of the first. In this case, we say that the first set is *more than* the second. For example, $\{\triangle, \bigcirc, \square\}$ is more than $\{\times, +\}$.

Finally, we might use up all the members of the first set before those of the second. In this case we say that the first set is *less than* the second. For example, $\{\triangle, \bigcirc, \square\}$ is less than $\{+, -, \times, \div\}$.

The most important fact about the operation of pairing is that the outcome does not depend on the order in which we pick out the members. Thus, if we pair the members of two sets and discover that they match, then we can be sure that if we shuffle the members of the first set and also shuffle the members of the second set, and then repeat the operation, the outcome will be the same. They will still match. Whether or not two sets match depends only on the sets and not on the way the members of the sets are arranged.

■ **Question 6** Which of these sets is more than the other: $A = \{$silver dollar, \$10 bill$\}$, $B = \{$penny, penny, penny$\}$?

There are some obvious but important properties of this relation between sets which we call matching.

1 It is always true that a set matches itself. Thus if $A = \{$Bill, Dick, Mary, Max$\}$, then the set $\{$Mary, Dick, Max, Bill$\}$ is the same set A and the sets can be matched by pairing each child in the first listing with himself in the second.

2 If A and B are any two sets and if A matches B, then B matches A.

This is true since the pairing of each member of A with a member of B can be considered equally well as a pairing of each member of B with a member of A.

3 If A, B, and C are any three sets, and if A matches B and B matches C, then A matches C.

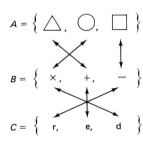

Figure 1.3

For example, in Figure 1.3, the upper set of arrows shows one way of matching A and B. The lower set of arrows shows a way of matching B and C. Now, start with a member of A, follow the upper arrow to a member of B and then follow the lower arrow to a member of C. We pair the member of A with this member of C.

■ **Question 7** List all the sets which can be formed from the members of the set $\{$Bill, Dick, Mary, Max$\}$ which match the set $\{$Larry, Stu, Bill$\}$.

When we paired off the members of A with those of B, it might have happened that we ran out of members of B before we had used up all the members of A, or vice versa. In the first case we said that A was *more than B.* Thus the set $\{\triangle, \bigcirc, \square\}$ was more than the set $\{\times, +\}$. We also say that $\{\times, +\}$ is *less than* $\{\triangle, \bigcirc, \square\}$. The set C in Figure 1.4 is more than the set D since in the one-to-one pairing of members illustrated we run out of members of D before we use up all the members of C.

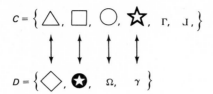

Figure 1.4

Just as there were some important properties of the matching relationship between sets, so there are some important properties of the "more than" relationship.

1   If A, B, and C are any three sets, and if A is more than B and B is more than C, then A is more than C.

This is illustrated in Figure 1.5.

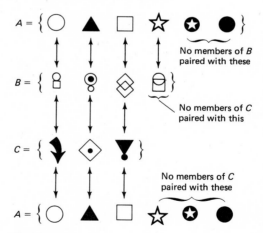

Figure 1.5

2   If A and B are any two sets such that A is more than B, and if C is any set which matches A and D is any set that matches B, then C is more than D.

This also is illustrated by a figure such as Figure 1.6. We can pair each member of D with a member of C by tracing through the pairings from a member of D to a member of B to a member of A to a member of C. But when all the members of D are thus paired, there are still some members of C, such as Z, which are not paired with any member of D. Thus C is more than D.

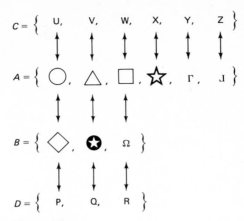

Figure 1.6

■ **Question 8** You put one textbook on each desk in the classroom. When the children sit down, one seat is empty. How does the set of texts compare with the set of children?

We do not need to spend much time now on the relationship "less than." It has properties corresponding exactly to those of "more than" since to say "B is less than A" is to say exactly the same thing as "A is more than B."

Finally, we notice that when we compare two sets to see whether or not they match, we do not need to pay any attention to what the members of the sets are like. It does not matter whether they are large or small. It does not matter what color they are. It does not matter what shapes they have. The set consisting of the Washington Monument, the largest elephant in the Bronx Zoo, and the quart can of green paint in my workshop matches the set consisting of the three pennies in my pocket.

Size, weight, shape, color, etc. are irrelevant when we talk about matching sets. And this is as it should be, because the answer to a "how many" question about a set does not depend on the size, weight, shape, or color of the members.

■ **Question 9** You have a set of six red balloons, each filled with helium. You now let all the helium out. Do you still have the same set of balloons?

PROBLEMS

1 Which of the following phrases define sets?
   (a) Great novels of the nineteenth century
   (b) Positive whole numbers greater than 12 but less than 30

(c)   Positive whole numbers greater than 12 but less than 10

(d)   Three consecutive large numbers

(e)   The beautiful girls in Watsonville

2   Is it possible for a set to have more than 100 members?

3   List the elements of each of the following sets whose descriptions are:

(a)   {the days of the week whose names begin with the letter W}

(b)   {months of the year whose names have less than six letters}

(c)   {whole numbers between 7 and 8}

(d)   {the capitals of Japan and England}

(e)   {the colors of the rainbow}

(f)   The members of the set of whole numbers between 5 and 11

(g)   The set of all states of the United States which begin with the letter A

4   List the set of all Presidents of the United States whose last names begin with the letter:

(a)   I

(b)   J

(c)   K

5   Which expression states that the letter y is an element of the set of letters in the word "Friday"?

(a)   y is an element of {Friday}

(b)   {y} is an element of {Friday}

(c)   y is an element of {F, r, i, d, a, y}

(d)   {y} is an element of {F, r, i, d, a, y}

6   If the following sets match, show a pairing. If they do not, tell which set is less than the other.

(a) $A = \{\square, \triangle, \star, \bigcirc\}$ (b) $C = \{\text{cow, tree, blimp}\}$

   $B = \{X, I, V, M, C\}$ $D = \{\text{dirigible, trunk, milk}\}$

7   Observe the sets in Figure P1.7.

(a)   Which pairs of sets match? Draw arrows to show the matching.

(b)   Which sets are less than A? Draw arrows to show the pairing, and show the members of A left over.

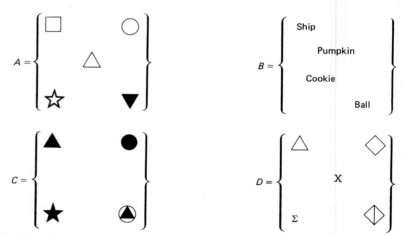

Figure P1.7

8   Divide these into groups of matching sets.

(a)   {a, b, c, d, e, f}   (b)   {$\square, \bigcirc, \odot, \triangle$}

(c)   {1, 2, 3, 4, 5, 6}   (d)   { }

(e)   {$\triangle, \diamondsuit, \square, \bullet, \star$}

9 Is there a matching between the states of the United States and cities (in the United States) of over 1 million population?

10 State why we do not necessarily have a one-to-one correspondence between the children mn the class and their first names.

11 Which of the following pairs of sets match?
 (a) {17} and {71}
 (b) {letters in the word "bundle"} and {n, d, b, l, u, e}
 (c) {*, q, Ø} and {q, *, Ø}
 (d) {zero} and {peacocks native to the North Pole}
 (e) {1, 2, 3, 4} and {a, b, c, d}
 (f) {are} and {era}
 (g) {M, i, s, p} and {the letters in the word "Mississippi"}

12 (a) When you buy a carton of a dozen eggs at a grocery store, do you have to count the eggs to tell if the dozen are there?
 (b) When the breakfast table has been arranged, need you count to tell if there are as many cups as saucers?

13 Suppose you are interested in finding out whether there are more girls than boys in a school auditorium. What could you do?

14 There are two pastures of goats. One pasture is separated from the other by a very swift river. The two primitive tribesmen who own each of these pastures want to know which has more goats. They have a raft which will transport one of the tribesmen but will not transport the goats. There are pebbles scattered around both pastures. What might they do?

15 Show three (or more) different ways of matching $A = \{\triangle, \bigcirc, \square\}$ and $B = \{r, e, d\}$.

16 Order the sets X, Y, Z in terms of "less than."
 $X = \{1, 2\}$ $Y = \{3, 4, 5, 6\}$ $Z = \{789\}$

17 Write the order of the following sets, beginning with the set that is the least.
 (a) $A = \{$letters of the alphabet in the word "peacock"$\}$
  $B = \{$letters of the alphabet in the word "letters"$\}$
  $C = \{$letters of the alphabet in the word "Mississippi"$\}$
  $D = \{$letters of the alphabet in the word "mathematics"$\}$
 (b) $A = \{1, 2, 3, 4\}$
  $B = \{2, 3, 5, 7, 11, 13\}$
  $C = \{a, b, c, d, e, f\}$
  $D = \{\ \}$

18 Which of the following pairs of sets match? For those that do not match, state which set is more than the other.
 (a) $A = \{$letters in the word "group"$\}$ and $B = \{g, o, p, r, u\}$
 (b) $C = \{23\}$ and $D = \{232\}$
 (c) $A = \{1, 2, 3, 4, 5\}$ and $B = \{c, d, e, f\}$
 (d) $B = \{c, d, e, f\}$ and $C = \{$oyster, walrus, carpenter$\}$
 (e) $A = \{1, 2, 3, 4, 5\}$ and $C = \{$oyster, walrus, carpenter$\}$

19 If A is more than B, and C is more than A, which set is more than either of the others?

20 If B is more than A, and C is more than A, which set is less than each of the others?

2. NUMBERS AND NUMERATION SYSTEMS

The members of the last two sets mentioned in the text of the previous chapter, the set A consisting of the Washington Monument, the largest elephant in the Bronx Zoo, and the can of green paint in my workshop, and the set B of the three pennies in my pocket, have little in common. However, the two *sets* do share a common property. They match each other. Also, they match each of these two sets:

$$C = \{\triangle, \bigcirc, \square\} \qquad D = \{r, e, d\}$$

so all four of these sets share the common property of matching each other.

There are many other sets with the same common property, i.e., sets such as $\{*, \diamondsuit, \mathbf{\bullet}\}$ or {boy, cow, dog}, which can also be matched with each of them.

NUMBER PROPERTY AND NUMBER

The property shared by all the sets which can be matched with the above sets is called the *number property* of these sets. For these particular sets, we choose to call the number property "three" and to write "3." All the sets which match the set $\{\bigcirc, \triangle, \square, +\}$ share a different number property which we call "four" and write as "4." Of course, many other sets of objects share this number property. The set {Joe, Jane, John, Jean} has the number property 4; the set {boat, house, car, garage, bank} has the number property 5.

In order to write this easily, we use $N(A)$ as an abbreviation for the phrase "the number property of A." Since this number property is a property shared by all the sets which match A, we can say that if A and B are sets which match, then they have the same number property and we can write $N(A) = N(B)$.

■ **Question 1** What do we call the number property of the empty set?

It is important to realize that the number property associated with $\{\triangle, \bigcirc, \square\}$ is something which has many different names. A Roman knew this, he called it *tres* and wrote III; a Chinese knows it, he calls it *sahn* and writes $\equiv$; a Frenchman and a German know it, they call it *trois* and *drei,* but write 3 as we do. The many names by which this number property may be called are *numerals.* But it is the number property shared by all the sets which match $\{\triangle, \bigcirc, \square\}$ which is the number itself. It is not the name, the numeral, which is important; it is the recognition that

A *number* is the common property shared by a collection of matched sets

which is the important idea we must get. Such numbers are called whole numbers and we see that they are connected fundamentally with sets of objects. The properties of these whole numbers will therefore follow naturally from the properties of sets.

ORDER AMONG THE WHOLE NUMBERS

When two sets are compared by a one-to-one pairing of their members, we find that A may match B, that A may be more than B, or that A may be less than B, and these are the only things that can happen. Corresponding to these three situations, we can say in the first case $N(A)$ and $N(B)$ are equal, $N(A) = N(B)$†; in the second case $N(A)$ is greater than $N(B)$, $N(A) > N(B)$; and in the third case $N(A)$ is smaller than $N(B)$, $N(A) < N(B)$. Note the symbols $>$ and $<$ to express the relationship "greater than" and "smaller than" between numbers.

How are we to decide between two numbers, say 4 and 6, as to which one is the greater? We choose a set with the number property 4, for example,

$$A = \{\triangle, \bigcirc, \square, +\}$$

and another with the number property 6, for example,

$$B = \{u, v, w, x, y, z\}$$

Now, when we pair the members of A with the members of B, we run out of elements of A first, so A is less than B and so $4 < 6$.

We can compare any two whole numbers in this way. Therefore we can now take any set of whole numbers, such as $\{6, 3, 2, 0, 4, 1, 5\}$, compare them in this fashion, and then *order* them by putting them in a row with the smaller ones to the left. Thus we soon find by comparing sets that $2 < 4$, and we already know $4 < 6$. So we order them 2, 4, 6. What about 3? Comparing a set of 3 with a set of 2 we find $2 < 3$, so 3 goes to the right of 2 but where in relation to 4 and 6? Comparing a set of 3 with a set of 4 we find $3 < 4$. Now we have 2, 3, 4, 6. In like manner we consider the numbers 5, 0, 1 and discover that the correct ordering is 0, 1, 2, 3, 4, 5, 6.

† We shall always use the equal sign in this way. "$X = Y$" means that X and Y are two names for the same thing.

Question 2 Which numeral is larger, 3 or 8?

SUCCESSIVE NUMBERS

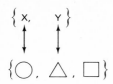

Figure 2.1

If we look at sets of 2 and 3 we see that in the pairing shown in Figure 2.1 the set of members left over after all the possible pairings is a set, {□}, with exactly one member. Any set, therefore, which is more than a set of 2 must either match a set of 3 or be more than a set of 3. This means that there is no whole number which is greater than 2 and also smaller than 3. We say that 3 is "one more than" 2.

Any number such as 6 is the number property of a set of matched sets. Take one such set. It is always possible to put another object in this set. Thus we can put ❂ into the set $A = \{\triangle, \bigcirc, \square, \star, \Omega, \Gamma\}$ and get the set $B = \{\triangle, \bigcirc, \square, ❂, \star, \Omega, \Gamma\}$. This is one of a collection of matched sets and to this collection is associated a number $N(B)$ which is one more than 6. Of course, this is the number we call 7. But the important thing to realize is that for any number we can always go through the same procedure and find another number which is one more than it. How can we possibly name all these numbers? This is a very real question to which many different answers have been given. We will return to this question shortly.

■ **Question 3** 2 is one more than one more than what?

COUNTING

Suppose a visitor from another country knew about numbers but did not know which number the English word "three" referred to. How could you explain? The solution is simple. Point to a set with three elements, perhaps this one:

$\{\triangle, \bigcirc, \square\}$

or this one:

$\{r, e, d\}$

Any three-member set will do. No one such set seems to be better than any others for this purpose. There seems to be no unique prototype for the number three, or for any other number for that matter.

However, there is one particular set which has a very useful arrangement when it comes to numbers. This is the set of whole numbers, except for 0, arranged in order starting with 1:

$N = \{1, 2, 3, 4, 5, 6, 7, 8, 9, \ldots\}$

The special fact about this set which makes it so useful is this:

> If we chop off the sequence just after a number, n, and throw away everything to the right of n, then the number property of what remains is exactly n.

For example, if we chop off everything to the right of 5, then what is left is the set

$\{1, 2, 3, 4, 5\}$

and it is easy to see that this set has five members. If we chop off everything to the right of 2, then we have left the set

$\{1, 2\}$

which has two elements.

■ **Question 4** Does N have a last element?

It is this fact that makes it possible to find out how many members there are in a set by counting them. Indeed, the operation of counting the elements of a set A is just this: An element of A is chosen and is paired with 1. (Actually, we usually touch the element and say the word "one" to indicate that they are paired.) Next, another element of A is chosen and is paired with 2, after which still another element of A is chosen and is paired with 3. This process of pairing is continued until the last element of A is reached and is paired with whatever number, call it n, has been reached in the sequence

$\{1, 2, 3, \ldots n, \ldots\}$

Now we chop off everything to the right of n, and we are left with the set

$\{1, 2, 3, \ldots, n\}$

The pairing operations we carried out show that this set matches A, so they both have the same number property. But the number property of $\{1, 2, 3, \ldots, n\}$ is just n, so the number property of A, the number of elements in A, is also n.

■ **Question 5** Starting with the last word and working forward, count the number of words in this sentence by pairing them with the whole numbers arranged in order. How many words are there? Which word did you pair with 10?

NUMBER SYMBOLS AND NUMERATION

So far in this chapter we have used either the name of or the symbol for these numbers:

zero	one	two	three	four	five	six	seven	eight	nine
0	1	2	3	4	5	6	7	8	9

Obviously, there are many more numbers than this. Think, for example, of the number of dollars in the U.S. national debt, and all the numbers from nine up to this number. What are we to do about names and symbols for all these numbers?

The particular names and symbols for the first ten numbers, zero to nine, are historical accidents, as a glance at any encyclopedia will show. Nevertheless, it would surely be unsatisfactory if each of the numbers after nine had its own individual name and symbol, chosen without regard for the names and symbols of other numbers. Our memories could not cope.

Fortunately, we have a simple but clever and efficient system for assigning symbols to numbers. (Our system for choosing names for

Figure 2.2

numbers is not as simple. We will take it up later.) This is called the Hindu-Arabic numeration system.[1] This system hinges on the choice of a particular number, ten (probably because each of us has ten fingers), called the *base* of the system, and a process of repeated counting, but never beyond ten. To illustrate, suppose we want to find the symbol for the number property of the set of marks[2] inside the box in Figure 2.2. We first count out ten of the marks and put them aside in a pile. (In our diagrams we will put them in a row and draw a line around them.) Then we have the situation shown in Figure 2.3.

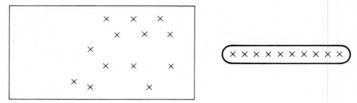

Figure 2.3

Now we start out again and count out another ten. Then we have what is shown in Figure 2.4.

Figure 2.4

Now start counting again. This time we will end at three when we run out of marks. Finally, and this is the key point of the system, we count the piles of ten. There are two of them.

Now, the rule is to write the symbol for the number of individual marks left over after the piles of ten were counted out, and then, to the *left*, write the symbol for the number of piles of ten:

2 3

[1] This system, based in part on some ideas of ancient Indian mathematicians and in part on ideas of Arabic mathematicians of the first millenium A.D., was perfected over 1,000 years ago. Numerous other numeration systems, all of them more complicated and less effective, were developed in the past. Some of them are still in use in some parts of the world. The interested reader should consult any standard encyclopedia.

[2] In this and the following illustrations we use sets of ×'s but this is merely for the convenience of the printer. In actual practice, we usually deal with sets of more important objects, such as people or dollars or bushels of corn.

These ×'s are the objects we are counting. These are not symbols for other objects. Thus {×, ×} has the number property 2, not 1.

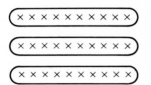

Figure 2.5

Figure 2.6

Figure 2.7

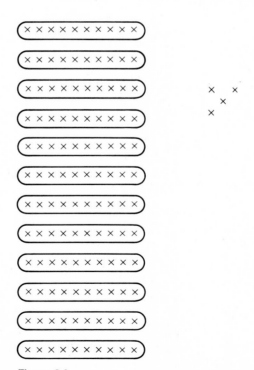

Figure 2.8

■ Question 6 The set of marks in Figure 2.5 has no piles of ten and has six individual marks. Why don't we write the symbol 06 for it?

Let us now look at some other examples, chosen to illustrate certain points about this numeration system. In each case, the set is displayed with the sets of ten already counted out and placed in rows. What is the symbol for the number of marks in the set shown in Figure 2.6?

Since there are six marks left over after counting out the set of ten, we write a 6 and then, to the left, a 1 to indicate just one pile of ten:

1 6

How about the set shown in Figure 2.7? In this case, after the three sets of ten have been counted out, there are no marks left, so we write a 0 and, to its left, a 3 for the three piles of ten:

3 0

How about the set in Figure 2.8? There is one set of ten and no marks left over:

1 0

There is a more complicated situation in Figure 2.9. When we count out the piles of ten, we reach ten before reaching the last pile.

Figure 2.9

What we do in this case is to put together ten piles of ten into a "superpile," and indicate this in our diagrams (see Figure 2.10) by putting a box around them.

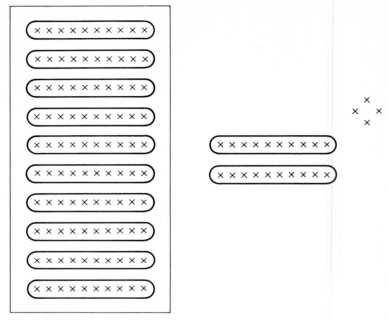

Figure 2.10

The number symbol for this set of marks is formed by writing a 4 for the marks left over, then, to the left, a 2 for the two piles of ten left over after the "superpile" was counted out, and finally, further to the left a 1 to indicate one "superpile":

1 2 4

As you remember, the name for the number property of ten piles of ten is "hundred."

■ **Question 7** What is the largest number whose symbol needs only two digits?

Consider the set in Figure 2.11. Here there are five left over after all piles of ten and the superpile have been counted out, no piles of ten, and one superpile, so the symbol is

1 0 5

Similarly, the symbol for this set, which contains exactly one hundred elements (see Figure 2.12), is

1 0 0

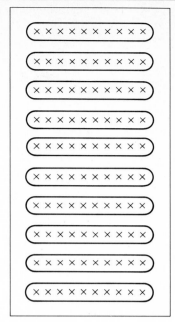

Figure 2.11

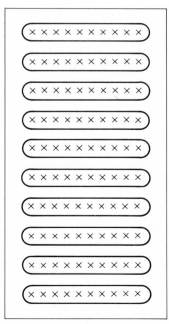

Figure 2.12

Finally, the number symbol for this set (Figure 2.13) is

3 2 1

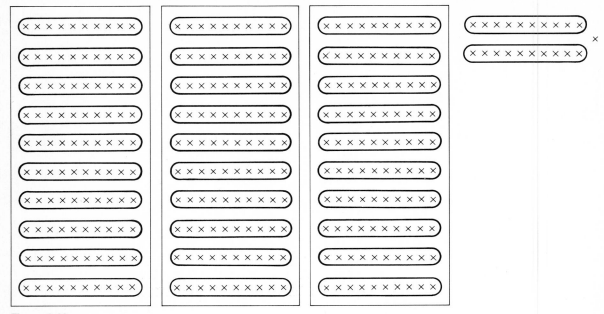

Figure 2.13

These examples are enough to remind the reader of the essential features of the Hindu-Arabic numeration system. Of course, when counting still larger sets, we can arrive at a case where there are as many as ten "superpiles" (hundreds). These we group into "super-superpiles," and the name for the number property of a super-super-pile is "thousand." As you remember, there are names for various "super- . . . -superpiles." For example, the name for the number property of "ten piles of ten piles of ten piles of ten piles of ten piles of ten" is "million," etc.

■ **Question 8** Look under the heading "number" in Webster's Unabridged Dictionary to find the name of this number:

1,000,000,000,000,000,000,000,000,000,000,000

PLACE VALUE This Hindu-Arabic numeration system is called a *place-value* system because when a particular one of the digits from 0 to 9 appears in the symbol for a larger number, the place where it appears in the symbol is important.

For instance, we are quite familiar with the fact that in the numeral 2,222, each digit 2 does not have the same "value." The "value" of each 2 is determined by its place or position in the numeral as a whole (see Figure 2.14).

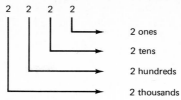

Figure 2.14

Or, we may convey the same idea in a slightly different way (see Figure 2.15).

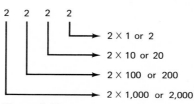

Figure 2.15

We frequently find it helpful to use an *expanded form* of notation illustrated by

$$2{,}222 = (2 \times 1{,}000) + (2 \times 100) + (2 \times 10) + (2 \times 1)$$

None of the notations used thus far has made explicit the important role of the base, ten, in determining the "place values." Each place to the left of the ones place in a numeral has associated with it a "value" that is ten times the "value" associated with the place immediately to its right. For the numeral 2,222, we can show this important idea as in Figure 2.16 or

$$2{,}222 = (2 \times 10 \times 10 \times 10) + (2 \times 10 \times 10) + (2 \times 10) + (2 \times 1)$$

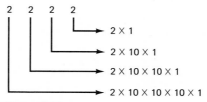

Figure 2.16

The importance of the zero symbol, 0, in connection with our place-value numeration system is reflected in numerals such as 2,220, 2,202, 2,022, 2,200, and 2,002. Without the zero symbol such numerals could not be distinguished readily from 222 (in the case of 2,220, 2,202, and 2,022) or from 22 (in the case of 2,200, 2,020, and 2,002). Without some symbol to denote "not any" in a particular place, a numeration system with a place-value principle would not be feasible. In fact, the relatively late invention of a symbol for "not any" (a symbol for the number pertaining to the empty set), was the reason for the relatively late creation of a place-value numeration system.

■ **Question 9** Write 2,002 in expanded form.

The chart in Figure 2.17 may be helpful in summarizing some of the ideas just discussed regarding our numeration system.

Millions	
Tens	*Ones*
10,000,000	1,000,000
10 × 10 × 10 × 10 × 10 × 10 × 10	10 × 10 × 10 × 10 × 10 × 10
	7

Thousands		
Hundreds	*Tens*	*Ones*
100,000	10,000	1,000
10 × 10 × 10 × 10 × 10	10 × 10 × 10 × 10	10 × 10 × 10
2	0	5

Units		
Hundreds	*Tens*	*Ones*
100	10	1
10 × 10	10 × 1	1
0	4	6

Figure 2.17

Consider the numeral 7,205,046 which we read as: "seven million, two hundred five thousand, forty-six." (Notice that the word "and" is *not* used in reading numerals for whole numbers. Otherwise, it would not be clear, for example, when we say "two hundred and five thousand" whether we mean "200 + 5,000" or "205,000".)

The numeral 7,205,046 means: 7 millions, 2 hundred-thousands, 0 ten-thousands, 5 thousands, 0 hundreds, 4 tens, 6 ones. Since 0 ten-thousands and 0 hundreds both result in zero, the words for them may be omitted. Thus, 7,205,046 means: 7 millions, 2 hundred-thousands,

5 thousands, 4 tens, 6 ones. We also may use an expanded notation form:

7,000,000 + 200,000 + 5,000 + 40 + 6

or

$(7 \times 1{,}000{,}000) + (2 \times 100{,}000) + (5 \times 1{,}000) + (4 \times 10) + (6 \times 1)$

or

$(7 \times 10 \times 10 \times 10 \times 10 \times 10 \times 10) + (2 \times 10 \times 10 \times 10 \times 10 \times 10) +$
$(5 \times 10 \times 10 \times 10) + (4 \times 10) + (6 \times 1)$

BASE THREE NUMERATION

Figure 2.18

The base of the Hindu-Arabic numeration system is the number ten, so it is called a *decimal* system. (The Latin word for tithe is *decima*.) However, other numbers could be, and in fact have been, used. It is instructive to glance briefly at one example, the case where the base is the number three.

Suppose, for example, that instead of grouping 14 objects as in Figure 2.18, we group them as one "superpile" of 3 sets of three, 1 set of three, and 2 ones (Figure 2.19). The *base three* symbol for this set would be

1 1 2

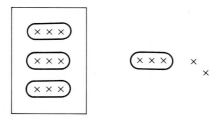

Figure 2.19

Let us use sets of one, two, three, . . . , eight objects to see how such a base three numeral system might be developed. This is done in the chart in Figure 2.20 which includes contrasting base ten numerals.

Note that in the decimal system, each set of ten objects is grouped as 1 ten and the number of these groups is indicated in the tens place. Thus, 23 is 2 tens and 3 ones, and the number of ones left ungrouped is given by the digit 3. The possible digits in the ones place are then any of the numerals 0, 1, 2, 3, . . . , 9. Similarly, groups of tens are regrouped into hundreds when there are 10 or more of these groups, groups of hundreds are regrouped into thousands when there are 10 or more of the hundreds, and so on. Thus, any digit in any place is one of the numerals 0, 1, 2, 3, . . . , 9. A similar analysis shows that any digit in the base three numeration system is one of the numbers 0, 1, 2 since any number of groups exceeding 2 would be regrouped into groups of the next larger size.

Set of Marks	Number of		Base Three Numeral	Base Ten Numeral
	Threes	Ones		
✕		1	1	1
✕ ✕		2	2	2
(✕ ✕ ✕)	1	0	10*	3
(✕ ✕ ✕) ✕	1	1	11	4
(✕ ✕ ✕) ✕ ✕	1	2	12	5
(✕ ✕ ✕) (✕ ✕ ✕)	2	0	20	6
(✕ ✕ ✕) (✕ ✕ ✕) ✕	2	1	21	7
(✕ ✕ ✕) (✕ ✕ ✕) ✕ ✕	2	2	22	8

Figure 2.20

■ **Question 10** If we were to go to base two numeration, what digits would we use (see Figure 2.20)?

We now face a problem. What, for instance, does the numeral "12" mean: "1 ten and 2 ones" or "1 three and 2 ones"? We commonly resolve this problem in the following way:

If we see the numeral "12," for example, we assume that it is written in base ten and understand it to mean "1 ten and 2 ones." This simply follows familiar convention.

If, on the other hand, we wish to write a numeral to convey a base three grouping such as "1 *three* and 2 ones" we agree to use the form "12_{three}." The subscript "three" indicates the base in which the numeral is written.

On occasion, when showing the base ten numeral for twelve, for instance, we may write "12_{ten}" instead of simply "12," just to be certain that there is no misunderstanding. Thus, we agree that

$$12 = 12_{ten}$$

However, be sure to keep clearly in mind that

$$12 \neq 12_{three}†$$

and that

$$12_{ten} \neq 12_{three}$$

In fact, it is true that

$$12_{ten} = 110_{three}$$

and that

$$12_{three} = 5_{ten}$$

† The symbol $\neq$ means "is not equal to."

Our base ten numeration system includes more than just two places, a tens place and a ones place. Likewise, a base three numeration system includes more than just a threes place and a ones place, as we have already seen. We now examine more carefully this extension of grouping by threes.

Question 11 Which is larger, 2_{ten} or 2_{three}?

We know that ninety-nine is the greatest whole number that can be named as a *two*-place numeral in our base ten numeration system: 99. The next whole number, 10 tens, or 1 hundred, necessitates a new place to the left of tens place. Thus, we name 10 tens or 1 hundred with the numeral shown in Figure 2.21.

Similarly, eight is the last whole number than can be named with a two-place numeral in a base three numeration system: 22. The next whole number, 3 threes, or nine, necessitates a new place to the left of the threes place. Thus, we name 3 threes or nine with the numeral shown in Figure 2.22.

The diagram in Figure 2.23 may help us interpret a numeral such as 212_{three}. Thus, 212_{three} is another name for 23_{ten}: $212_{three} = 23$.

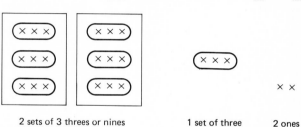

2 sets of 3 threes or nines   1 set of three   2 ones

Figure 2.23

The place values associated with a base three numeration system follow the same pattern as do the place values associated with a base ten numeration system, as shown in the chart shown in Figure 2.24.

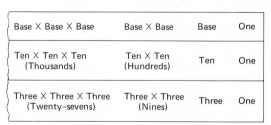

Base × Base × Base	Base × Base	Base	One
Ten × Ten × Ten (Thousands)	Ten × Ten (Hundreds)	Ten	One
Three × Three × Three (Twenty-sevens)	Three × Three (Nines)	Three	One

Figure 2.24

Thus, the numeral 2021_{three} may be interpreted as:

$2021_{three} = (2 \times 3 \times 3 \times 3) + (0 \times 3 \times 3) + (2 \times 3) + (\underline{1} \times 1)$

$2021_{three} = (2 \times 27) + (0 \times 9) + (2 \times 3) + (1 \times 1)$

$2021_{three} = 54 + 0 + 6 + 1$

$2021_{three} = 61$ (that is, 61_{ten})

Figure 2.21 shows:

1 0 0

ones
tens
10 tens or hundreds

Figure 2.21

Figure 2.22 shows:

1 0 0

three
ones
threes
3 threes or nines

Figure 2.22

PROBLEMS

1 Name the number property of the sets:
  (a) $\{\square, \triangle, \bigcirc, \circlearrowright\}$
  (b) $\{243\}$
  (c) $\{\text{zero}\}$
  (d) $\{\text{letters in the word "deeded"}\}$
  (e) $\{\text{the number of vowels in "bureau"}\}$
  (f) $\{\text{whole numbers less than 1}\}$

2 What numbers are associated with the following sets?
  (a) $\{\ \ \}$ (b) $\{\blacktriangle, \square\}$ (c) $\{\blacktriangle, \square, \bullet, \triangle, \bigcirc\}$
  (d) $\{\int\}$ (e) $\{\bigcirc, \bullet, \odot, \ominus, \bar{\bigcirc}, \ominus, \wp, \breve{\bigcirc}, \bar{\bigcirc}, \wp\}$

3 Write number sentences using appropriate symbols to show the relationship that exists among these sets.
  $A = \{\triangle, \bigcirc, \boxtimes, \Gamma, \Psi\}$ $B = \{a, b, c, d\}$ $C = \{p, q, r\}$

4 If a is the successor of b, and b is the successor of c, is it true that a is the successor of c?

5 What whole number is not the successor of any whole number?

6 Group in tens the elements in each of the sets in Figure P2.6 and write the decimal numeral for the number of elements in the set.

(a)   (b)   (c)   (d)

Figure P2.6

7 Write the base ten numeral for each of these expressions.
  (a) 7 hundreds, 4 tens, 9 ones
  (b) 8 thousands, 3 hundreds, 6 ones
  (c) $2,000 + 700 + 50 + 1$
  (d) $40,000 + 6,000 + 80 + 3$
  (e) $(5 \times 1,000) + (0 \times 100) + (2 \times 10) + (4 \times 1)$
  (f) $(7 \times 10,000) \times (6 \times 100) + (9 \times 1)$
  (g) $(8 \times 10 \times 10 \times 10) + (4 \times 10 \times 10) + (3 \times 1)$
  (h) $(9 \times 10 \times 10 \times 10 \times 10) + (5 \times 10 \times 10 \times 10) + (6 \times 10)$

8 Express each of these base ten numerals in three ways as shown in the illustrative example below.

EXAMPLE:

$4,257 = 4,000 + 200 + 50 + 7$
$4,257 = (4 \times 1,000) + (2 \times 100) + (5 \times 10) + (7 \times 1)$
$4,257 = (4 \times 10 \times 10 \times 10) + (2 \times 10 \times 10) + (5 \times 10) + (7 \times 1)$

  (a) 6,184   (b) 7,350   (c) 40,702

9 From the list below write all the letters which are beside correct names for 467.

(a) Four hundred sixty-seven (b) Forty-six and seven more
(c) 46 tens and 7 (d) 40 hundreds + 67 ones
(e) 300 + 160 + 7 (f) Seven plus four hundred
(g) 400 + 60 + 7 (h) 300 + 150 + 7
(i) 467 tens

10 Answer Yes or No.
(a) 3,729 is 37 tens plus 29 ones
(b) 10 hundreds plus 40 tens plus 9 ones is the same as one thousand forty-nine
(c) 5,000 + 500 + 1 = 5,501
(d) 36 hundreds + 1 ten + 18 ones = 3,628
(e) 734 = 600 + 120 + 24

11 In each ring write = or > or < so that the sentence will be true.
(a) 7,000 + 600 + 50 ○ 7,000 + 60 + 5
(b) $(3 \times 1{,}000) + (8 \times 100) + (4 \times 1)$ ○ 3,840

12 Write each of these in expanded form:
(a) 1,020,304 (b) 12
(c) 10 (d) 10,000

13 What is the meaning attached to the digit 4 in each numeral?
(a) 345 (b) 435 (c) 354

14 The sum of the digits of a two-digit number is 10. If the digits are interchanged, the number thus formed is 36 less than the original number. Find the number.

15 Consult an encyclopedia or other source to find our names for the powers of one thousand (billion, trillion, etc.), then write our word form for this number:

600,570,107,004,230,500,005

16 Write the numerals from 8 to 26 in base three.

17 Rename the following using a base three numeral system:
(a) 31 (b) 50 (c) 81

18 Express each of these base three numerals as a base ten numeral.
(a) 112_{three} (b) 1112_{three}
(c) 1002_{three} (d) 2001_{three}

19 What is the next number?
(a) 20_{three} (b) 21_{three} (c) 22_{three}
(d) 202_{three} (e) 20020_{three}

20 On planet X-101, the pages in books are numbered in order as follows:

I, ∟, △, □, ⧅, ⊠, I–, II, I∟, I△, I□, I⧅, I⊠, ∟—, ∟I, etc.

What seems to be the base of the numeration system these people use? Why? How would the next number after ∟I be written? Write numerals for numbers from □— to ⧅△.

3. GEOMETRY—ANOTHER WAY OF LOOKING AT THE WORLD

The first sentence in Chapter 1 reads: "Mathematics consists of a number of special ways of looking at the world we live in." In the first two chapters we found that looking at the world in terms of sets was helpful if we were faced with a "how many" question.

There is much more to be said about sets, numbers, and "how many" questions. However, we will now turn aside for a while to examine another way of looking at the world, a way which is helpful when we are faced with a "What shape is it?" question. This way of looking is called geometry.

GEOMETRIC FIGURES

Ball

Ice cream cone

Box

Can

Figure 3.1

Geometry is also a system of ideas based on our experiences with physical objects. For example, from our experiences with various boxes we have developed an idea of the shape and form of a box. This idea we have of a box is called a *geometric figure* and it is this figure, not the box, which is studied in geometry. In other words, geometry is developed as a *mathematical model* of our experience with physical space.

In addition to these geometric ideas we have in our minds, derived from concrete objects, such as boxes, we also need some way of representing them on the pages of books such as this. We can do this by means of sketches of objects, as in Figure 3.1, or we can use drawings such as those in Figure 3.2.

(a)

(b)

(c)

Figure 3.2

■ **Question 1** What does Figure 3.3 suggest?

Figure 3.3

The drawings in Figure 3.2 are examples of typical representations of three-dimensional geometric figures. There may be some difficulty in visualizing the three-dimensional nature of the figures since the drawings are restricted to two dimensions. The broken lines are included to aid perception. They represent parts of the figures which would not be visible from this vantage point.

In each of these figures, the "inside" may or may not be filled. The object identified by Figure 3.2a might represent a wooden block or a brick. The object identified by Figure 3.2c might represent a bowling ball or a ball bearing. Hollow physical objects which can be associated with the same figures are a shoe box (including the lid), an empty oatmeal box with the lid on, and a balloon.

In order to learn about geometric figures, we need to start with some of the simplest and most basic of them. We need to agree on names for them so that we can discuss them among ourselves. We need also to see how these basic figures are related to each other. The rest of this chapter will be devoted to this task.

Figures 3.2a to c can be abstracted from numerous physical objects which are available to any of us. For our purposes in developing some basic concepts and vocabulary, we will concentrate only on Figure 3.2a.

This "box" (a more technical name is: *rectangular prism*) is made of six flat surfaces which are called faces of the prism. A *face* of a geometric figure is a flat surface of the figure.

Where two faces meet is an *edge*. Each face of this figure has a boundary consisting of four edges. The "skeleton" of the prism is made up of twelve edges. The figure in Question 1 has nine edges.

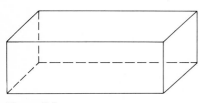

Figure 3.4

■ **Question 2** Does the figure represented in Figure 3.3 have more faces than edges?

POINTS One other characteristic which we wish to identify in the above solid is that it has "corners" where three edges come together. Each is a *vertex* (plural: vertices) of the prism, and is an example of a point.

The basic ingredient of all geometric configurations is what is called a *point*. A point may be thought of as a precise location. Points are represented by dots on a paper or as the end of a sharply pointed pencil, but these are just visual aids to assist us in conceptualizing the nature of a point. The dot made on a sheet of paper is merely an attempt to mark the idealized geometric entity, the "point." In fact, the dot covers not one point but an infinite number (in this instance more

●

A poor
representation
of a point

●

A better
representation
of a point

Figure 3.5

than can be counted). Viewed under some magnifying device such as
a microscope, any dot is clearly seen to cover many locations. Hence
no device can be used to mark a point accurately (see Figure 3.5).

Also, we note that a point is thought of as a fixed location. A point
does not move. If the dot made on a sheet of paper were erased, the
location previously marked by this dot still would remain. Again, if the
sheet of paper were moved to some other place, the point originally
marked by the dot would remain fixed. Perhaps a more graphic
demonstration of the permanency of the geometric point is given by
the demolition of a building. The points occupied by each corner of
each room remain unchanged. The difference is that they are no
longer represented by the physical objects called the corners. They
would now have to be described by some set of directions leading to
the location, such as, for example, 10 feet north of some marked point
and then 12 feet up. Finally, think of a pencil held in some position.
Its tip represents a geometrical point. If the pencil is moved, its tip
now represents a different geometrical point.

■ **Question 3** Is it possible for one point to be entirely inside another
point?

It is customary to name points by letters, just as we did sets, and to
say "the point A" or "the point B." We write the letter we have assigned to a particular point next to the dot which represents the point.

Once the geometrical meaning of point is understood, we are then
prepared to envision *geometrical space* or simply *space*.

Space is the set[1] of all points.

Since points can be thought of as represented by locations, space
may be thought of as represented by *all possible* locations.

CURVES
If we take two points A and B in space and think of moving the tip of a
pencil from the location of A to the location of B, the path traced by
the pencil tip will give a pretty good idea of a curve from A to B.
There are, of course, many paths which might have been taken besides the one we did select. Each path consists of a set of points and
the set of all points on such a path is called a *curve*. Each of the five
paths represented in Figure 3.6 is called a curve. Even the one
usually called straight is a special case of a curve. (Note that this
mathematical use of the word "curve" is not the same as the ordinary
use.)

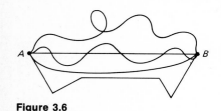

Figure 3.6

■ **Question 4** Is there anything outside of space?

Now think of the particular curve which may be *represented* by a
string tightly stretched between A and B (see Figure 3.7). This special curve is called a *line segment* or a *straight line segment*.

[1] Note that the sets dealt with in Chapters 1 and 2 were usually finite. Sets of
points, in geometry, are usually infinite, i.e., have more elements than can be
counted.

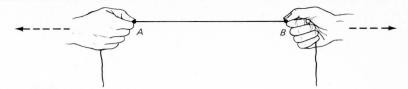

Figure 3.7

Another representation of a line segment would be the pencil mark, drawn with a ruler and pencil, connecting two points. The curve includes the points A and B which are called the *endpoints* of the line segment. The line segment can be thought of as the line of sight between A and B, and can be described as the most direct path. The line segment exists, of course, independently of any of its representations. For example, if the stretched string were removed, the line segment would remain since it is a set of locations. The symbol for the line segment determined by the points A and B is $\overline{AB}$. The fact that we say "*the* line segment $\overline{AB}$" implies that there is only one such segment.

If $\overline{AB}$ is extended in both directions along the line of sight so that it does not stop at any point, the result is a *straight line*. Its symbol is $\overleftrightarrow{AB}$. For brevity a straight line is called simply a *line*. Note that a line is a set of points, a particular set of points whose properties we can describe but which cannot wholly be represented in a figure because of its indefinite length. We have to use our imagination to conceive of the unlimited nature of the line.

■ **Question 5** How many endpoints does a line have?

POINTS AND LINES We can see that through a point A many lines can be drawn; in fact so many that we cannot possibly count them all. In other words, there are infinitely many lines passing through a single point A.

If a second point B is given different from A, there is always one line which passes through both A and B but more importantly, there is only one such line. We say it is *determined* by the two points.

If two distinct points are given, there is exactly one line which contains both points.

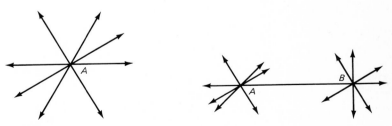

Figure 3.8

PLANES Perhaps a more familiar concept is that of the set of points we call a *plane*. Once again we do not define a plane, but we describe its properties and its relationship to lines and points and thus attempt to get a good idea of what is meant by it. Any flat surface such as the wall of a room, the floor, the top of a desk, or a door in any position suggests the ideas of a part of a plane in mathematics. But as with a line, a plane is thought of as being unlimited in extent. A plane is represented by a picture or a drawing of a flat surface, but this is only a representation of a part of a plane. An ever-growing table top provides a better and better representation. Also, we must remind ourselves that we are only representing certain sets of points in space. If the table were removed, the set of points (locations) does not change.

A plane of which the table top is a partial representation is the set of all points of all the lines obtained by extending the line segments with endpoints in the table top. In this way it is clear that a plane contains more lines than can be counted, in other words, infinitely many lines. Moreover, if two points of a line are contained in a plane, then the entire line is contained in the plane.

■ **Question 6** How many planes in space contain any particular point?

INTERSECTIONS OF SETS

Before going on, we need to consider another idea about sets. There is a standard way of creating a third set from two given sets. Suppose that one set of St. Nick's reindeers consists of Dasher, Dancer, Prancer, and Vixen. Suppose another set consists of Prancer, Vixen, Comet, Cupid, and Donder. Then we have two sets

R = {Dasher, Dancer, Prancer, Vixen}
M = {Prancer, Vixen, Comet, Cupid, Donder}

These sets overlap: Prancer and Vixen are members of both. In fact, common members of two sets suggest a natural set of elements— namely, the set consisting of all members that the sets have in common. Associated with two given sets then, is the set whose members are simultaneously elements of both given sets. This new set is called the *intersection* of the two sets, and is denoted by the symbol "∩". Thus,

$R \cap M$ = {Prancer, Vixen}

Note that $M \cap R$ is also {Prancer, Vixen}. It is obvious that if Prancer and Vixen are members common to R and M, then these same elements are members that are common to M and R.

The operation of intersection of sets is *commutative*.

That is to say, under the intersection operation, the *order* of intersection is immaterial.

■ **Question 7** What is the intersection of {Prancer, Vixen} and the empty set?

Now since the intersection of two sets is a set, we may consider the possibility of intersecting this set with yet another set. To illustrate, we have the two sets R and M listed above and also the set

$T = \{\text{Dasher, Blitzen, Prancer}\}$

We saw that

$R \cap M = \{\text{Prancer, Vixen}\}$

We can also work out that the intersection of $R \cap M$ with T is

$(R \cap M) \cap T = \{\text{Prancer}\}$

As before, the parentheses around R and M indicate the grouping of these sets to form the first intersection. Thus, intersection of sets may be formed successively one upon another. So we may ask the question about the result of grouping these same three sets differently. The question then might be: "How does $R \cap (M \cap T)$ compare with $(R \cap M) \cap T$?" To answer this, first observe that

$M \cap T = \{\text{Prancer}\}$

Therefore,

$R \cap (M \cap T) = \{\text{Dasher, Dancer, Prancer, Vixen}\} \cap \{\text{Prancer}\}$
$= \{\text{Prancer}\}$

This clearly gives the same result that we obtained above for $(R \cap M) \cap T$. In general, we have the *associative* property under intersection:

> For sets A, B, and C, it is true that
> $(A \cap B) \cap C = A \cap (B \cap C)$

and so we may simplify both of these expressions by dropping the parentheses:

$(A \cap B) \cap C = A \cap (B \cap C) = A \cap B \cap C$

■ **Question 8** Let $A = \{\text{C, L, S, W}\}$, $B = \{\text{L, S, V, W}\}$, $C = \{\text{C, S, V, W}\}$, and $D = \{\text{C, L, S, W}\}$. What is $(A \cap B) \cap (C \cap D)$? What is $[(D \cap A) \cap B] \cap C$?

The sets we have used to illustrate these ideas about the intersection of sets were all finite. But the ideas still make sense when applied to infinite sets of points, such as lines, line segments, planes, etc., as we shall now see.

POINTS, LINES, AND PLANES Let us look at the way points, lines, and planes can be fitted together in space. First we recall that through any two different points there is exactly one line. On the other hand, suppose we have two different lines and suppose there is a point which belongs to both of the lines. You can see that there is no other point which belongs to both of the lines.

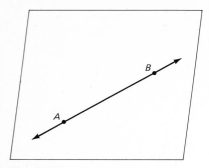

Figure 3.9

If two distinct lines intersect, their intersection consists of exactly
one point.

We think of a plane as being "flat." If any two points in the plane
are selected, they determine a line. Where are all the other points of
this line? The straightness of the line and the flatness of the plane
suggest that they lie in the plane. Thus:

If two different points lie in a plane, the line determined by the
points lies in the plane.

In Figure 3.10 the surface of a tin can is *not* a plane since $\overleftrightarrow{AB}$ cuts
through the space inside the can and does not lie wholly on the sur-
face.

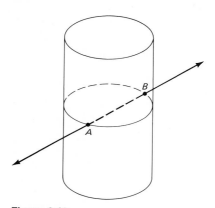

Figure 3.10

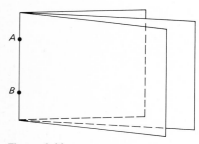

Figure 3.11

Can there be more than one plane containing two different points?
Think of two points at the hinges of a door. The door, which represents part of a plane, can swing freely, and therefore there must be
many planes through the two given points (Figure 3.11). Another way
to think of it would be to think of a 3 by 5 card held at opposite corners
between the thumb and third finger. The card can spin freely and in
each position represents a portion of a plane.

Through two points in space, and hence through a line in space,
there are many possible planes.

■ **Question 9** How many lines are there in a plane?

It seems that two different points determine a line. How many are
needed to determine a plane? By "determine" we mean that there is
at least one plane containing the given points and not more than one.

The card mentioned above held between thumb and third finger can
no longer spin if a third finger is extended. The tips of the three
fingers determine a plane. A three-legged stool always sits firmly on
the floor, while a badly made four-legged table may wobble. These

Figure 3.12

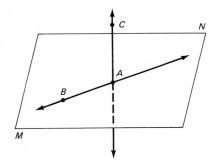

Figure 3.13

illustrate the idea that three points can fix a plane. Of course, they must not lie on the same line since, for instance, if there were another point C on $\overline{AB}$ in Figure 3.11, all three planes would contain it as well as A and B. So:

> Any three points not in the same line determine one and only one plane.

Two intersecting lines also determine a plane since we can pick two points in one line and the third one in the other line. In the same way a line and a point which is not in it determine one and only one plane.

■ **Question 10** Can you think of four points which are not in the same plane?

If the intersections of lines and planes are considered, three more interesting properties can be noted.

Two different lines may not intersect at all, but if they do, what would their intersection be? A single point. Just think of two pencils and how they may be held.

If a line and plane intersect, their intersection must be only a single point unless the whole line lies in the plane (see Figure 3.13). These ideas are summarized in the following properties:

> If two different lines in space intersect, their intersection is one point.

> If a line and a plane intersect, their intersection is either one point or the entire line.

A more difficult thing to see, perhaps, is how two planes intersect. Of course, the floor and ceiling of an ordinary room represent portions of two planes which would not intersect at all. The floor and a flat wall seem to intersect in a straight line. In fact, consideration of any two planes which have some points in common will indicate that they must meet in a straight line. This is indeed true.

> If two different planes intersect, their intersection is a straight line.

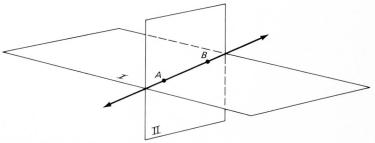

Figure 3.14

It is possible, of course, for two planes not to intersect at all, in which case we say that they are *parallel.* The same is true for a line and a plane (see Figure 3.15).

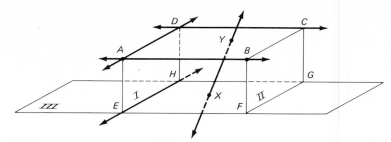

Figure 3.15

In this figure, planes *I* and *II* are parallel whereas planes *I* and *III* intersect in $\overleftrightarrow{EH}$. $\overleftrightarrow{AB}$ intersects plane *I* point *A* but $\overleftrightarrow{AB}$ is parallel to plane *III*. Two planes always either intersect or are parallel. It is possible, however, for two lines not to intersect and yet not to be parallel to each other either. This is the case with $\overleftrightarrow{XY}$ and $\overleftrightarrow{EF}$. Such lines are said to be *skew* and they never lie in the same plane. We speak of two lines as being parallel only if they *lie in the same plane and do not intersect.* $\overleftrightarrow{AD}$ and $\overleftrightarrow{EH}$ are parallel.

■ **Question 11** Can you think of three lines located in space so that any pair of them are skew?

SEPARATION

There is another very important idea closely connected with the relationship between points, lines, and planes, and that is the idea of *separation.* What is involved may be made clear by some examples and illustrations. Think of a plane represented by the wall of a room. This plane separates space into two sets of points, those in front of the wall and those behind it. In this sense we think of any plane as having two sides and we say that a plane separates space, that is, it divides the points of space into three sets, one of which is the plane itself while the other two are the points on each side of the plane. By convention we refer to these latter two sets of points as *half-spaces.* Note that the separating plane does not lie in either half-space. Later on we shall use "half-plane" and "half-line" in a similar sense. Points are in the same half-space if they are on the same side of the plane. If points such as *A* and *B* in Figure 3.16 are on the same side of the plane, then there always exist curves connecting *A* and *B* which do not intersect the plane. In particular $\overline{AB}$ does not intersect the plane. On the other hand, if points such as *A* and *C* are on opposite sides of the plane, then any curve connecting *A* and *C*, in particular $\overline{AC}$, must intersect the plane. (Note that while the segment $\overline{AB}$ does not intersect the separating plane in Figure 3.16 the line $\overleftrightarrow{AB}$ may very well do so.)

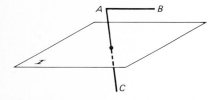

Figure 3.16

■ **Question 12** How many lines can there be in a half-space?

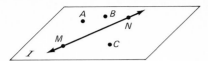

Figure 3.17

In the same way, if we consider a certain plane and a line in that plane, the line separates the points of the plane into two *half-planes*. Thus, in Figure 3.17, $\overleftrightarrow{MN}$ separates plane I into two half-planes such that A and B lie in the same half-plane but A and C lie in different half-planes. The line $\overleftrightarrow{MN}$ does not lie in either half-plane. There exist curves in plane I connecting A and B which do not intersect $\overleftrightarrow{MN}$. $\overline{AB}$ is such a curve. On the other hand, every curve in plane I connecting A and C must intersect $\overleftrightarrow{MN}$. We see that $\overline{AC}$ does in particular.

In the same manner, a point in a line separates a line. In Figure 3.18, P separates line ℓ into two *half-lines* such that A and B lie in the same half-line, but A and C lie in different half-lines. The point P does not lie in either half-line. Again $\overline{AB}$ does not intersect the separating point P while $\overline{AC}$ does.

Figure 3.18

The three cases are much the same idea applied in different situations. Thus:

Any plane separates space into two *half-spaces*.

Any line of a plane separates the plane into two *half-planes*.

Any point of a line separates the line into two *half-lines*.

Figure P3.2

PROBLEMS

1 Draw a representation of a geometric figure shaped like the pyramids of Egypt. How many vertices does it have?

2 How many faces does Figure P3.2 contain?

3 Which of the figures in Figure P3.3 have edges but no vertices?

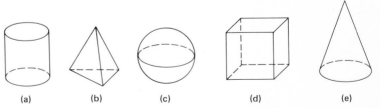

(a)   (b)   (c)   (d)   (e)

Figure P3.3

4 Let $A = \{1, 3, 5\}$, $B = \{1, 5, 9\}$, and $C = \{1, 4, 6, 8\}$. Find:
(a) $A \cap B$ (b) $A \cap C$ (c) $B \cap C$
(d) $A \cap (B \cap C)$ (e) $(A \cap B) \cap C$

5 Let V be the set of odd whole numbers. Let W be the set of whole numbers less than 20. Let X be the set of whole numbers divisible by 5.
(a) $V \cap W = \{\ ?\ \}$ (b) $W \cap X = \{\ ?\ \}$ (c) $(V \cap W) \cap X = \{\ ?\ \}$

6 For each set A,
(a) $A \cap A = ?$ (b) $(A \cap A) \cap A = ?$ (c) $A \cap \{ \ \} = ?$

7 Draw a line. Label three points of the line A, B, and C with B between A and C.
(a) What is $\overline{AB} \cap \overline{BC}$? (b) What is $\overline{AC} \cap \overline{BC}$?

8 Draw a horizontal line. Label four points on it P, Q, R, and S in that order from left to right. Name two segments
(a) whose intersection is a segment.
(b) whose intersection is a point.
(c) whose intersection is empty.

9 Let A and B be two points. Is it true that there is exactly one segment *containing* A and B? Draw a figure explaining your answer.

10 Draw two segments $\overline{AB}$ and $\overline{CD}$ for which $\overline{AB} \cap \overline{CD}$ is empty but $\overleftrightarrow{AB} \cap \overleftrightarrow{CD}$ is one point.

11 Draw two segments $\overline{PQ}$ and $\overline{RS}$ for which $\overline{PQ} \cap \overline{RS}$ is empty but $\overleftrightarrow{PQ}$ is $\overleftrightarrow{RS}$.

12 Draw a segment. Label its endpoints X and Y. Is there a pair of points of $\overline{XY}$ with Y between them? Is there a pair of points of $\overleftrightarrow{XY}$ with Y between them?

13 How many lines can be drawn through four points, a pair of them at a time, if the points lie
(a) in the same plane? (b) not in the same plane?

14 May four different points, no three of which lie in the same straight line, lie in one plane?

15 Suppose four points are not all in the same plane.
(a) How many different planes contain at least three of them?
(b) How many different lines contain at least two of them?

16 (a) Suppose P, Q, and R are three distinct points and are all in each of two different planes. What can be said about P, Q, and R?
(b) Suppose points P, Q, and R are in only one plane. What can be said about the line containing P and Q?

17 Consider the sketch of the outline of a house in Figure P3.17. Think of the lines and planes suggested by the figure. Name lines by a pair of points and planes by three points. Name:
(a) a pair of parallel planes.
(b) a pair of planes whose intersection is a line.
(c) three planes that intersect in a point.
(d) three planes that intersect in a line.
(e) a line and a plane which do not intersect.
(f) a pair of parallel lines.
(g) a pair of skew lines.
(h) three lines that intersect in a point.

18 Consider a piece of paper as a representation of a plane.
(a) Does a segment separate the plane? A line?
(b) Into how many parts do two intersecting lines separate a plane? Two parallel lines?

19 (a) Does a half-plane separate space? Does a line?
(b) Into how many parts do two intersecting planes separate space?

20 If A and B are points on the same side of the plane M (in space), must $\overline{AB} \cap M$ be empty? Can $\overleftrightarrow{AB} \cap M$ be empty?

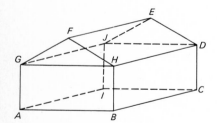

Figure P3.17

4. SOME FAMILIAR GEOMETRIC FIGURES

PLANE CURVES Let us go back to the idea of a curve in the previous chapter and see what more we can learn from it. A curve from A to B is any set of points which can be represented by a pencil point which started at A and moved around to end up at B. Such a curve might wander around in space and cross and recross itself many times. We want to confine our attention to fairly simple curves and so we impose some restrictions. The first is that our pencil point stays in the same plane. Such curves are called *plane curves*. We can represent them by figures we draw on a sheet of paper.

Line segments are examples of such curves. Curves may or may not contain portions that are straight. (In everyday language we do not usually use the term "curve" in this same sense, but it is convenient to do so in mathematics.) In Chapter 3 we looked at some of the basic geometric figures. We learned their names and we learned something about the way they can be related. There are still a few more basic ideas to study.

Figure 4.1c is an important type of curve called a *broken-line* curve. It consists of a number of line segments joined at their endpoints.

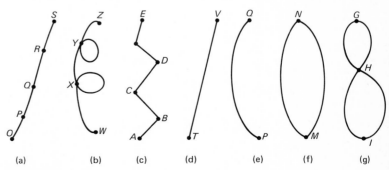

Figure 4.1

We need a way to distinguish curves such as those shown in Figure 4.1f and g from other curves. These are called closed curves. A *closed curve* is a plane curve whose representation can be drawn without retracing and with the pencil point stopping at the same point from which it started.

If a closed curve does not intersect itself at any point, we call it a *simple closed curve.*

■ **Question 1** Can the intersection of two curves be a line segment if neither of the curves is a broken-line curve?

Figure 4.1f is a simple closed curve. We could speak of going around the curve and, when we do, we pass through each point just once (except, of course, the starting point).

Curves a, d, and e in Figure 4.2 are closed curves.

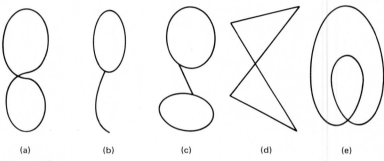

(a)　　　(b)　　　(c)　　　(d)　　　(e)

Figure 4.2

Each curve in Figure 4.3 is a simple closed curve, but not one of those in Figure 4.2 is.

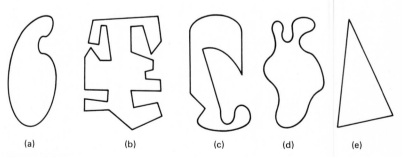

(a)　　　(b)　　　(c)　　　(d)　　　(e)

Figure 4.3

SEPARATION From looking at representations of simple closed curves we might guess that any such curve separates the plane into two parts, one of which might be called the interior of the curve and the other the exterior. As a matter of fact this is true and is a very important property

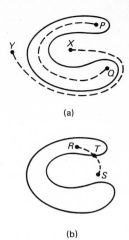

(a)

(b)

Figure 4.4

of a simple *plane* closed curve, but one which may not hold if the curve is drawn on another surface. Every simple plane closed curve separates the plane in which it lies into two parts, the interior of the curve and the exterior of the curve. The curve itself does not belong to either part.

When a line separated a plane into two half-planes we could join any two points in the same half-plane by a segment which did not intersect the line. In like manner, any two points in the interior of the simple closed curve can be joined by a curve which does not intersect it. The same is true for two points in the exterior as illustrated in Figure 4.4a.

Again, if point R is in the interior and point S in the exterior of the curve, we can never join R and S by a plane curve which does not intersect the given simple closed curve. See Figure 4.4b. The interior of any simple closed curve, together with the curve, is called a *region*. The curve is the *boundary* of the region. Note that the boundary of the region is part of the region.

■ **Question 2** Must the intersection of a line and a region always be a line segment?

POLYGONS If a simple closed curve consists of three or more line segments, it is called a *polygon*. Note that a curve can consist of three or more segments without being a polygon. (See Figure 4.1c.) Polygons have special names according to how many line segments are involved. Those with three segments are called *triangles*, with four, *quadrilaterals*. After that, the names are made up of the Greek word for the appropriate number followed by the syllable "gon." Thus, *penta*gon, *hexa*gon, *octa*gon, and *deca*gon are the names for the polygons with 5, 6, 8, or 10 sides, respectively (see Figure 4.5).

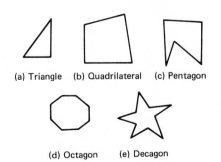

(a) Triangle   (b) Quadrilateral   (c) Pentagon

(d) Octagon   (e) Decagon

Figure 4.5

The simplest of the polygons is the triangle. Let A, B, and C be three points, not all on the same line. The triangle ABC, written as $\triangle ABC$, consists of $\overline{AB}$, $\overline{AC}$, and $\overline{BC}$. Each of the points A, B, and C is a *vertex* of the triangle. The segments $\overline{AB}$, $\overline{BC}$, and $\overline{CA}$ are called the *sides* of the triangle. It should be noted that the triangle consists of just the three line segments and specifically does not include the interior.

■ **Question 3** Could a polygon have as many as 1,000 sides?

CONVEX POLYGONS

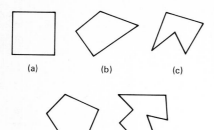

(a)   (b)   (c)

(d)   (e)

Figure 4.6

If we consider quadrilaterals and other polygons with more than three sides, we soon see that two quite different situations are possible. These are illustrated in Figure 4.6 where the polygons a, b, and d surely have a different character from polygons c and e.

A polygon is *convex* if for any two points A and B on the polygon, $\overline{AB}$ contains besides A and B only points in the interior or on the boundary of the polygon. Usually, it is convex polygons that we are interested in (see Figure 4.7).

Simple convex polygon

Simple nonconvex polygon

Neither simple nor convex—not a polygon

Figure 4.7

■ **Question 4** Is every triangle convex?

We will come back to these ideas in a later chapter. There is very little more that is interesting that we can say now about geometric figures. We need some new ideas first, and we will start to develop one of them in the rest of this chapter.

COMPARING LINE SEGMENTS

Nowhere yet in these two chapters have we looked at any properties of geometric figures which needed the idea of size or measure. But now, in order to get a better idea of the familiar figures, such as triangles, we need to be able to compare two segments so as to say whether the first is longer than the second, or the second is longer than the first, or whether neither statement is correct. This concept of "longer than" is much the same as the concept "more than" used in comparing two numbers in Chapter 1. At that time, we used the word "equals" and the symbol "=" to mean that we had two names for the same number. Thus, $8 - 5 = 12 \div 4$ because both $8 - 5$ and $12 \div 4$ are names for the number 3.

In all our work we are going to reserve the word "equals" and the symbol "=" for this idea, that is to say, if A, B, C, etc., are names for points, $A = B$ means that A and B are both names for precisely the same point. $\overline{AB} = \overline{CD}$ means that either $A = C$ and $B = D$ or that $A = D$ and $B = C$ so that we are talking about the same segment.

In Figure 4.8, while $\overline{AB}$ and $\overline{CD}$ seem to have the same length, it is *not* true that $\overline{AB} = \overline{CD}$ since A is not the same point as either C or D.

Figure 4.8

■ **Question 5** Is the set of all four-sided triangles equal to the set of all three-sided quadrilaterals?

CONGRUENCE OF SEGMENTS

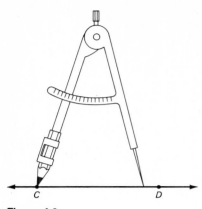

Figure 4.9

We come now to the problem: How do we compare two segments $\overline{AB}$ and $\overline{CD}$? Since A and B are fixed locations in space, we cannot move them around. It is possible, however, to have a representation of the segment $\overline{AB}$ on a piece of paper, or as a stretched, taut piece of string, or as the segment between the tips of a compass. If there is a similar representation for $\overline{CD}$, we can physically compare the two representations. Although we cannot move the segments, we can move their representations. Suppose the tips of a compass are used to represent $\overline{AB}$. Place one tip of the compass on C and see where the other tip falls on the line $\overleftrightarrow{CD}$. If it falls between C and D, we say that $\overline{CD}$ is longer than $\overline{AB}$. If the tip of the compass falls on D, $\overline{AB}$ is *congruent* to $\overline{CD}$. If it falls beyond D, $\overline{AB}$ is longer than $\overline{CD}$ (see Figure 4.9).

It is perhaps a bit difficult to see why we want to say that segments are "congruent" instead of using the more familiar word "equal." But strictly speaking the familiar usage "two lines are equal" is inaccurate on two counts. First of all, it is clearly "line segments" rather than "lines" that are being referred to. Secondly, and more to the point here, what is really being said is that the two segments have the same "length" or "measurement." The "length of a line segment" is a *number* which measures the segment in a sense which will be discussed in Chapter 10, and to say that two lengths are equal is to say that the same number is the measure of each. But, without this idea at all, it makes perfectly good sense to say that $\overline{AB}$ is congruent to $\overline{CD}$, with the meaning attached to this statement in the previous paragraph, and this is all we need at the moment.

■ **Question 6** Are any two points congruent?

Notice that this method of comparing two segments is not the same as, but is similar to, the method of comparing two sets in Chapter 1. If we tried to match the elements of set P with those of set Q and ran out of elements of P before we did those of Q, so that we matched the elements of P with those of only part of Q, we said Q was "more than" P. It was also possible for P to be more than Q or for P to "match" Q. These relationships between two sets roughly correspond to the relationships between two segments when they are compared as we did a moment ago.

For brevity in writing, symbols are needed to represent these relationships between segments. Since the symbol "=" has been reserved to indicate two different names for the same thing, whether it be a number, a set, a segment, a concrete object, or an idea, we need a new symbol for congruence. To indicate that $\overline{AB}$ and $\overline{CD}$ are congruent we write $\overline{AB} \cong \overline{CD}$. If $\overline{CD}$ is longer than $\overline{AB}$, we see that $\overline{AB}$ is congruent to a part of $\overline{CD} : \overline{CE}$ in Figure 4.10. We write this $\overline{CD} > \overline{AB}$ (or $\overline{AB} < \overline{CD}$). Note that this use of symbols ">" and "<" indicates a comparison relationship between line segments. In

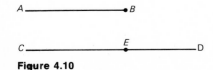

Figure 4.10

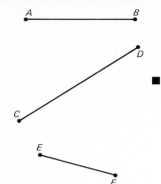

Figure 4.11

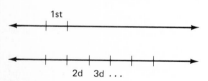

Figure 4.12

THE NUMBER LINE

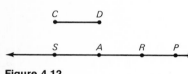

Figure 4.13

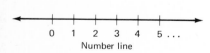

Number line

Figure 4.14

Chapter 2 we used the same symbols to indicate a comparison relationship between two numbers. We should keep carefully in mind the difference between the two usages. Usually it will be easy to see when we are talking about segments and when we are talking about numbers.

■ **Question 7** If one polygon has more sides than another, are its sides longer?

When we compare two segments $\overline{AB}$ and $\overline{CD}$ by the method outlined above, only three things can happen: The tip of the compass, representing B, can fall between C and D, or on D, or beyond D. Accordingly we find that one and only one of the three statements $\overline{AB} < \overline{CD}$, $\overline{AB} \cong \overline{CD}$, and $\overline{AB} > \overline{CD}$ will be true.

If we compare $\overline{AB}$ with two other segments $\overline{CD}$ and $\overline{EF}$, as shown in Figure 4.11, we may find that $\overline{CD} > \overline{AB}$ and $\overline{AB} > \overline{EF}$. If we now compare $\overline{CD}$ and $\overline{EF}$, we will always find that $\overline{CD} > \overline{EF}$. Thus: If $\overline{CD} > \overline{AB}$, and $\overline{AB} > \overline{EF}$, then $\overline{CD} > \overline{EF}$.

Suppose we have a segment $\overline{CD}$ and a point A fixed on a line $\overleftrightarrow{AP}$, as in Figure 4.12. Using the tips of a compass to represent $\overline{CD}$, we can place one tip on A and with the other tip determine two points of $\overleftrightarrow{AP}$, say R and S, one on each side of A so that $\overline{AR} \cong \overline{CD}$ and $\overline{AS} \cong \overline{CD}$. Thus we can determine two segments with endpoint A on line $\overleftrightarrow{AP}$ congruent to $\overline{CD}$. If, as in Figure 4.12. $\overline{SA} \cong \overline{AR}$ and A is between S and R, we say that point A *bisects* $\overline{SR}$.

■ **Question 8** Are there triangles in which no pair of sides are congruent?

Congruent segments give us a way of relating numbers with points on a line. Given any two points on a line, a segment is determined. We can continue to mark off points, one after another, so that each segment is congruent to the first (Figure 4.13). The points may be labeled 0, 1, 2, 3, 4, . . . , in the order of the whole numbers. Although one can assign these labels from right to left, conventionally we proceed from left to right. When the line is vertical we usually proceed upwards. When points are labeled thus, the numbers associated with the points are called the *coordinates* of the points, and the line together with its coordinates is called a *number line* (Figure 4.14).

The number line thus gives us a way of matching the set of endpoints of congruent segments and the set of whole numbers. That is, each endpoint is associated with one and only one whole number, and each whole number is associated with one and only one endpoint of the congruent segments on the line. This device is quite useful for us. It enables us to visualize the order of numbers by the position of corresponding points on the line. We will later connect operations in arithmetic with operations on the number line.

■ **Question 9** What point on the number line bisects the segment from 2 to 8?

CONCLUSION In these two chapters we have had only an introduction to a geometric way of looking at the world we live in. There is much more to geometry than this. There are many more interesting and useful aspects of geometry, and we will take a look at some of them in later chapters now that we have become acquainted with the basic ideas and terminology.

As a final comment on this introduction, we wish to emphasize that this geometric way of looking at the world and the arithmetic way, discussed in the first two chapters, are not entirely separate. The number line provides a link between them. This idea, of a number line, will be strengthened and deepened as we proceed, and will eventually turn out to be probably the most important unifying idea of all the mathematics we will encounter.

PROBLEMS

1 Which of the curves in Figure P4.1 are simple closed curves?

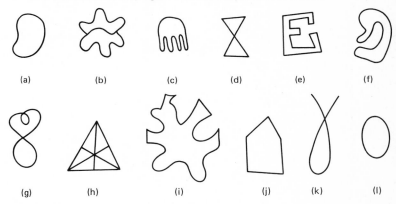

(a) (b) (c) (d) (e) (f)

(g) (h) (i) (j) (k) (l)

Figure P4.1

2 Which of the curves in Figure P4.2 are simple?

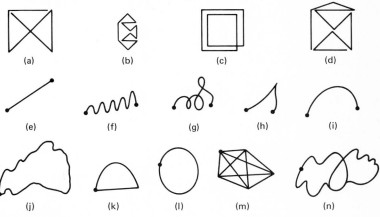

(a) (b) (c) (d)

(e) (f) (g) (h) (i)

(j) (k) (l) (m) (n)

Figure P4.2

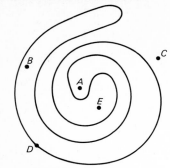

Figure P4.4

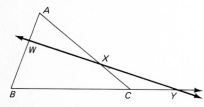

Figure P4.5

3 If the letters of the alphabet are printed in block type, which letters indicate simple closed curves?

A B C D . . .

Which are simple but not closed? Which separate the plane?

4 In Figure P4.4 can a curve be drawn from A to B without crossing the given curve? From A to C? From C to E? State the reason for each case.

5 Refer to Figure P4.5.
  (a) What is $\overline{YW} \cap \triangle ABC$?
  (b) Name the four triangles in the figure.
  (c) Which of the labeled points, if any, are in the interior of any of the triangles?
  (d) Which of the labeled points, if any, are in the exterior of any of the triangles?
  (e) Name a point on the same side of $\overleftrightarrow{WY}$ as C and one on the opposite side.

6 Refer to Figure P4.5.
  (a) Label a point P not in the interior of any of the triangles.
  (b) Label a point Q inside two of the triangles.
  (c) Label a point R in the interior of $\triangle ABC$ but not in the interior of any of the other triangles.

7 Draw two simple closed curves whose interiors intersect in three separate parts.

8 Which of the simple closed curves in Figure P4.8 are pictures of polygons?

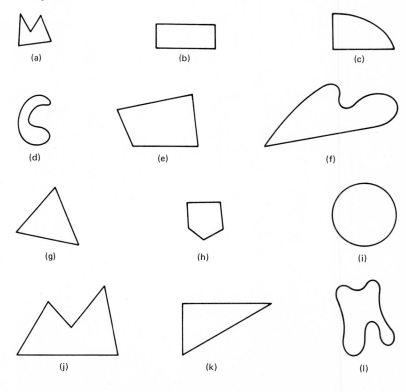

(a)

(b)

(c)

(d)

(e)

(f)

(g)

(h)

(i)

(j)

(k)

(l)

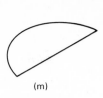

(m)

(n)

(o)

Figure P4.8

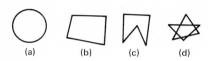

(a)

(b)

(c)

(d)

(e)

(f)

(g)

(h)

Figure P4.9

9 Which figures in Figure P4.9 are pictures of quadrilaterals?

10 Certain regions in the plane are indicated in each part of Figure P4.10. For each say whether or not it is convex.

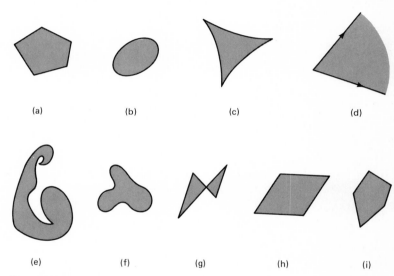

(a)

(b)

(c)

(d)

(e)

(f)

(g)

(h)

(i)

Figure P4.10

11 Is the intersection of two plane convex polygons also convex?

12 If possible, make sketches in which the intersection of two triangles is
   (a) the empty set.
   (b) exactly two points.
   (c) exactly four points.
   (d) exactly five points.

13 Which of the following segments in Figure P4.13 are congruent?

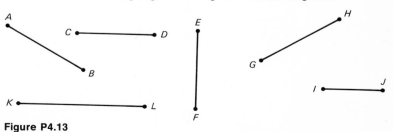

Figure P4.13

14 In each of the following pairs of segments in Figure P4.14 determine if $\overline{AB} < \overline{CD}$, $\overline{AB} > \overline{CD}$, or $\overline{AB} \cong \overline{CD}$.

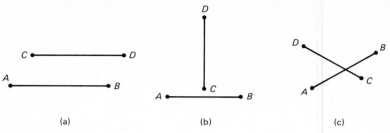

(a)   (b)   (c)

Figure P4.14

15 If $\overline{AB} \cong \overline{CD}$ and $\overline{CD} \cong \overline{EF}$, is it true that $\overline{AB} \cong \overline{EF}$? Is $\overline{CD} \cong \overline{AB}$?

16 If $\overline{AB} > \overline{CD}$ and $\overline{AB} > \overline{EF}$, is it true that $\overline{CD} > \overline{EF}$? Is $\overline{CD} > \overline{AB}$?

17 See Figure P4.17.

Figure P4.17

    (a) Name all the line segments that have two of these four points for end points. How many of these segments are there?

    (b) Which is the shortest of the line segments? The longest?

18 Each set of three vertices of a cube determines a triangle. For example, in Figure P4.18 points A, B, C determine a triangle whose two sides $\overline{AB}$ and $\overline{BC}$ are of the same length.

    (a) Find three vertices that determine a triangle in which no two sides have the same length.

    (b) Find three vertices that determine a triangle in which three sides have the same length.

19 If one number is greater than another, what do you know about their positions on the number line?

20 What is the smallest whole number represented on the number line?

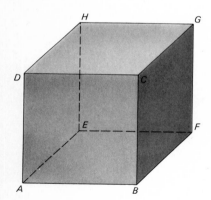

Figure P4.18

5.

ADDITION AND SUBTRACTION

In Chapter 1, it was stated that the ideas presented there about sets and about numbers would be helpful in answering "how many" questions. But you have not yet been shown just how they can be used. Some new ideas are needed before this can be done.

In general, in order to find an answer to a "how many" question, it is necessary to carry out some operations, or computations, on some numbers. Therefore, we need to look at these operations. We will start with the simplest one, addition. We saw that our ideas of whole numbers were developed from some ideas (e.g., pairing, matching) about sets of concrete objects. It will come as no surprise, then, that the concept of addition of whole numbers is derived from a particular kind of operation on sets.

SET UNION We have already seen one operation, that of forming the intersection, which creates a third set from two given sets. There is another equally important operation, called *join*, which creates a new set from two given sets. The join or *union* of two sets A and B, denoted by $A \cup B$, is the set consisting of all the things which are members of A or members of B, or members of both. Thus, if we have these two sets:

$R = \{$Dasher, Dancer, Prancer, Vixen$\}$

$M = \{$Prancer, Vixen, Comet, Cupid, Donder$\}$

then

$R \cup M = \{$Dasher, Dancer, Prancer, Vixen, Comet, Cupid, Donder$\}$

Note that in this case the two sets had some members in common, Prancer and Vixen. This is not necessary. We can still join two sets that have nothing in common or, as we shall say, are *disjoint*.

■ **Question 1** What is $A \cup B$ if $A = B$?

Suppose you have a set A consisting of one jelly bean, one gumball, and one chocolate drop. Also, you have a set B consisting of an apple and a banana. Then

$A \cup B = \{$ jelly bean, gumball, chocolate drop, apple, banana $\}$

You could have formed this set of goodies by first putting the set of candies into a bowl and then adding the set of fruit to the bowl. Or, you could have done it in the opposite order, thus forming the set $B \cup A$. Obviously, you end up with the same set of objects in the bowl whether you put the candies in first or the fruit in first.

This illustrates that the operation of forming the union of sets is *commutative:*

$A \cup B = B \cup A$

You will remember that the other set-forming operation we discussed, intersection, is also commutative.

There are many actual situations in which commutativity holds. Taking 3 red marbles from a sack, 4 green marbles from a second sack, and putting these together into a third sack is another illustration of a commutative operation. Taking 4 green marbles from the second sack and 3 red marbles from the first to put together into the third sack would net the same result.

■ **Question 2** What is $\{ \ \} \cup A$?

On the other hand, there are situations where the results do depend on the order in which the operation is carried out. For example, applying a coat of red paint on top of a coat of green paint gives a different visual effect than reversing this procedure. Therefore, it is pertinent to point out commutativity when it does occur.

There is another property of the operation of joining that we wish to point out. Suppose that in addition to the set A of candy and the set B of fruit, you also have a set C consisting of one bottle of strawberry pop. Suppose also that you have already formed the set $A \cup B$ by putting the candy and the fruit together in a bowl. Since $A \cup B$ is a set, we can form the union of it and the set C:

$(A \cup B) \cup C$

where the parentheses around $A \cup B$ indicate that the union of A and B was formed first and that then the union of that set with C was formed.

Now we could have formed the final set in a different order. We could have formed the union of B and C and then formed the union of A with that set:

$A \cup (B \cup C)$

But when we look in the bowl at the end of either of these processes, we see the same things: three pieces of candy, two pieces of fruit, and a bottle of pop. This is an illustration of the *associative* property of joining:

$$(A \cup B) \cup C = A \cup (B \cup C)$$

■ **Question 3**   If

$$A = \{1, 2, 3\} \qquad B = \{0, 1, 4\} \qquad C = \{0, 1, 8\}$$

write $(A \cup B)$, $(A \cup B) \cup C$, $(B \cup C)$, and $A \cup (B \cup C)$.

ADDITION OF WHOLE NUMBERS

Now we are ready to define the operation of addition. This is an operation which assigns to any pair of whole numbers another whole number. The rule for this operation is:

> If a and b are two whole numbers, take two disjoint sets A and B, with $N(A) = a$ and $N(B) = b$. The *sum* of these two numbers a and b is $N(A \cup B)$ and is denoted by $a + b$:
>
> $$a + b = N(A \cup B)$$

Thus, if we want to know what $3 + 2$ is, we could go back to our set A of 3 pieces of candy and our set B of 2 pieces of fruit. When we count the members of $A \cup B$, we find 5 objects, so $3 + 2 = 5$.

Notice how important it is that A and B be disjoint. Suppose we want to find out what $3 + 3$ is. If we were careless, we might choose these two sets:

$$A = \{\square, \bigcirc, \triangle\} \qquad B = \{\times, \bigcirc, \triangle\}$$

Of course, $N(A) = 3$ and $N(B) = 3$. But

$$A \cup B = \{\square, \times, \bigcirc, \triangle\}$$

and $N(A \cup B) = 4$, which is certainly not the same as $3 + 3$. The trouble here is that A and B are not disjoint. They have the members $\bigcirc$ and $\triangle$ in common.

■ **Question 4**   What does the definition tell us about $5 + 0$?

Suppose two different people tried to find the sum of 3 and 2. Would they get the same answer? They might choose different sets for A and B, so there is a question. To illustrate, suppose one person chose the sets of candies and of fruit, as above, while another person

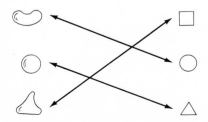

Figure 5.1

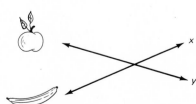

Figure 5.2

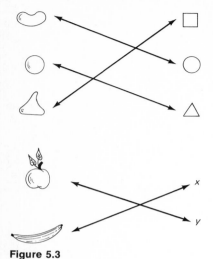

Figure 5.3

chose set $A' = \{\Box, \bigcirc, \triangle\}$ and set $B' = \{x, y\}$. Since $N(A') = 3$, and $N(B') = 2$, the second person is following the rule correctly.

Now, the second person would say that $3 + 2 = N(A' \cup B')$. We can see that in this case the second person gets 5 for the sum, just as did the first person. This will always be so, as we can easily see. Since $N(A) = 3 = N(A')$, the sets A and A' must match, because if they did not match they would have different number properties. The arrows in Figure 5.1 show one way of matching A and A'. Also, B and B' must match, since $N(B) = N(B')$. The arrows in Figure 5.2 show one way of matching them.

But now (Figure 5.3) we can use the arrows for A and A' together with the arrows for B and B' to show a matching for $A \cup B$ and $A' \cup B'$.

Obviously, if we had other numbers for a and b, we could still use the same trick of using the arrows which show a matching of A and A' together with the arrows which show a matching of B and B' to show a matching of $A \cup B$ with $A' \cup B'$. Since these unions match, $N(A \cup B) = N(A' \cup B')$. We get the same result for $a + b$ no matter which sets we choose for A and B as long as A and B are disjoint and $N(A) = a$ and $N(B) = b$.

It is important to remember that addition is an operation on two numbers, while joining is an operation on two sets. We put together or join two sets to form a third set, while we add two numbers to get a third number.

PROPERTIES OF ADDITION

Since addition of two numbers is defined in terms of the joining of two sets, we may determine the properties of addition by considering the properties of joining sets. The first important property of addition is simply that we can always do it. That is, if we add two whole numbers, we always get a whole number. Technically, we say that the set of whole numbers is *closed* under the operation of addition.

What other properties of addition can we get from our definition that $a + b$ is the number of members in $A \cup B$? To find $b + a$ we have to consider the number of members of $B \cup A$. We already know that $B \cup A$ is precisely the same set as $A \cup B$. Therefore it is true that for any whole numbers a and b,

$$a + b = b + a$$

We name this property the *commutative property of addition*, or we say "addition is commutative."

If addition is an operation on two numbers, how can we add three numbers a, b, and c? We add two of them together and then to the result we add the third number. Is it important which two we add first? Going back to the definition of addition in terms of the union of sets, we find that $(a + b) + c$ is the number of members in the set $(A \cup B) \cup C$, while $a + (b + c)$ is the number of members in the set $A \cup (B \cup C)$. Since these two sets are identical, the number we get either way is the same:

$(4 + 6) + 7 = 10 + 7 = 17$ but also $4 + (6 + 7) = 4 + 13 = 17$

This property is called the *associative property of addition,* or we say, "addition is associative."

That is, for any three numbers a, b, and c,

$$(a + b) + c = a + (b + c)$$

■ **Question 5** Which properties of addition are illustrated by the following?

1 $18 + 11 = 11 + 18$
2 $8 + (7 + 3) = (8 + 7) + 3$
3 $9 + (8 + 7) = 8 + (7 + 9)$

Repeated use of the associative and commutative properties enables us to group numbers for addition in any order we wish.

There is one whole number which plays a special role with respect to addition and that is the number zero. If we join the empty set to any set A, we still have set A. Therefore we see that $0 + a = a + 0 = a$. Since the addition of zero to any number leaves that number identical, we say that zero is the *identity element* with respect to addition. Zero is the only number with this interesting property.

ADDITION ON THE NUMBER LINE

Another useful way of thinking about addition is to consider it with respect to the representation of numbers on the number line. Draw a number line with the point 1 lying to the right of the point 0 (see Figure 5.4).

Figure 5.4

To add the number 3 to the number 4 start from 0, move 4 units to the right to the number 4, then move 3 more units to the right from the number 4. We stop at 7, so $4 + 3 = 7$. Although we are using the specific numbers 3 and 4 for our example, this process will work for any numbers a and b and illustrates the closure property (see Figure 5.5).

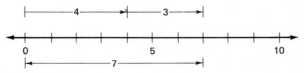

Figure 5.5

The commutative property may be illustrated on the number line (see Figure 5.6).

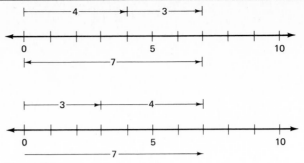

Figure 5.6

The associative property can also be illustrated on the number line, though the process is a bit more involved. Figure 5.7a shows $(3 + 5) + 4$ by first showing $(3 + 5)$, then taking the result and adding 4 to it. Figure 5.7b, on the other hand, shows $3 + (5 + 4)$ by first showing $(5 + 4)$, then taking that result and adding 3 to it. The broken-line segments show the result of $(5 + 4)$ being moved down to be added to 3 (see Figure 5.7).

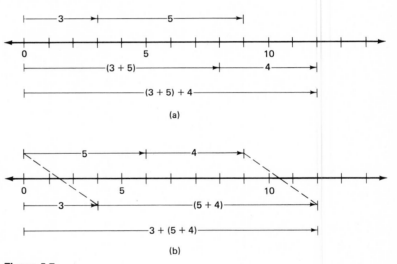

(a)

(b)

Figure 5.7

■ **Question 6** Illustrate $(1 + 0) + 2 = 1 + (0 + 2)$ on the number line.

AN ADDITION TABLE The reader should now obtain 18 conveniently small objects, such as matches, wooden blocks, jelly beans, or pennies, and use appropriate sets of these to work out the sum $a + b$ for each pair of whole numbers a and b, both of which are no greater than 9. The results (which you knew before you started, because you learned them in grade school and never forgot them) can be arranged as in Table 5.1.

To use this table to find the sum of two numbers, a and b, first find the row labeled, at the left-hand end, with the number a. Then move

Table 5.1

+	0	1	2	3	4	5	6	7	8	9
0	0	1	2	3	4	5	6	7	8	9
1	1	2	3	4	5	6	7	8	9	10
2	2	3	4	5	6	7	8	9	10	11
3	3	4	5	6	7	8	9	10	11	12
4	4	5	6	7	8	9	10	11	12	13
5	5	6	7	8	9	10	11	12	13	14
6	6	7	8	9	10	11	12	13	14	15
7	7	8	9	10	11	12	13	14	15	16
8	8	9	10	11	12	13	14	15	16	17
9	9	10	11	12	13	14	15	16	17	18

along this row to the right until you come to the column headed with the number b. The number in the a row and the b column is the number $a + b$.

The numbers in the top row illustrate that $0 + a = a$. Those in the left-hand column illustrate that $a + 0 = a$. The commutative property is illustrated by the fact that the same number appears in the a row and b column as in the b row and a column.

Of course, by working with more objects, it would be possible to work out more sums and to expand the table to more rows and more columns. In fact, you can extend the addition table as far as you want, say to 1,000 or even 1 million rows and columns, if only you are willing to devote enough time to the task. However, very few people would be able to put up with the boredom of extending the addition table to 1,000 rows and columns. (This would allow us to read off the sum of any two whole numbers neither of which is greater than 999.) Therefore we should ask if there are easier ways for working out sums.

■ **Question 7** Use the table to find $[(1 + 2) + 3] + 4$.

AN ADDING MACHINE One way would be to construct a simple adding machine. This consists of a V-shaped trough, running right and left on the table in front of us, with a number line at the top of the far side. The trough should be closed at the 0 point and should be slightly raised at the right end so that a marble put in the trough will roll as far as it can to the left (see Figure 5.8).

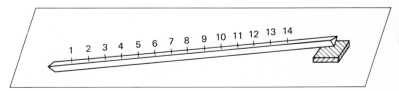

Figure 5.8

The number line should be drawn so that the marks are, say, 1 inch apart. Finally, we need a supply of marbles 1 inch in diameter.

Now suppose we want to add the whole numbers a and b. We first drop marbles in the trough, one at a time, until the last one just comes up to the a mark on the number line. Then there are a marbles in the trough. Pour them into an empty container. Next, drop more marbles into the trough until there are b of them there.

Finally, drop the set of a marbles in, to the right of the set of b marbles already there, one at a time, until they are all in. The mark on the number line at the right-hand side of the marble farthest to the right is the number $a + b$.

Such an adding machine will always work. However, it too has drawbacks. If you wanted to be sure, for example, that you could always add any pair of whole numbers no greater than 1 million, you would need a trough over 30 miles long.

Fortunately, thanks to our place-value system of numeration, there is an easy way of finding the sum of any two whole numbers, no matter how large they are. We will study this in the next chapter.

■ **Question 8** If you stretched along the equator a trough adding machine that could add any two numbers less than 1 billion, how far would it go?

SUBSETS

Another important and useful operation on numbers is called *subtraction*. It is closely related to the addition operation and, in particular, the concept of subtraction is derived, as is addition, from set operations. First we need two more set concepts.

We say that a set B is a *subset* of a set A if every member of B is also a member of A. Thus, for example, the set of all U.S. states which touch the Atlantic Ocean is a subset of all the U.S. states which are east of the Mississippi River. Similarly, the set of all even whole numbers is a subset of the set of all whole numbers.

No matter what set A you may think of, you will have to agree that every member of A is indeed a member of A, and so we have to agree that any set A is a subset of itself.

We might as well agree also that the empty set is a subset of any set whatsoever. If B is some set and if you want to convince someone that the empty set is *not* a subset of B, then you have to point to a member of the empty set which is not a member of B. That is hard to do.

The standard symbolism to indicate that B is a subset of A is:

$$B \subset A$$

Now suppose that we have a set A and one of its subsets, B, as in Figure 5.9. Then another subset of A is automatically created: the set of all members of A which are *not* members of B. We call this subset of A the *complement* of B in A (sometimes it is called the *remainder* set of B) and we use the symbol $A \sim B$ to stand for this new subset of A (see Figure 5.9).

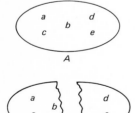

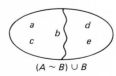

Figure 5.9

We notice that every member of A is either in B or is not in B, so

$$B \cup (A \sim B) = A$$

Also, nothing can be in B and at the same time not in B, so
$$B \cap (A \sim B) = \{\quad\}$$

■ **Question 9** If B is a subset of A, what is $A \cap B$ and what is $A \cup B$?

SUBTRACTION

Now we are ready to use these two ideas of subset and complement to define the operation of subtraction. Suppose then that a is a particular whole number and that b is another whole number, with $b \le a$. Choose any set A for which $N(A) = a$. Then choose a subset B of A for which $N(B) = b$. Now we define the number $a - b$, the result of subtracting b from a, by

$$a - b = N(A \sim B)$$

For example, if $a = 5$ and $b = 2$, we can choose A to be the set

$$A = \{\bigcirc, \wedge, \square, \star, \mathscr{E}\}$$

Next we can choose B to be the subset

$$B = \{\wedge, \star\}$$

Then

$$A \sim B = \{\bigcirc, \square, \mathscr{E}\}$$

Now our definition tells us that

$$5 - 2 = N(A \sim B) = 3$$

Note that if we made a different choice for B, for example,

$$B = \{\square, \mathscr{E}\}$$

the result would be the same. Also, if we had chosen a different set A, for example $A = \{V, W, X, Y, Z\}$, and any two-member subset of this set as B, the result would still be the same.

■ **Question 10** If $A \cup B = A$, must B be a subset of A?

Before we go on, let us ask how we know that we can find a subset B of A with $N(B) = b$. That turns out to be easy. Go back to the counting set we discussed in Chapter 2:

$$N = \{1, 2, 3, \ldots\}$$

Both a and b are whole numbers, so they both appear somewhere in this set. In the case where $b < a$, b appears to the left of a:

$$N = \{1, 2, 3, \ldots, b, \ldots, a, \ldots\}$$

Now since $N(A) = a$, A matches what is left of N when we chop off everything to the right of a:

A matches $\{1, 2, 3, \ldots, b, \ldots, a\}$

The pairing of A and $\{1, 2, 3, \ldots, b, \ldots, a\}$ automatically pairs the members of $\{1, 2, 3, \ldots, b\}$ with some of the members of A, and these form a subset B of A. But $N(B) = N(\{1, 2, 3, \ldots, b\}) = b$, so we have found the subset of A we were looking for.

Figure 5.10 is an illustration of this. A is the set of shapes. By pairing the elements of A with elements of N, we find that $N(A) = 8$. Now suppose that $b = 4$. The shapes paired with 1, 2, 3, and 4 then form a subset B of A with $N(B) = 4$.

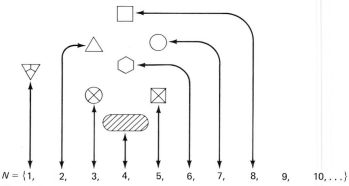

$N = \{1, \quad 2, \quad 3, \quad 4, \quad 5, \quad 6, \quad 7, \quad 8, \quad 9, \quad 10, \ldots\}$

Figure 5.10

Of course, if $b = a$, the situation is much simpler since we can choose B to be the same as A.

PROPERTIES OF SUBTRACTION

Now let us observe some properties of the operation of subtraction. First, suppose that we have two disjoint sets A and B, with $N(A) = a$ and $N(B) = b$. Form their union, $A \cup B$. Obviously, B is a subset of $A \cup B$. Therefore we can consider the complement of B in $A \cup B$. But this is just A itself.

Now, $N(A \cup B) = a + b$, by the definition of addition of whole numbers. Also, $N((A \cup B) \sim B) = (a + b) - b$, by the definition of subtraction. Finally, $(A \cup B) \sim B = A$, so

$$N((A \cup B) \sim B) = N(A) = a$$

Putting this all together, we see that

$$(a + b) - b = a$$

■ **Question 11** Could you convince a skeptic that $(30{,}000 + 2) - 2 = 30{,}000$?

Next, suppose that B is a subset of another set A, with $N(B) = b < N(A) = a$. We have already seen that $B \cup (A \sim B) = A$, so

$$N(B \cup (A \sim B)) = N(A) = a$$

But B and $(A \sim B)$ are disjoint, so the definition of addition tells us that

$$N(B \cup (A \sim B)) = N(B) + N(A \sim B)$$

$N(B) = b$ and the definition of subtraction tells us that $N(A \sim B) = a - b$. Putting all this together gives us:

$$b + (a - b) = a$$

Finally, since addition is commutative,

$$(a - b) + b = a$$

These two results tell us that if we start with a number a and first add and then subtract a number b, or first subtract and then add the number b, we get back to the number that we started from. In this sense, subtraction undoes addition, and vice versa, so we often say that these two operations are *inverse* to each other.

Let us take another look at the result:

$$b + (a - b) = a$$

This tells us that if we have two numbers a and b, with $b \leq a$, then the number $a - b$ is just that number that, when added to b, gives us a. We will see in the next chapter how we can use this idea to find the answers to complicated subtraction problems.

Going back to addition, when we have three numbers a, b, and c, and when

$$b + c = a$$

we often say that b and c are *addends* and that a is the *sum*. In an addition problem, we are told what the addends are and we are asked to find the sum.

Now suppose we are given a and b and are asked to find $a - b$. The result

$$b + (a - b) = a$$

tells us that we have been given one addend, b, and the sum, a, and have been asked to find the other addend, $a - b$. For this reason, we can say that the operation of subtraction is the same as finding the missing addend in an addition problem.

■ **Question 12** What is the missing addend (indicated by the box □) in each of these problems?

1 $5 + \square = 8$
2 $\square + 10 = 20$
3 $7 + \square = 107$

There are two more properties of subtraction that are easy to see. It is easy to see that

$$A \sim \{ \ \} = A$$

and so

$$a - 0 = a$$

no matter what number a we start with.

Also, it is easy to see that, for any set A,

$$A \sim A = \{\ \ \}$$

Consequently, for any number a, we have

$$a - a = 0$$

We noticed above that addition is commutative:

$$a + b = b + a$$

What about subtraction? The answer is easy to see. Subtraction is *not* commutative. We can subtract b from a, if $b < a$. But then the definition of subtraction does not apply to the case $b - a$, and so we cannot write

$$b - a = a - b$$

because $b - a$ has no meaning.

Neither is subtraction associative. For example,

$$(13 - 5) - 2 = 8 - 2 = 6$$

but

$$13 - (5 - 2) = 13 - 3 = 10$$

SUBTRACTION ON THE NUMBER LINE

If we consider subtraction with respect to the representation of numbers on the number line, we can illustrate many of its important properties.

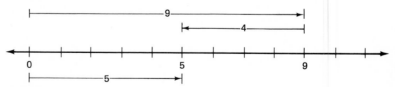

Figure 5.11

What is the answer to $9 - 4$? We start on the number line at 9 and take away, or move to the left, 4 units thus arriving at 5 which is our answer (see Figure 5.11). We know that $9 - 4$ is the number n for which $4 + n = 9$. This concept can be shown on the number line as follows where we see that $n = 5$ by bringing back the arrow representing n so that it starts at 0 (see Figure 5.12).

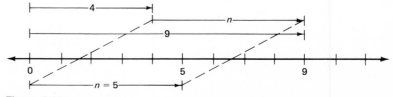

Figure 5.12

We illustrated in Figure 5.7 the use of the number line to show the associative property of addition. We have also seen that subtraction does not have the associative property. In Figures 5.13 and 5.14 these problems are worked out on number lines. Figure 5.13a shows that $13 - 5 = 8$ and this result is used in Figure 5.13b to get the answer 6.

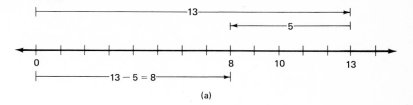

(a)

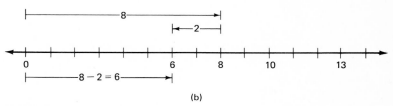

(b)

Figure 5.13

Similarly Figure 5.14a shows that $5 - 2 = 3$ and this result is used in 5.14b to get the answer 10.

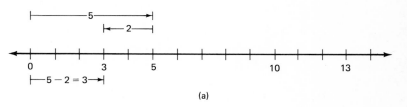

(a)

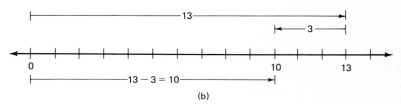

(b)

Figure 5.14

■ **Question 13** Use a number line to illustrate that

$$(9 - 6) - 3 \neq 9 - (6 - 3).$$

A SUBTRACTION TABLE Earlier in this chapter we constructed an addition table. We could use the same set of 18 small objects to work out the subtraction table shown in Table 5.2.

Table 5.2

−	0	1	2	3	4	5	6	7	8	9
0	0	×	×	×	×	×	×	×	×	×
1	1	0	×	×	×	×	×	×	×	×
2	2	1	0	×	×	×	×	×	×	×
3	3	2	1	0	×	×	×	×	×	×
4	4	3	2	1	0	×	×	×	×	×
5	5	4	3	2	1	0	×	×	×	×
6	6	5	4	3	2	1	0	×	×	×
7	7	6	5	4	3	2	1	0	×	×
8	8	7	6	5	4	3	2	1	0	×
9	9	8	7	6	5	4	3	2	1	0
10		9	8	7	6	5	4	3	2	1
11			9	8	7	6	5	4	3	2
12				9	8	7	6	5	4	3
13					9	8	7	6	5	4
14						9	8	7	6	5
15							9	8	7	6
16								9	8	7
17									9	8
18										9

An easier way to construct this table would be to go back to Table 5.1. For example, if we wanted to find the number for the row labeled 13 and the column labeled 6, we could remember that $13 - 6$ is the number which, when added to 6, gives 13. If we look along the row labeled 6 in Table 5.1, we find that 13 appears in the column labeled 7, so 7 goes in the 13 row, 6 column of Table 5.2.

We can use this table to look up $a - b$ when someone gives us a and b. We go to the row labeled, at the left-hand end, with the number a. Next, we go to the right along this row until we reach the column which has the number b for the label at the top. Then the number in the a row and the b column is the number $a - b$.

The cells in the upper right-hand corner have been ×ed out since there are no such numbers. No whole number can be the result of subtracting a whole number from a smaller whole number.

The cells in the lower left-hand corner could easily be filled in, but we really do not need to do so. We will see in the next chapter that as long as we know the results in this table we can work out any subtraction problem, no matter how large and complicated.

The adding machine, consisting of a long trough with a number line on one rim, can also be used as a subtraction machine. If we want to use it to find the value of $a - b$, we first put a marbles in the trough and let them roll as far as possible to the left. Next we insert a thin metal wedge between the two marbles which touch each other at the b mark on the number line. Then, of course, there are b marbles to the left of the wedge. We use the wedge to hold back the marbles to the right while we take out all the marbles to the left. Then we remove the wedge, let all the marbles roll to the left, and the number at the right of the last marble will be $a - b$.

■ **Question 14** Use the table to find $[(15 - 6) - 1] - 5$.

SOME "HOW MANY" QUESTIONS

We have claimed that the set way of looking at the world, the idea of the number property of a set, and the operations of addition and subtraction all help us to find the answers to "how many" questions. To illustrate, here are four simple problems that are typical of what might be found in an elementary school arithmetic text.

1 Algernon always eats 4 scones at tea and Cuthbert always eats 3. How many scones should the maid put on the tea table so each of the two men will have just enough?

2 Yesterday there were 23 lettuce plants in my garden. Last night rabbits ate 17 of them. How many lettuce plants were left in my garden this morning?

3 Lower Slobovia has 7 extinct volcanoes. The state of Washington has 4. How many more extinct volcanoes does Lower Slobovia have than Washington?

4 Douglas has 5 hats. Eddie has only 2. How many more hats does Eddie have to get to have as many as Douglas?

Problem 1 immediately suggests addition. Let A be Algernon's set of scones and let B be Cuthbert's. Obviously A and B are disjoint. What the maid needs to bring to the tea table is $A \cup B$, and the number property of this set is $4 + 3$. Figure 5.15 shows a diagram of the problem.

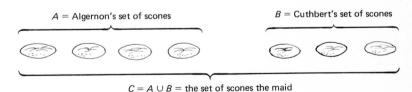

A = Algernon's set of scones B = Cuthbert's set of scones

$C = A \cup B$ = the set of scones the maid needs to put on the tea table

Figure 5.15

Problem 2 immediately suggests subtraction. Set A is the set of lettuce plants in my garden yesterday. Set B is the subset of plants eaten during the night. What is left this morning is the complement of B in A, and $N(A \sim B) = 23 - 17$. Figure 5.16 is an illustration of this problem.

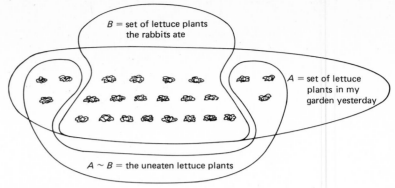

Figure 5.16

Problem 3 is a little different. It suggests subtraction, but the sets are not arranged properly. If A is the set of Lower Slobovian extinct volcanoes and B the set of Washington State's extinct volcanoes, then B is not a subset of A (in fact A and B are disjoint) and so we are not set up for subtraction (see Figure 5.17).

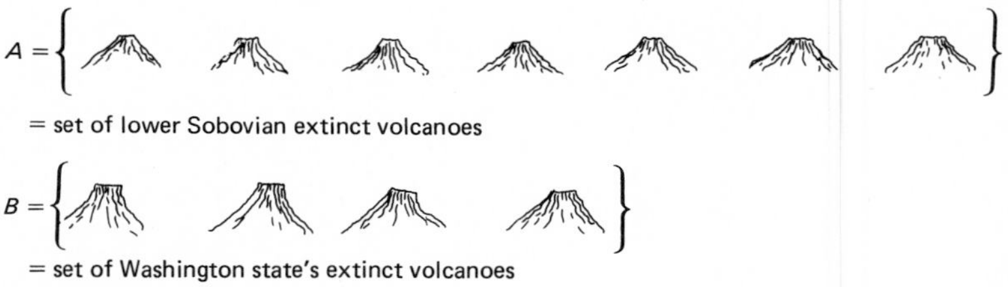

$A = \left\{ \text{ } \right\}$

= set of lower Sobovian extinct volcanoes

$B = \left\{ \text{ } \right\}$

= set of Washington state's extinct volcanoes

Figure 5.17

However, we can look at it this way. Let us start pairing the Washington extinct volcanoes with the Lower Slobovian ones. We know we will run out of the Washington ones first. The instant we do we will have specified a subset B^* of A which just matches B (see Figure 5.18).

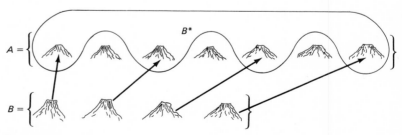

Figure 5.18

This means that

$$N(B^\star) = N(B) = 4$$

Now $A \sim B^*$ consists of those Lower Slobovian extinct volcanoes not paired with any of the Washington extinct volcanoes, so $N(A \sim B^*)$ tells us how many more Lower Slobovia has. But $N(A \sim B^*) = N(A) - N(B^*) = 7 - 4$.

Problem 4 also suggests subtraction but, as in Problem 3, is not in standard form. We can let A be the set of hats which Douglas has and B the set of hats which Eddie has (see Figure 5.19). A and B are disjoint, while to have the arrangement for subtraction we should have B a subset of A.

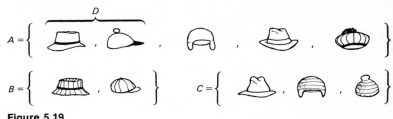

Figure 5.19

However, think of it this way. Eddie is going to get another set of hats, C, so that $B \cup C$ will match A. We want to know $N(C)$. We can match Eddie's original set of hats, B, with a subset D of Doug's set, as in Figure 5.19. If the additional set of hats, C, that Eddie is going to obtain will bring him up even with Doug, then C will have to match $A \sim D$.

Now we know that $N(A \sim D) = N(A) - N(D) = 5 - N(D) = 5 - N(B) = 5 - 2 = 3$. So $N(C) = 3$, and we have found that Eddie has to find 3 more hats to match Doug's set.

These are, of course, rather simple problems, but they do illustrate the various ways in which addition and subtraction can be used to answer "how many" questions.

■ **Question 15** Would you use addition or subtraction to answer the question "How many more days until Christmas?"

PROBLEMS

1 For each pair of sets given below, find $A \cup B$ and $A \cap B$.
   (a) $A = \{a, b, c, d, e\}$, $B = \{c, e, f\}$
   (b) $A = \{c, e, f\}$, $B = \{a, b, c, d, e\}$
   (c) $A = \{a, b, c\}$, $B = \{a, b, c\}$
   (d) $A = \{a, b, c\}$, $B = \{d, e\}$
2 For each of the following problems, state whether C represents the union or the intersection.
   (a) $A = \{1, 2, 3, 4\}$, $B = \{2, 4, 5, 9\}$, $C = \{2, 4\}$
   (b) $A = \{1, 2, 3\}$, $B = \{4, 5, 6, 7\}$, $C = \{1, 2, 3, 4, 5, 6, 7\}$
   (c) $A = \{1, 2, 3\}$, $B = \{4, 5, 6, 7\}$, $C = \{\ \ \}$
   (d) $A = \{1, 2, 3, 4, 5\}$, $B = \{1, 2, 3\}$, $C = \{1, 2, 3\}$
   (e) $A = \{1, 2, 3, 4\}$, $B = \{1, 2, 3, 4\}$, $C = \{1, 2, 3, 4\}$
   (f) $A = \{1, 2, 3, 4\}$, $B = \{\ \ \}$, $C = \{\ \ \}$
   (g) $A = \{1, 2, 3, 4\}$, $B = \{\ \ \}$, $C = \{1, 2, 3, 4\}$

(h) $A = \{$stockholders of Donner Imports, Inc.$\}$
$B = \{$stockholders of Blitzen Exports, Ltd.$\}$
$C = \{$stockholders of both corporations$\}$

3   If A is a set, what is $A \cup A$?

4   Let $A = \{2, 4, 6, 8, 10\}$, $B = \{3, 6, 9\}$, $C = \{3, 5, 7, 9\}$.
(a)   Write the set $(A \cap B) \cup (A \cap C)$. Is this set equal to $A \cap (B \cup C)$?
(b)   Write the set $(A \cup B) \cap (A \cup C)$. Is this set equal to $A \cup (B \cap C)$?

5   Let $n(A)$ denote the number of elements in set A. For example, if $A = \{a, b, c\}$, then $n(A) = 3$. Which of the following statements is always true?
(a) $n(A) + n(B) = n(A \cup B)$ (b) $n(A) + n(B) < n(A \cup B)$
(c) $n(A) + n(B) > n(A \cup B)$ (d) $n(A) + n(B) \leq n(A \cup B)$
(e) $n(A) + n(B) \geq n(A \cup B)$

6   Give an example of two sets, A and B, containing 3 and 4 elements, respectively, whose union contains 5 elements. How many elements are in $A \cap B$?

7   Using the addition table, verify the associative property for:
(a) $2 + 3 + 4$ (b) $3 + 5 + 2$ (c) $8 + 1 + 7$
Why can you not similarly verify this principle for $3 + 4 + 7$?

8   Make up a table for the basic addition facts, base 3. (Three rows and three columns.)

9   What properties of addition of whole numbers are illustrated by each of the following statements?
(a) $5 + 7 = 7 + 5$ (b) $3 + 0 = 3$
(c) $8 + (6 + 4) = (8 + 6) + 4$ (d) $8 + (9 + 7) = 8 + (7 + 9)$
(e) $0 + 18 = 18$ (f) $14 + (9 + 7) = (9 + 7) + 14$

10   Draw number lines to show the following addition examples.
(a) $3 + 6 = 9$ (b) $4 + 5 = 9$
(c) $(3 + 6) + 7 = 16$ (d) $3 + (6 + 7) = 16$

11   For each set in the left column, choose the sets from the right column which are subsets of it.
(a) $\{x, y, z\}$ (1) $\{\ \}$
(b)   The set of even numbers   (2) $\{2, 7\}$
(c)   The set of letters in the word "deer"   (3) $\{r, e, d\}$
(d)   The set of even numbers which are   (4) $\{0\}$
      odd numbers   (5) $\{2\}$

12   See Figure P5.12.

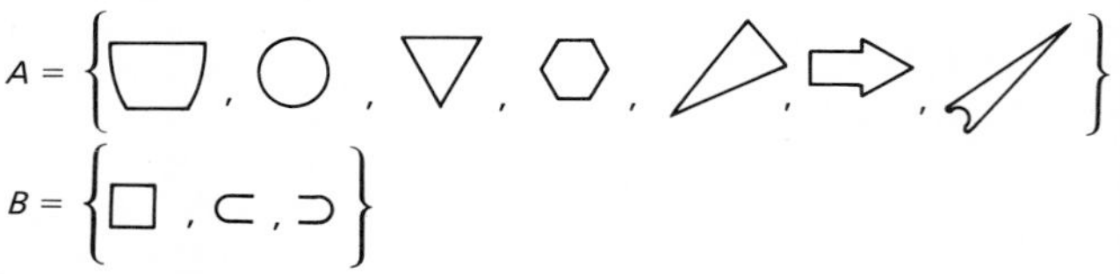

Figure P5.12

Join to B a set C, disjoint from B and A, such that $B \cup C$ matches A.

13

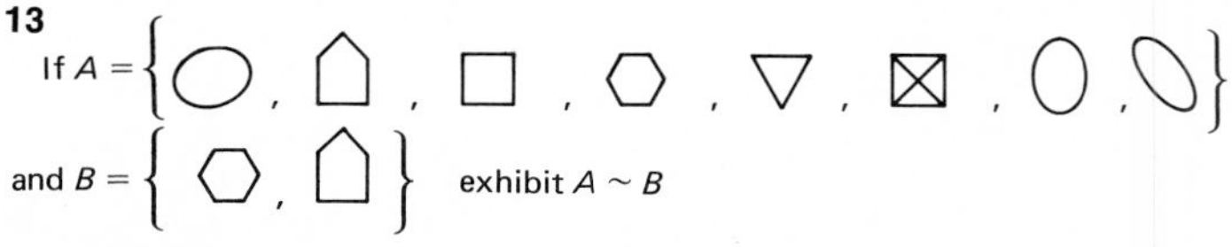

Figure P5.13

14 If $A = \{1, 2, 3\}$, $B = \{1, 2, 3, 4, 5, 6\}$, and $C = \{1, 2, 3, 4, 5, 6, 7\}$:
    (a) What is $C \sim A$?   (b) What is $C \sim B$?   (c) What is $B \sim A$?

15 If A is a subset of the empty set, what do you know about A?

16 (a) What does $A \sim (A \sim B)$ equal if $B \subset A$?
    (b) Simplify $A \sim [A \sim (A \sim B)]$.

17 Given the set B and subsets C, F, and G,

$$B = \{1, 2, 3, 4, 5, 6, 7, 8\} \qquad C = \{2, 6, 7, 8\}$$
$$F = \{1, 2, 4, 6\} \qquad\qquad G = \{1, 2, 3, 5, 8\}$$

List the elements of each of the following sets.
    (a) $C \cap F$ (b) $B \sim C$ (c) $B \sim (G \cap F)$
    (d) $B \cup C$ (e) $(B \sim F) \cap (B \sim G)$ (f) $(C \cup F) \cap G$
    (g) $C \cup (B \sim C)$

18 What operation is the inverse of adding 7 to any number? What is the inverse of subtracting 8?

19 Show a representation on the number line which illustrates the fact that $10 - 3 = 7$. Use the same figure to illustrate the idea that $10 = 7 + 3$.

20 Tell what number x is, given that:
    (a) $67 + x = 67$ (b) $92 - x = 0$ (c) $74 - x = 74$
    (d) $x - 43 = 0$ (e) $x - 0 = 99$ (f) $87 + (x - 12) = 87$
    (g) $96 - (x + 96) = 0$ (h) $x + x = 0$

6. TECHNIQUES OF ADDITION AND SUBTRACTION

In the preceding chapter we used sets to develop the ideas of addition and subtraction, and to develop the properties of these two operations. For example, if A is a set with 5 members and B is a disjoint set with 3 members, then we can count the members of $A \cup B$ and find that $5 + 3$ is 8. Once we know that $5 + 3 = 8$, then we can see that $8 - 5 = 3$.

But we also saw that we would have to spend a lot of time finding, for example, $928 + 487$ or $765 - 123$ if we had to rely just on the definitions of addition and subtraction. Now we are going to see how we can work these larger problems more quickly. In fact, you will see that all it takes to work any addition or subtraction problem with whole numbers, no matter how complicated, is

1 an understanding of our decimal place-value numeration system.
2 the addition table on page 53 and the subtraction table on page 60.

THE ADDITION PROCESS Let us start with a simple example: $42 + 37$. Figure 6.1 is a diagram of these two numbers. If we let A be the set of marks representing 42 and B the set of marks representing 37, then $42 + 37$ will be the number of members of the set $A \cup B$.

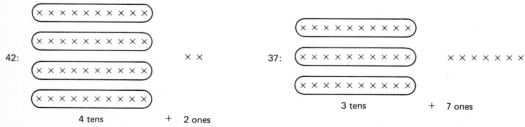

Figure 6.1

It is easy to join A and B to get the set shown in Figure 6.2.

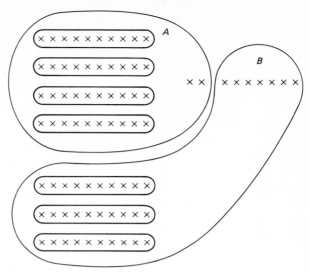

Figure 6.2

Now we can count the ones and we find there are 9 of them. Next, we can count the piles of ten and we see there are 7 of them. Therefore

$$42 + 37 = 79$$

If we were too lazy to count the ones, we could have looked up $2 + 7$ in our addition table, and we could also have looked up $4 + 3$.

■ **Question 1** Use sets of marks like those above to add $31 + 22 + 13$.

EXPANDED NOTATION AND ADDITION

Another way of looking at this problem skips the pictures of the sets A and B, and goes straight to the expanded notation for the two numbers.

$42 = 4$ tens $+ 2$ ones
$37 = 3$ tens $+ 7$ ones

Now, we add the tens and get 7 tens. Then we add the ones and get 9 ones.

$$
\begin{aligned}
42 &= 4 \text{ tens} & + 2 \text{ ones} \\
37 &= 3 \text{ tens} & + 7 \text{ ones} \\
\hline
42 + 37 &= (4 + 3) \text{ tens} + (2 + 7) \text{ ones} \\
&= 7 \text{ tens} & + 9 \text{ ones}
\end{aligned}
$$

"CARRYING"

Let us now add 27 and 35. This time we have (2 tens $+ 7$ ones) $+$ (3 tens $+ 5$ ones) which may be illustrated as in Figure 6.3.

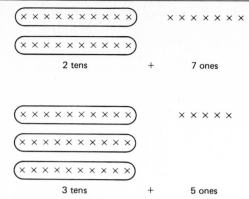

Figure 6.3

We put these groups together in Figure 6.4.

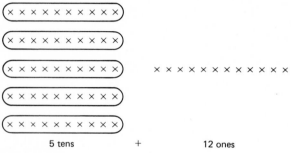

Figure 6.4

In Figure 6.5, we *regroup* the 12 ones and get another set of 1 ten and 2 ones.

1 ten   +   2 ones

Figure 6.5

Then, in Figure 6.6, we add 5 tens + 1 ten + 2 ones.

An older term for "regroup" was "carry." Instead of saying "We write down the 2 and carry the 1," we now say "We regroup the 12 into 1 ten and 2 ones."

Doing the same problem in terms of expanded notation instead of sets, we would write

$$27 = 2 \text{ tens} + 7 \text{ ones}$$
$$35 = 3 \text{ tens} + 5 \text{ ones}$$

$$27 + 35 = 5 \text{ tens} + 12 \text{ ones}$$
$$= 5 \text{ tens} + (1 \text{ ten} + 2 \text{ ones})$$
$$= 6 \text{ tens} + 2 \text{ ones}$$
$$= 62$$

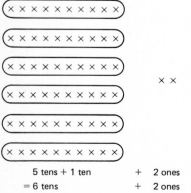

5 tens + 1 ten   +   2 ones
= 6 tens   +   2 ones
= 62

Figure 6.6

Notice that going from

5 tens + (1 ten + 2 ones)

to

6 tens + 2 ones

we used the associative property of addition.

■ **Question 2** Use expanded notation to add 19 + 28 + 37.

We worked each of these problems in two ways. For the first way, we put as many marks on the paper as we needed to represent each of the two numbers. For the second way, we made use of the expanded form of our decimal place-value way of representing numbers. Obviously, the second way is much quicker.

In our next example, we will not even bother with the set representation of the two numbers, but will go straight to the expanded notation for them.

When we add 68 and 57 we get 11 tens and 15 ones which we rewrite as follows:

$$68 = 6 \text{ tens} + 8 \text{ ones}$$
$$57 = 5 \text{ tens} + 7 \text{ ones}$$
$$11 \text{ tens} + 15 \text{ ones} = (1 \text{ hundred} + 1 \text{ ten}) + (1 \text{ ten} + 5 \text{ ones})$$
$$= 1 \text{ hundred} + 2 \text{ tens} + 5 \text{ ones}$$
$$= 125$$

In another way of working this problem, we switch from vertical to horizontal form. We write

$$68 + 57 = (60 + 8) + (50 + 7)$$
$$= (60 + 50) + (8 + 7) \quad \text{use of associative property and commutative property}$$
$$= 110 + 15$$
$$= (100 + 10) + (10 + 5)$$
$$= 100 + (10 + 10) + 5 \quad \text{use of associative property}$$
$$= 100 + 20 + 5$$
$$= 125$$

Thus, for example, whenever we add 7 groups of one kind to 5 groups of the same kind, we get 12 groups which we write as 1 group of the next kind plus 2 groups. It is this 1 which we can think of as "carrying" over to the next group.

■ **Question 3** In this problem

```
  54
+86
  10
  13
 140
```

What does the 13 stand for?

ADDING LARGER NUMBERS

Precisely the same process is used in adding three or more numbers. Once again the associative and commutative properties of addition are important. Thus, $563 + 787 + 1,384$ can be thought of as follows:

$$563 = \qquad 500 + 60 + 3 = \qquad (5 \times 100) + (6 \times 10) + (3 \times 1)$$
$$787 = \qquad 700 + 80 + 7 = \qquad (7 \times 100) + (8 \times 10) + (7 \times 1)$$
$$1,384 = 1,000 + 300 + 80 + 4 = (1 \times 1,000) + (3 \times 100) + (8 \times 10) + (4 \times 1)$$

and the sum

$$563 + 787 + 1,384$$

$$= (1 \times 1,000) + (15 \times 100) + (22 \times 10) + (14 \times 1)$$
$$= (1 \times 1,000) + [(1 \times 1,000) + (5 \times 100)] +$$
$$\qquad\qquad\qquad [(2 \times 100) + (2 \times 10)] + [(1 \times 10) + (4 \times 1)]$$
$$= [(1 \times 1,000) + (1 \times 1,000)] + [(5 \times 100) + (2 \times 100)] +$$
$$\qquad\qquad\qquad\qquad [(2 \times 10) + (1 \times 10)] + (4 \times 1)$$
$$= (2 \times 1,000) + (7 \times 100) + (3 \times 10) + (4 \times 1)$$
$$= 2,000 + 700 + 30 + 4$$
$$= 2,734$$

ABBREVIATING THE PROCESS

After a certain amount of experience, it becomes clear that we can abbreviate the process. Let us go back to our first example, $42 + 37$. Skipping over the set representation of the two numbers to the expanded notation form, we had

$$
\begin{array}{rll}
42 = & 4 \text{ tens} & + 2 \text{ ones} \\
37 = & 3 \text{ tens} & + 7 \text{ ones} \\
\hline
42 + 37 = & (4 + 3) \text{ tens} & + (2 + 7) \text{ ones} \\
= & 7 \text{ tens} & + 9 \text{ ones}
\end{array}
$$

First, we can abbreviate this to

$$
\begin{array}{rll}
42 = & 40 & + 2 \\
37 = & 30 & + 7 \\
\hline
42 + 37 = & (40 + 30) & + (2 + 7) \\
= & 70 & + 9 \\
= & 79 &
\end{array}
$$

We can abbreviate this even more and write

$$
\begin{array}{l}
42 \\
\underline{37} \\
9 \qquad \text{sum of ones} \\
\underline{70} \qquad \text{sum of tens} \\
79
\end{array}
$$

In the same way, the problem $563 + 787 + 1,384$ that we worked out above can be abbreviated to

$$
\begin{array}{rrrr}
500 + & 60 + & 3 \\
700 + & 80 + & 7 \\
1,000 + & 300 + & 80 + & 4 \\
\hline
1,000 + & 1,500 + & 220 + & 14
\end{array}
$$

This can be written with partial sums indicated as

```
   563
   787
 1,384
    14     sums of ones
   220     sums of tens
 1,500     sums of hundreds
 1,000     sums of thousands
 2,734
```

and the operation is still further abbreviated to

```
①②①
   563
   787
 1,384
 2,734
```

Finally, by omitting even the carry-over numerals we get

```
   563
   787
 1,384
 2,734
```

■ **Question 4** Is there anything wrong with adding the tens before adding the ones? For example:

```
  14
  53
  76
  13
  13
 143
```

It is important to realize that each of these versions of the addition process is just as valid as any of the others. If we wanted, we could always do addition problems by means of the marble machine described in the preceding chapter or by making sets of marks on the page as we did in the first example in this chapter. If we use either of these processes, and if we are not careless, we will get the right answer even though it may take us a long time.

If we use the expanded notation method, we spend a lot less time counting out marbles or marks on the paper. However, we do have to remember how our decimal place-value system works and, in particular, how we regroup whenever we have ten or more in a set. In other words, if we are going to use expanded notation instead of counting objects, we have to understand decimal place value well enough to "carry" correctly.

If we want to abbreviate even more, as we did in the last example above, then we are trading off a complete written account of the problem against the mental arithmetic involved in remembering to "write the 4 and carry the 1, write the 3 and carry the 2, write the 7 and carry the 1."

All these variations are equally valid, and the particular variation an individual wishes to use is a matter of personal preference.

ALGORITHMS

Another observation about these addition procedures is that each of them is an *algorithm*[1], a step-by-step procedure, which could be carried out by anyone and which is guaranteed to produce the desired result. Let us illustrate this by writing out in detail an algorithm for adding two numbers each of which has no more than two digits. Side by side, we will repeat the exercise of finding 27 + 35.

STEP 1:
Use the addition table to find the sum of the two right-hand digits. $7 + 5 = 12$

STEP 2:
Write down the right-hand digit of the sum.   2

STEP 3:
(a)   If the result of Step 1 was no more than 9, use the addition table to find the sum of the two left-hand digits. Write this sum to the left of the result of Step 2.
(b)   If the result of Step 1 was 10 or more, use the addition table to find the sum of 1 and the left-hand digit of the first number. $1 + 2 = 3$
Then use the table again to find the sum of the above digit and the left-hand digit of the second number. Write this last $3 + 3 = 6$
result to the left of the result of Step 2.   62

Finis

By now, if you really understand how our Hindu-Arabic numeration system works, you could write out similar directions for adding larger numbers, but as long as you understand what is going on it is not necessary to do so.

A final observation about these algorithmic procedures is that they are not enough. Even though an individual might memorize these algorithms and be as efficient at adding as any computing machine, what good will it do if he does not know whether adding is the right thing to do in a particular problem? Knowing which arithmetic operation to apply is just as important as being able to do the operation.

■ **Question 5** Write an algorithm for adding three single-digit numbers.

THE SUBTRACTION PROCESS

Let us now turn to the process of subtraction, and let us start with the example 49 − 17. Examine this problem from the point of view of the

[1] After the ninth century Arabian mathematician Al-Kworesmi who wrote the first book on arithmetic algorithms.

definition of subtraction in terms of complements or remainder sets. We can take for our set A a collection of 49 marks arranged as in Figure 6.7.

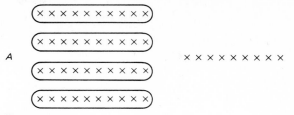

Figure 6.7

Now we need to pick a subset B of A which contains 17 members. Then the number of members of the remainder set $A \sim B$ will be $49 - 17$.

There are many ways to choose B. One of them is shown in Figure 6.8. But when we choose B this way, the remainder set $A \sim B$ is not easy to count. Some of the original bundles of ten have been broken up, and only pieces of them are in $A \sim B$.

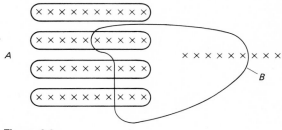

Figure 6.8

It is much better if we choose B so as to include either all of a bundle of ten or none of it. Figure 6.9 shows one way.

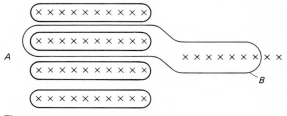

Figure 6.9

Now it is easy to count the remainder set $A \sim B$. It can be done in two steps. Looking at the right-hand side above, we see that the number of units in the remainder set is $9 - 7 = 2$. Looking at the left-hand side above, we see that the number of bundles of ten in the remainder set is $4 - 1 = 3$. Therefore the number of members in the remainder set is 32.

EXPANDED NOTATION AND SUBTRACTION

We could do the same problem using expanded notation

49 = 4 tens + 9 ones

17 = 1 ten   + 7 ones

Now, using our subtraction table if necessary, we can see that when we subtract 7 ones from 9 ones, we are left with 2 ones, and when we subtract 1 ten from 4 tens, we have 3 tens left, so

49 − 17 = 32

Figure 6.10

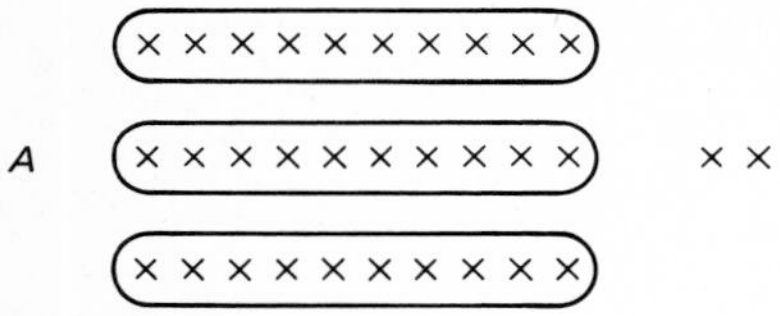

Now, let us examine in the same way another problem: 32 − 17. We can pick A to be a set of 32 marks (see Figure 6.10). We need to pick a subset B with 17 members, that is, 1 bundle of ten and 7 units. But A has only 2 units, so we will have to use some of the members of A in the bundles of ten. As we saw above, it is best if we use only whole bundles. Therefore, we will take 1 of the bundles of ten in A and put it in with the units (see Figure 6.11).

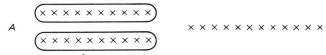

Figure 6.11

Now it is easy to see how we can pick a convenient subset B which has 17 members (Figure 6.12).

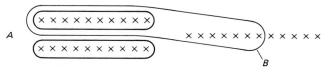

Figure 6.12

Now it is easy to count the remainder set $A \sim B$. The number of units is 12 − 7 = 5 and the number of tens is 2 − 1 = 1. Therefore 32 − 17 = 15.

Let us repeat this problem, 32 − 17, using expanded notation rather than sets.

32 = 3 tens + 2 ones

But we can rewrite 32 as

32 = 2 tens + 1 ten + 2 ones
   = 2 tens + 12 ones

next

17 = 1 ten + 7 ones

Now, using the subtraction table if necessary, we see that 32 − 17 is

1 ten + 5 ones = 15

A shortcut way of writing this is:

32
17
15

■ **Question 6** Use a set of marks to work this problem: $102 - 35$

Here is another example written out in two ways. The problem is to calculate $68 - 49$.

| *Horizontal form* | *Vertical form* |

STEP 1:
68 and 49 are written in expanded form:

$68 - 49 = (60 + 8) - (40 + 9)$

$68 = (60 + 8)$
$49 = (40 + 9)$

STEP 2:
68 is regrouped as $50 + 18$ because, looking ahead, we see the need for more ones in the ones place because $(8 - 9)$ cannot be computed with whole numbers.

$68 - 49 = (50 + 18) - (40 + 9)$

$50 + 18$
$40 + \ 9$

STEP 3:
Now we rearrange so as to subtract 9 from 18 and 40 from 50.

$68 - 49 = (50 - 40) + (18 - 9)$
$\qquad = 10 \qquad + \ 9$
$\qquad = 19$

$50 + 18$
$40 + \ 9$
$10 + \ 9 = 19$

This explanation is long and wordy, but the actual computation is fairly brief.

Here is another example, with the vertical form written in a more abbreviated way.

| *Horizontal form* | *Vertical form* |

$68 - 42 = (60 + 8) - (40 + 2)$
$\qquad = (60 - 40) + (8 - 2)$
$\qquad = 26$

68
−42
26

SUBTRACTING LARGER NUMBERS We may subtract larger numbers, of course, simply by extending the principles and procedures used with smaller numbers. Consider, for instance, subtracting 276 from 523.

Since we cannot subtract 6 ones from 3 ones nor 7 tens from 2 tens, regrouping is required. We may write:

5 hundreds + 2 tens + 3 ones = 5 hundreds + (1 ten + 1 ten) +
3 ones
= 5 hundreds + (1 ten + 10 ones) +
3 ones
= 5 hundreds + 1 ten + 13 ones
= (4 hundreds + 1 hundred) +
1 ten + 13 ones
= (4 hundreds + 10 tens) + 1 ten +
13 ones
= 4 hundreds + 11 tens + 13 ones

Ordinarily this procedure is simply indicated by

5 hundreds + 2 tens + 3 ones = 4 hundreds + 11 tens + 13 ones

We may now complete the problem 523 − 276 by writing:

$$\begin{array}{l} 5 \text{ hundreds} + 2 \text{ tens} + 3 \text{ ones} = 4 \text{ hundreds} + 11 \text{ tens} + 13 \text{ ones} \\ 2 \text{ hundreds} + 7 \text{ tens} + 6 \text{ ones} = 2 \text{ hundreds} + 7 \text{ tens} + 6 \text{ ones} \\ \hline 2 \text{ hundreds} + 4 \text{ tens} + 7 \text{ ones} = 247 \end{array}$$

or we may write

$$\begin{array}{l} 500 + 20 + 3 = 400 + 110 + 13 \\ 200 + 70 + 6 = 200 + 70 + 6 \\ \hline 200 + 40 + 7 = 247 \end{array}$$

or we may use an equation form, such as

$$\begin{aligned} 523 − 276 &= (500 + 20 + 3) − (200 + 70 + 6) \\ &= (400 + 110 + 13) − (200 + 70 + 6) \\ &= (400 − 200) + (110 − 70) + (13 − 6) \\ &= 200 + 40 + 7 \\ &= 247 \end{aligned}$$

We eventually may shorten such algorithms to the form

$$\begin{array}{r} ④ ⑪ ⑬ \\ 5 2 3 \\ -2 7 6 \\ \hline 2 4 7 \end{array}$$ or simply $$\begin{array}{r} 523 \\ -276 \\ \hline 247 \end{array}$$

■ **Question 7** Work this problem, 1,001 − 456, showing all the borrowings.

**ANOTHER SUBTRACTION
ALGORITHM** There is another algorithm which we could use in subtraction. We saw in the preceding chapter that subtraction can be thought of as the process of finding the missing addend in an addition problem for which we are given the sum and the other addend:

$$b + (a − b) = a$$

Let us use this in the problem 49 − 17. Let us write this as if it were an addition problem with a missing addend. We use the symbol $\bigcirc$ to show where this addend should go.

17
⌣
―――
49

Now, we ask ourselves what number added to 7 gives 9. We remember (or find by looking in the table on page 53) that is is 2, so we write it under the 7.

17
⟨2⟩
―――
49

Next, we ask what we have to add to 1 to get 4. The answer is 3, so we write it in.

17
⟨32⟩
―――
49

Now, since $17 + 32 = 49$, $49 - 17 = 32$, and we have finished the subtraction problem.

Let us try another problem, one which would require "borrowing" or regrouping if we used the algorithm illustrated before. Try $32 - 17$. We write this as

17
⌣
―――
32

Now, we ask what we add to 7 to get 2. There is no such number. But we now remember that when we add 5 to 7 we get 12, which ends in 2. So we write the 5 under the 7 and remember that we have to carry the 1.

17
⟨5⟩
―――
32

Now, what do we add to $1 + 1$ in order to get 3? The answer is 1, so we write it in.

17
⟨15⟩
―――
32

We see that $17 + 15 = 32$, so $32 - 17 = 15$.

The same process can be used for larger numbers. Try $342 - 187$.

187
⟨155⟩
―――
342

Just as in the case of addition, all these variations of the algorithms are equally valid. You may use whichever one you prefer.

■ **Question 8** Use this algorithm to work the problem 1,001 − 456.

PROBLEMS

1 Use sets of marks to work each of these:
(a) 25 + 32   (b) 27 + 38   (c) 77 + 38
2 Use sets of marks to work each of these:
(a) 35 − 22   (b) 83 − 27   (c) 101 − 228

Here is an addition problem, 42 + 36, worked in four different ways:

(1) 4 tens + 2 ones   (2) 40 + 2
    3 tens + 6 ones   30 + 6
    7 tens + 8 ones = 78   70 + 8 = 78

(3) 42   (4) 42
    36   36
     8   78
   70
   78

3 Work this problem in each of these four ways: 58 + 17
4 Work this problem in each of these four ways: 777 + 964
5 Work this problem in each of these four ways: 5,678 + 9,753

Here is a subtraction problem, 69 − 24, done in three different ways:

(1) 6 tens + 9 ones   (2) 60 + 9
    2 tens + 4 ones   20 + 4
    4 tens + 5 ones = 45   40 + 5 = 45

(3) 69
    24
    45

6 Work this problem in each of these three ways: 81 − 35
7 Work this problem in each of these three ways: 446 − 168
8 Work this problem in each of these three ways: 43,201 − 7,068
9 Work this problem in each of these three ways: 10,000 − 11
10 Work each of Problems 6 to 9 by the "missing addend" method.
11 Use expanded notation to work this problem:
   342
   174
 +418

12 In this problem, what does the 23 stand for? The 28?
   5,481
   4,737
    896
    938
     22
     23
    28
     9
 12,052

13 You could subtract from left to right. For example, you could work the
problem 49 − 17 by these steps:

$49 - 10 = 39$

$39 - 7 = 32$

Work each of these problems in the same way:

(a) $57 - 33$ (b) $59 - 15$ (c) $543 - 357$

14 A student makes the same mistake in each of these problems:

48	35	274	417
+ 9	+ 9	+ 96	+226
111	71	811	661

If he makes the same mistake in this problem, what will his answer be?

$$\begin{array}{r} 244 \\ + \ 56 \\ \hline \end{array}$$

15 A student makes the same kind of mistake in each of these problems:

48	35	274	417
− 9	− 9	− 96	−226
49	36	288	291

If he makes the same mistake, what answer will he record for this problem?

$$\begin{array}{r} 244 \\ - \ 56 \\ \hline \end{array}$$

16 In each of these exercises, each capital letter stands for one of the digits 0, 1, 2, . . . , 9. Find which digit each letter stands for.

$$\begin{array}{r} 63 \\ - \ A \\ \hline B8 \end{array} \qquad \begin{array}{r} 47 \\ + \ D \\ \hline CC \end{array} \qquad \begin{array}{r} EE \\ + \ F \\ \hline 8\,0 \end{array} \qquad \begin{array}{r} 7G \\ -GH \\ \hline 1\,4 \end{array}$$

17 Fill in this table for addition in base three.

18 Work each of these base three addition problems using sets of marks and then using expanded notation and the addition table:

(a) $11 + 10$ (b) $22 + 22$ (c) $201 + 112$

19 Fill in this table for subtraction in base three, using the table in Problem 17.

20 Work each of these base three subtraction problems using sets of marks and then using expanded notation and the subtraction table:

(a) $22 - 10$ (b) $20 - 12$ (c) $101 - 22$

+	0	1	2	10
0				
1				
2				
10				

−	0	1	2
0		×	×
1			×
2			
10			
11			

7. MULTIPLICATION AND DIVISION

In addition to the two operations of addition and subtraction on whole numbers, there are two others that are useful in answering certain "how many" questions. These are multiplication and division. As is the case with addition and subtraction, these two operations have to do with sets of objects.

MULTIPLICATION

Let us look at two "how many" questions to see how the idea of multiplication arises.

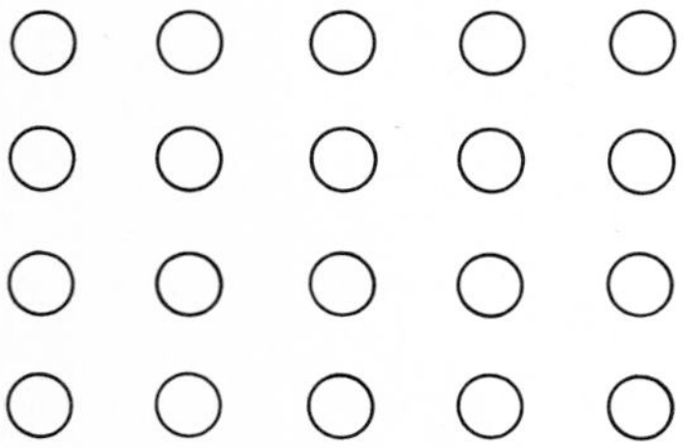

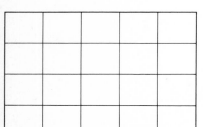

Figure 7.1

1 If I put five coins in my piggy bank each day, how many coins will be in the bank at the end of four days?
2 Four varieties of plants are each tested with five different kinds of fertilizers. A plot of ground is needed to test each plant with each fertilizer. How many plots are needed?

Let us draw a sketch for the first question. We can draw a row of five coins for each of the four days (Figure 7.1). Once we have this sketch, we can see that one way to find the answer to Question 1 is to count the number of pictures of coins in the sketch.

We can also make a sketch for Question 2. We can think of the plots laid out in rows, one for each variety of plant. Since there are five kinds of fertilizer, there will have to be five plots in each row. As soon as we see the sketch in Figure 7.2, we can see that we can answer Question 2 by counting the number of little squares in the sketch.

Figure 7.2

■ **Question 1** Make a sketch for the question "How many coins will I save in five days if I put four in the bank each day?"

There are many questions which lead to sketches similar to those in Figures 7.1 and 7.2. It is convenient to introduce a special name for

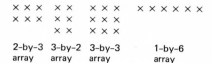

2-by-3 3-by-2 3-by-3   1-by-6
array   array   array   array

Figure 7.3

this kind of sketch. An *array* is a collection of objects arranged in rows and columns, with the same number of objects in each row and with the same number of objects in each column. The objects may be real things, such as coins or plots of ground, or they may be pictures of real objects, or they may just be marks on the paper.

If the number of rows is *a* and the number of objects in each row is *b*, we call it an *a*-by-*b* array. Both the sketches above are 4-by-5 arrays. Some other arrays are illustrated in Figure 7.3.

Now we can specify what the operation of multiplication is. The process of *multiplying* the whole number *a* by the whole number *b* is that of forming an *a*-by-*b* array and counting the number of elements in the array. The numbers *a* and *b* are called the *factors,* and the number of elements in the array is called the *product,* which we write as $a \times b$.

Questions 1 and 2 both led to 4-by-5 arrays and the number of elements in each array was 20, so

$$4 \times 5 = 20$$

A look at the other arrays above, and some simple counting, tells us that

$2 \times 3 = 6$ $3 \times 3 = 9$
$3 \times 2 = 6$ $1 \times 6 = 6$

■ **Question 2**   Do the stars on the United States flag form an array?

There are many ways of counting the elements of a 4-by-5 array. One way is to note that a 4-by-5 array is the union of 4 disjoint sets, each set having 5 members. Consequently, 4×5 can be computed by the successive addition

4 addends

$\overbrace{5 + 5 + 5 + 5}$

that is, 5 is used as an addend 4 times. (This is sometimes referred to as the repeated addition description of multiplication.) It is also true that the array is the union of 5 disjoint sets, each set having 4 members, as in Figure 7.4.

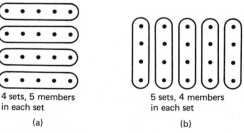

4 sets, 5 members   5 sets, 4 members
in each set   in each set
(a)   (b)

Figure 7.4

Thus 4×5 can be computed by the successive addition of 4 as an

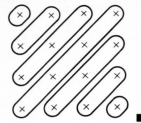

Figure 7.5

addend 5 times

5 addends

$$\overbrace{4+4+4+4+4}$$

But of course any way of counting the elements of the array is as good as any other way.

■ **Question 3** If a 4-by-4 array is partitioned as in Figure 7.5, what addition does it suggest?

PROPERTIES OF MULTIPLICATION

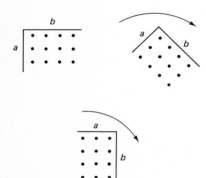

Figure 7.6

When we discussed addition, we found that we could discover some of its properties by looking at sets. We can do the same for multiplication by looking at arrays, which are particular arrangements of sets.

The first thing we notice about multiplication is that we can always do it. No matter what pairs of whole numbers a and b we are given, we can always construct an a-by-b array and then count the number of elements in the array. The result, of course, is another whole number. Therefore we can say that the set of whole numbers is *closed* under multiplication.

From the study of various arrays another property of multiplication of whole numbers may be discovered. In general, if a and b are whole numbers, $a \times b = b \times a$. This is the *commutative property of multiplication*. This property may be seen from the definition directly, for an a-by-b array can be changed into a b-by-a array simply by rotating through 90 degrees (see Figure 7.6). Commuting the order of factors does not alter the product ($3 \times 4 = 4 \times 3$). Commutativity is pedagogically quite important because, for example, to some people the product of 3 and 41 seems simpler than the product of 41 and 3. This property also reduces the number of multiplication facts to be remembered since $a \times b$ and $b \times a$ may be learned simultaneously, and the property is useful in simplifying calculations.

Given three numbers a, b, and c, it is not immediately obvious whether any meaning may be attached to $a \times b \times c$. One may note that if a and b are whole numbers, $(a \times b)$ is a single whole number. Now this whole number may be paired with c to obtain a product. This is indicated by $(a \times b) \times c$. Or alternately, one may note that $(b \times c)$ is a whole number, and may be paired with a to obtain a product which may be described as $a \times (b \times c)$. For example,

$$(3 \times 2) \times 4 = 6 \times 4 = 24$$

and

$$3 \times (2 \times 4) = 3 \times 8 = 24$$

In general, it is true that $(a \times b) \times c = a \times (b \times c)$, so it does not matter how the pairings are grouped; a unique number is assigned to the triple (a, b, c) as the product. This freedom of grouping is similar to the associative property of addition, and is called the *associative property of multiplication*. That is, for any three whole numbers a, b, and c,

$$(a \times b) \times c = a \times (b \times c)$$

This is, in fact, the property which permits $a \times b \times c$ to be written without parentheses. The same procedure used above also permits multiplication to be extended to more than three factors.

■ **Question 4** If you add a column of 47 eighty-fours, and then a column of 84 forty-sevens, how will the two answers compare?

The physical model of a box made up of cubical blocks with dimensions a by b by c may be used to illustrate the associativity of multiplication (Figure 7.7).

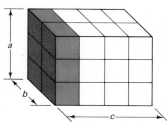
$a \times b$ blocks in each vertical
slice; c vertical slices

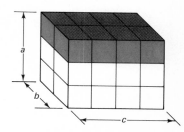
$b \times c$ blocks in each horizontal
slice; a horizontal slices

Figure 7-7

Since multiplication has the commutative and the associative properties, multiplication has the same flexibility in grouping and rearranging of the factors as addition has in the grouping and rearranging of the addends. For example, for

$25 \times 15 \times 3 \times 4$, it can be seen that the product of 25 and 4 is 100, and the product of 3 and 15 is 45; consequently $25 \times 15 \times 3 \times 4 = 4{,}500$.

This flexibility amounts to a sort of "do-it-whichever-way-we-want" principle in any problem where only multiplications are involved.

THE IDENTITY ELEMENT FOR MULTIPLICATION

The number 1 occupies, with respect to multiplication, the same position that 0 occupies with respect to addition. Notice that

$1 \times 3 = 3 \times 1 = 3$
$1 \times 5 = 5 \times 1 = 5$
$1 \times 6 = 6 \times 1 = 6$
$1 \times 8 = 8 \times 1 = 8$

It is true that $1 \times n = n$ for all numbers n because a 1-by-n array consists of only one row having n members, and therefore the entire array contains exactly n members. Since $1 \times n = n$, the number 1 is called the *identity* element for multiplication (Figure 7.8).

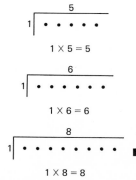

$1 \times 5 = 5$

$1 \times 6 = 6$

$1 \times 8 = 8$

Figure 7.8

■ **Question 5** Is there any number which could replace N on both sides of this sentence and make it false?

$(2 \times N) \times (3 \times 4) = 2 \times (N \times 12)$

| **MULTIPLICATION PROPERTY OF 0** | The number 0, besides playing the role of the identity element for *addition*, also has a rather special property with respect to *multiplication*. The number of members in a 0-by-n array (that is, an array with 0 rows, each having n members) is 0 because the set of members of this array is empty. Similarly, the set of members of an array of n rows, each of them having 0 members, is empty. Thus, for any number n, |

$$0 \times n = n \times 0 = 0$$

What has been done so far shows that multiplication, as with addition, is an operation on the whole numbers which has the properties of closure, commutativity, and associativity. There is a special number 1 that is an identity for multiplication just as 0 is an identity for addition. Moreover, 0 plays a special role in multiplication for which there is no corresponding property in addition.

| **THE DISTRIBUTIVE PROPERTY** | We have seen that multiplication may be described by repeated addition. Aside from this, there is another important property that links the two operations. This property is called the *distributive property of multiplication over addition*. The distributive property states that if a, b, and c are any whole numbers, then |

$$a \times (b + c) = (a \times b) + (a \times c)$$

The distributive property may be illustrated by considering an a-by-$(b + c)$ array (see Figure 7.9).

Figure 7.9

It is true that this array is formed from an a-by-b array and an a-by-c array (Figure 7.10). Consequently, the number $a \times (b + c)$ of members in the large array is the sum of $(a \times b)$ and $(a \times c)$, the numbers of members of the subsets. That is, $a \times (b + c) = (a \times b) + (a \times c)$.

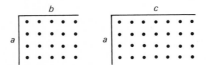

Figure 7.10

■ **Question 6** Is $0 \times (2 + 3) = (0 \times 2) + (0 \times 3)$?

Since multiplication is commutative, both the "left-hand" and the "right-hand" distributive properties hold, that is,

Left-hand: $a \times (b + c) = (a \times b) + (a \times c)$

Right-hand: $(b + c) \times a = (b \times a) + (c \times a)$

These distributive properties are demonstrated in Figure 7.11.

Left hand:

$$3 \times (5 + 8) = (3 \times 5) + (3 \times 8)$$

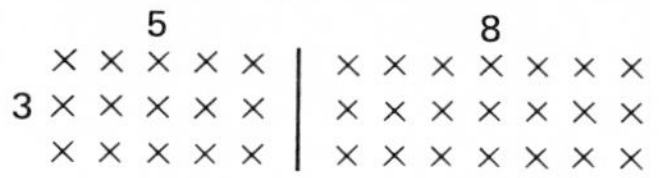

```
             5                    8
     × × × × ×  |  × × × × × × × ×
 3 × × × × ×    |  × × × × × × × ×
     × × × × ×  |  × × × × × × × ×
```

Right hand:

$$(4 + 3) \times 2 = (4 \times 2) + (3 \times 2)$$

```
             2
           × ×
           × ×
       4   × ×
           × ×
          ─────
           × ×
       3 × ×
           × ×
```

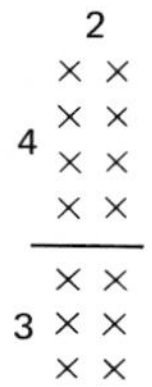

Figure 7.11

Sometimes we use the distributive property to replace a product by the sum of two products:

$$3 \times 13 = 3 \times (10 + 3) = (3 \times 10) + (3 \times 3)$$
$$= \quad 30 \quad + \quad 9 \quad = 39$$

```
× × × × × × × ×
× × × × × × × ×
× × × × × × × ×
× × × × × × × ×
× × × × × × × ×
× × × × × × × ×
× × × × × × × ×
```

Figure 7.12

and sometimes to replace two products by a single product:

$$(5 \times 4) + (5 \times 6) = 5 \times (4 + 6)$$
$$= 5 \times \quad 10 \quad = 50$$

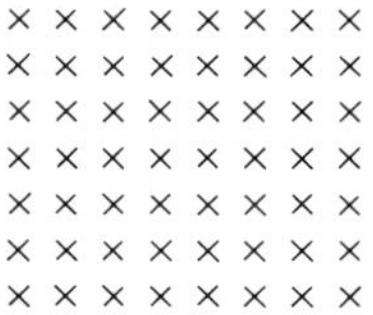

Sometimes we need to consider the product of two numbers each of which is written as a sum, for example, $(3 + 4) \times (2 + 6)$, which is of course the same as 7×8. We can draw an array for this (Figure 7.12). In Figure 7.13 divide the array into two arrays. This tells us that

$$7 \times 8 = (7 \times 2) + (7 \times 6)$$

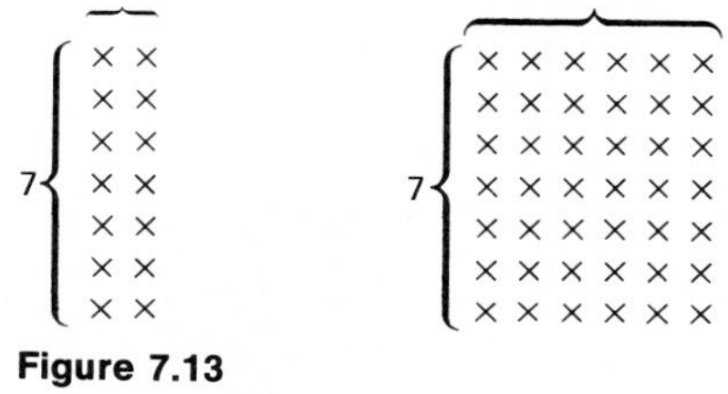

```
       2                    6
     ⎧ × ×              ⎧ × × × × × ×
     ⎪ × ×              ⎪ × × × × × ×
     ⎪ × ×              ⎪ × × × × × ×
   7 ⎨ × ×            7 ⎨ × × × × × ×
     ⎪ × ×              ⎪ × × × × × ×
     ⎪ × ×              ⎪ × × × × × ×
     ⎩ × ×              ⎩ × × × × × ×
```

Figure 7.13

We can divide each of these arrays into two (see Figure 7.14). The two arrays on the left of Figure 7.14 tell us that

$$7 \times 2 = (3 \times 2) + (4 \times 2)$$

The two arrays on the right of Figure 7.14 tell us that

$$7 \times 6 = (3 \times 6) + (4 \times 6)$$

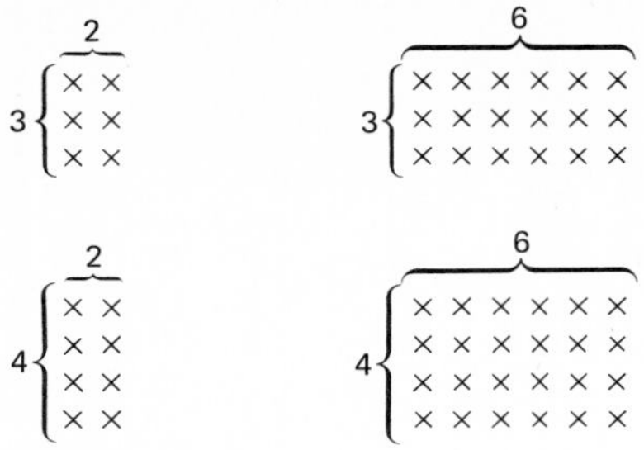

Figure 7.14

Figure 7.15

Putting this all together, we see that

$$7 \times 8 = (3 + 4) \times (2 + 6) = (3 \times 2) + (4 \times 2) + (3 \times 6) + (4 \times 6)$$

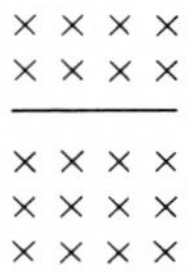

■ **Question 7** Is Figure 7.15 a correct diagram for $(2 + 3) \times (4 + 0)$?

Of course, we could have divided the original array by a horizontal rather than vertical cut at the first step, but the end result would still be the same four small arrays.

Now we could have done this without resorting to the arrays but by using the distributive property more than once. First we think of $(3 + 4)$ as a single whole number [the a in $a \times (b + c)$] and use the distributive property to write

$$(3 + 4) \times (2 + 6) = [(3 + 4) \times 2] + [(3 + 4) \times 6]$$

Next, we use the right-hand form of the distributive property twice to write

$$(3 + 4) \times 2 = (3 \times 2) + (4 \times 2)$$

and

$$(3 + 4) \times 6 = (3 \times 6) + (4 \times 6)$$

Putting all these steps together, we have

$$(3 + 4) \times (2 + 6) = (3 \times 2) + (4 \times 2) + (3 \times 6) + (4 \times 6)$$

A curious person might ask if the distributive property works the other way, with the $+$ and the $\times$ interchanged.

That is, is it always the case that

$$a + (b \times c) = (a + b) \times (a + c)$$

This is false. For example, if $a = 1$, $b = 3$, and $c = 2$,

$$a + (b \times c) = 1 + (3 \times 2) = 1 + 6 = 7$$

but

$$(a + b) \times (a + c) = (1 + 3) \times (1 + 2) = 4 \times 3 = 12$$

So it cannot be stated that $a + (b \times c)$ is always equal to $(a + b) \times (a + c)$.

■ **Question 8** Is it always the case that

$$a \times (b + c + d) = (a \times b) + (a \times c) + (a \times d)$$

Another related question might be asked as to whether multiplication distributes over subtraction; that is, whether it is true that

$$a \times (b - c) = (a \times b) - (a \times c)$$

For the values $a = 1$, $b = 3$, $c = 2$,

$$a \times (b - c) = 1 \times (3 - 2) = 1 \times 1 = 1$$

and

$$(a \times b) - (a \times c) = (1 \times 3) - (1 \times 2) = 3 - 2 = 1$$

Other examples may be tried and it will turn out that in every instance multiplication does distribute over subtraction, subject to the restriction of course that $b \geq c$, otherwise $b - c$ is not defined.

A MULTIPLICATION TABLE By constructing appropriate arrays and counting their members, a multiplication table can be prepared which records in convenient form all the results of multiplying together numbers that have only one digit (see Figure 7.16).

×	0	1	2	3	4	5	6	7	8	9
0	0	0	0	0	0	0	0	0	0	0
1	0	1	2	3	4	5	6	7	8	9
2	0	2	4	6	8	10	12	14	16	18
3	0	3	6	9	12	15	18	21	24	27
4	0	4	8	12	16	20	24	28	32	36
5	0	5	10	15	20	25	30	35	40	45
6	0	6	12	18	24	30	36	42	48	54
7	0	7	14	21	28	35	42	49	56	63
8	0	8	16	24	32	40	48	56	64	72
9	0	9	18	27	36	45	54	63	72	81

Figure 7.16

Note that the first row and first column can be filled in, without counting any arrays, by remembering the multiplication property of 0. Also, the second row and second column can be filled in by using the fact that 1 is the identity element for multiplication. Once the elements on and above the diagonal from upper left to lower right have been filled in, the rest of the table can be completed by remembering that multiplication is commutative.

■ **Question 9** Prepare a multiplication table for numbers written in base three.

**MULTIPLICATION ON
THE NUMBER LINE**

Through the interpretation of multiplication as repeated addition, multiplication may be illustrated on the number line. For example, 3×4 means 3 addends, each addend being 4. That is,

$$3 \times 4 = 4 + 4 + 4$$

Therefore, this may be represented by three successive arrows as shown in Figure 7.17.

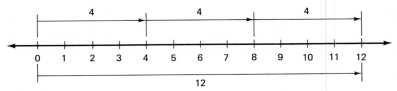

Figure 7.17

On the other hand, 4×3 means 4 addends of 3. The representation on the number line is as in Figure 7.18.

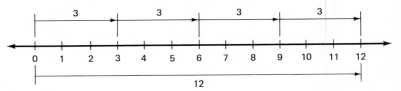

Figure 7.18

As we can see, the representations in Figures 7.17 and 7.18 are different; however, both of these yield the same result. By combining these two in a single diagram, we illustrate the commutative property of multiplication.

When more than two factors are involved, this too may be illustrated. For example, to show $(2 \times 3) \times 4$, we have the diagram in Figure 7.19. Likewise, $2 \times (3 \times 4)$ may be shown by obtaining two (3×4) "arrows" and abutting them. By combining the diagrams for $(2 \times 3) \times 4$ and $2 \times (3 \times 4)$, associativity may be illustrated.

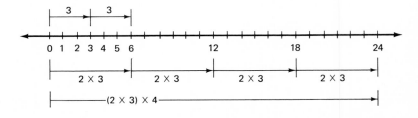

$$(2 \times 3) \times 4 = 24$$

Figure 7.19

■ **Question 10** On the number line, draw an illustration of

$$2 \times (4 + 3) = (2 \times 4) + (2 \times 3)$$

DIVISION In order to determine the unknown number n which is the product $4 \times 5 = n$ of the two known factors 4 and 5, we counted the number of members in a 4-by-5 array—that is, in an array of 4 rows with 5 members in each row.

An associated problem is to start with 20 objects and ask how many rows there will be if this set of 20 members is arranged in an array with 5 to a row. In this particular case, the answer is 4.

There are many problems where the number relationship can be recognized as that of a known product and one known factor. Thus, if 75 tulip bulbs are to be planted in equal rows of 15 each, how many rows will there be? The number sentence to express this relationship is

$$n \times 15 = 75$$

Division is the operation of finding an unknown factor in a multiplication problem when the product and one factor are known.

The normal symbol for the operation of division is ÷. Thus, 8 ÷ 2 is the unknown factor, if there is one, which when multiplied by 2 gives the product 8. It is also the number of columns in an array of 8 objects arranged in 2 rows.

Of course, since multiplication is commutative, 8 ÷ 2 can be thought of as the number of rows in an array containing 8 objects arranged in 2 columns, as in Figure 7.20b.

(a)   (b)

Figure 7.20

As another example, 17 ÷ 4 is the number n, if any, for which $n \times 4 = 17$. A few trials will suffice to indicate that there is no such whole number. Also, the attempt to arrange an array of 17 elements in equal columns of 4 members each is doomed to failure. As a matter of fact, there is no a-by-b array with 17 members except the 1-by-17 or the 17-by-1 array.

■ **Question 11** The multiplication table in Figure 7.16 can also be used to divide. Illustrate by showing how to use the table to find 48 ÷ 6.

Thus we see that if a and b are known whole numbers, $a \div b = n$ and $a = b \times n$ (or $a = n \times b$) are two number sentences which say the same thing. This "missing factor" concept in division parallels the "missing addend" concept in subtraction where it is noted that if a and

b are known whole numbers, $a - b = n$ and $a = b + n$ are two number sentences which say the same thing. Accordingly, if $b \neq 0$, *division* may be defined as follows:

$$a \div b = n \qquad \text{if and only if} \qquad a = b \times n$$

(Why $b = 0$ is to be ruled out in division will be discussed later.)

PROPERTIES OF DIVISION

In the same way that subtraction is the inverse of addition, division by n may be thought of as the inverse of multiplication by n. Thus,

$$(8 \times 3) \div 3 = 8 \qquad \text{and} \qquad (17 \times 4) \div 4 = 17$$

However, caution must be exercised in thinking about multiplication as the inverse of division because it is true that $(15 \div 3) \times 3 = 15$, while $(8 \div 3) \times 3$ is meaningless since $8 \div 3$ is not a whole number. This is similar to the caution we must exercise in this "doing and undoing" process with subtraction; thus, while

$(15 - 3) + 3 = 15$ is perfectly acceptable

$(5 - 13) + 13$ is meaningless

since $(5 - 13)$ is not a whole number.

■ **Question 12** How can you tell from the multiplication table that there is no such whole number as $48 \div 7$?

Many examples may be given to show that the whole numbers are not closed under division. For example, while $6 \div 3 = 2$, $3 \div 6$ is not a whole number. These same two examples show that $6 \div 3 \neq 3 \div 6$; hence the operation is not commutative. To see that division is not associative, again many examples may be produced. One such example is the following:

$(12 \div 6) \div 2 = 2 \div 2 = 1$

but

$12 \div (6 \div 2) = 12 \div 3 = 4$

The different results obtained for $(12 \div 6) \div 2$ on the one hand, and for $12 \div (6 \div 2)$ on the other, shows that, in general, it is not true that $(a \div b) \div c = a \div (b \div c)$.

So far, division with respect to whole numbers has revealed itself as an operation that does not have the properties of closure, commutativity, and associativity. To free ourselves from the impression that not much can be said about this operation, we need to consider only the important notion that division by b is the inverse of the operation of multiplication by b. That is, $(a \times b) \div b = a$, as long as $b \neq 0$.

THE DISTRIBUTIVE PROPERTY

Recall that the distributive property relates multiplication to addition and subtraction. In a limited way, division also has a distributive property, but care is needed in using it.

If $(b + c) \div a$ is a whole number, and if $b \div a$ and $c \div a$ are whole numbers, then it is true that

$$(b + c) \div a = (b \div a) + (c \div a)$$

This is what we mean when we say that division has a limited distributive property; that it has only the right-hand distributive property, and only when $(b \div a)$ and $(c \div a)$ are defined. For example:

$$(15 + 24) \div 3 = 39 \div 3 = 13$$

and

$$(15 \div 3) + (24 \div 3) = 5 + 8 = 13$$

and thus we see that the two results are the same; that is,

$$(15 + 24) \div 3 = (15 \div 3) + (24 \div 3)$$

■ **Question 13** For which whole numbers n is it true that $(12 \div n) \times n = 12$?

On the other hand,

$20 \div (2 + 5) = 20 \div 7$ is not a whole number, whereas

$(20 \div 2) + (20 \div 5) = 10 + 4 = 14$ is a whole number

So $20 \div (2 + 5) \neq (20 \div 2) + (20 \div 5)$

In general, then, $a \div (b + c) \neq (a \div b) + (a \div c)$, but it is true that $(b + c) \div a = (b \div a) + (c \div a)$, provided $b \div a$ and $c \div a$ have meaning. Many examples may also be produced to confirm that division has the right-hand distributive property over subtraction, provided each of the indicated subtraction and divisions has meaning; that is,

$$(b - c) \div a = (b \div a) - (c \div a)$$

provided $b - c$, $b \div a$, and $c \div a$ are whole numbers.

As an example, we can use $(24 - 15) \div 3$ and $(24 \div 3) - (15 \div 3)$ to illustrate this point.

THE ROLE OF 0 AND 1 IN DIVISION The operation of division was connected to the operation of multiplication by the statement that

$$a \div b = n \qquad \text{if and only if} \qquad a = b \times n$$

Since 0 and 1 played special roles in multiplication, it may be appropriate to pay particular attention to the two numbers in division.

If $a = 0$ and b is any number not zero, then $0 \div b$ is that number n, if there is one, such that $0 = b \times n$. From the multiplication facts, $0 = b \times n$ is certainly true if $n = 0$. So,

$$0 \div b = 0 \qquad \text{if} \qquad b \neq 0$$

■ **Question 14** For what numbers is it the case that

$$a \div (b - c) = (a \div b) - (a \div c)$$

The above is true for $a = 0$ and $b \neq 0$. Now consider $a = 0$ and

$b = 0$; this is the case, $0 \div 0$. By the definition of division, $0 \div 0$ is equal to that number n, if there is one, for which it is true that $0 = 0 \times n$. But by the special multiplication property of 0, $0 \times n$ is equal to 0 for *any whole number n whatever*. Thus

0 ÷ 0 is an ambiguous symbol

The case of $a \div 0$, where $a \neq 0$, is still another situation. This must be equal to that number n such that $a = 0 \times n$. But this is impossible for any number a not equal to 0 since we must have $a = 0 \times n$ and $0 \times n$ is always equal to 0. For this reason, $a \div 0$ is meaningless for $a \neq 0$; that is,

a ÷ 0 is undefined

Notice that $0 \div b = 0$ if $b \neq 0$, but that $a \div 0$ is either ambiguous or meaningless, depending on whether or not $a = 0$. In either case, division by zero is impossible. Thus, 0 plays a very special role with respect to division—a role that is not understood clearly by many. In summary,

$a \div b$ is ambiguous if $a = 0$ and $b = 0$

$a \div b$ is meaningless if $a \neq 0$ and $b = 0$

$a \div b$ is zero if $a = 0$ and $b \neq 0$

■ **Question 15**   If $a \neq 0$ and $b \neq 0$, can $a \div b = 0$?

Using the definition it can be seen that for any whole number b ($b \neq 0$), $1 \div b$ is not a whole number at all unless $b = 1$, while $a \div 1 = a$ for any whole number a. Consequently,

$a \div 1 = a$ for any whole number a

$1 \div b$ is not a whole number unless $b = 1$

In the sense that $a \div 1 = a$, the number 1 acts somewhat like an identity element for division. Unlike the identity element for multiplication in which, for *any a,* $1 \times a = a \times 1$, the number 1 is limited to acting as an identity element for division if it is to the right of the symbol ÷.

INCOMPLETE ARRAYS   We have found that to determine $a \div b$, we may sometimes enlist the aid of a physical model in the form of an array. For example, to determine $35 \div 5$, we put the 35 elements in 5 rows and we find that this can be done with exactly 7 elements in each row.

On the other hand, if an attempt were made to determine $37 \div 5$, there is no whole number n such that $37 = 5 \times n$; hence $37 \div 5$ is not defined in the set of whole numbers. However, an approach may be made guided by the procedure in setting up an array as before. Filling out 5 rows with the 37 elements, it can be seen that 1 element in each row requires 5 elements; 2 elements in each row requires 10 elements; 3 in each row, 15 elements, etc., until 7 elements have been displayed in each of 5 rows. The array now employs 35 of the 37 elements, with 2 elements left over. This situation can be expressed by the number sentence

$$37 = (5 \times 7) + 2$$

Essentially what has been done (see Figure 7.21) is to set up as large an array as possible which has 5 rows and observe the number of elements left over. If no element remains undisplayed, then the number of columns obtained is precisely the missing factor n in $b \times n = a$; otherwise, the process is said to yield a remainder.

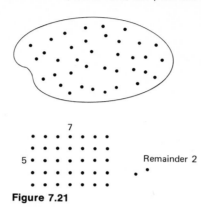

Figure 7.21

■ **Question 16** Is there any value for a such that $a \div 1$ leads to an incomplete array?

From the physical model that had helped to determine the missing factor, the procedure has shifted to a *process* of obtaining a *quotient* and *remainder*, using this model as a guide. This process is then applied to such questions as $37 \div 5$ for which it may be known that there is no missing factor among the whole numbers. This is done by expressing 37 as $(7 \times 5) + 2$, where 7 is the quotient. In general, this is done by expressing a as $(q \times b) + r$, where q is the quotient and r is the remainder.

Note that we first wrote $37 = (5 \times 7) + 2$ and then $37 = (7 \times 5) + 2$. Since $5 \times 7 = 7 \times 5$, the two are equivalent; that is,

$$37 = (5 \times 7) + 2 = (7 \times 5) + 2$$

**DIVISION ON THE
NUMBER LINE**

We can illustrate division using the number line by partitioning a segment into congruent subsegments. For example, to illustrate $6 \div 3$, we can partition a 6-unit segment into 3 congruent subsegments, each of which is congruent to the segment from 0 to 2 (see Figure 7.22). Thus, this partition conveys the concept $6 \div 3 = 2$. Clearly, this is associated with the representation of multiplication on the line in which three 2-unit arrows or 2-unit segments are abutted, resulting in a 6-unit arrow or a 6-unit segment. The association may be thought of as: One operation is the inverse of the other, or, from the point of view that

$$6 \div 3 = 2 \qquad \text{if and only if} \qquad 6 = 2 \times 3$$

Figure 7.22

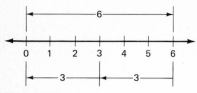

Figure 7.23

■ **Question 17** Use a number line to illustrate: $(8 \div 2) \div 2$

Another method of illustrating division on the number line is related to considering division in terms of repeated subtraction. This concept will be discussed in further detail in Chapter 8 when the division techniques are discussed. We can indicate here, however, this use of the number line in order to compare with the use shown above. Beginning with 6, we ask: "How many times can 3 be subtracted?" Corresponding to this, we can show division using the number line as in Figure 7.23. In this case, since subtraction is performed twice, $6 \div 3 = 2$.

PROBLEMS

1 Figure P7.1 shows five arrays.

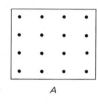

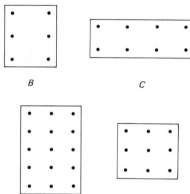

Figure P7.1

(a) Three Stanford Frisbee teams are to play three Berkeley Frisbee teams. Which array above shows the number of games that can be arranged if each Stanford team plays each Berkeley team just once?

(b) Mr. Rhodes is buying a two-tone car. Colors available are: red, yellow, green, and blue. Which array shows the various possible results? If Mr. Rhodes insists that the car must be two-toned, how many choices does he have?

(c) Three children play violin and two other children play piano. Which array above shows how many duets can be played so that each violinist plays once with each pianist?

(d) There are two roads from Palo Alto to San Jose, and four roads from San Jose on to Watsonville. Which array shows how many routes there are from Palo Alto?

2 A combination lock has two dials, one bearing all the letters of the alphabet, the other bearing all the digits. How many combinations are possible?

3 The middle section of an auditorium seats 28 to a row, and each side section seats 11 to a row. What is the capacity of this auditorium if there are 20 such rows?

4 How many possible arrays of 24 dots could you make? (Draw them if necessary.)

5 If $ab = 0$, what conclusions can you make concerning possible values of a and b?

6 A marching band always forms an array when it marches. The leader likes to use many different formations. Aside from the leader, the band has 59 members. The leader is trying very hard to find one more member. Why?

7 A standard deck of 52 cards is to be dealt to 7 players. Each player is to be dealt 5 cards. Write a number sentence telling how many cards are to be dealt and how many cards are to be left in the deck. How many more cards may each player be dealt?

8 Make each of the following a true statement illustrating the distributive property.
(a) $3 \times (4 + \underline{\hspace{1cm}}) = (3 \times 4) + (3 \times 3)$
(b) $2 \times (\underline{\hspace{1cm}} + 5) = (2 \times 4) + (\underline{\hspace{1cm}} \times 5)$
(c) $13 \times (6 + 4) = (13 \times \underline{\hspace{1cm}}) + (13 \times \underline{\hspace{1cm}})$
(d) $(2 \times 7) + (3 \times \underline{\hspace{1cm}}) = (\underline{\hspace{1cm}} + \underline{\hspace{1cm}}) \times 7$

9 Evaluate the following:
(a) $4 \times (7 + 3)$ (b) $(4 \times 7) + 3$
(c) $(5 + 9) \times 7$ (d) $5 + (9 \times 7)$

10 Use the commutative and associative properties to get the answer quickly by "picking and choosing" appropriate combinations:
(a) $5 \times 4 \times 3 \times 2 \times 1$ (b) $125 \times 7 \times 3 \times 8$
(c) $250 \times 14 \times 4 \times 2$ (d) $(513 \times 2) \times 50$
(e) $(11 \times 5) \times 20$

11 Rewrite each mathematical sentence below as a division sentence. Find the unknown factor.
(a) $n \times 5 = 20$ (b) $p \times 4 = 28$ (c) $n \times 1 = 6$

12 (a) Display an array to show $28 \div 7$.
(b) Illustrate $28 \div 7$ by a partitioning that is other than an array.

13 No division table was displayed in this chapter. State how you could use the multiplication table to find $a \div b$, if it exists, whenever $a \leq 81$ and $b \leq 9$.

14 For each of the following division expressions, write a quotient and remainder sentence.
(a) $20 \div 3$ (b) $20 \div 4$ (c) $3 \div 7$

15 Earl and Bill have the same number of candy bars. Earl divides his candy bars equally in 3 boxes, while Bill divides his equally in 5 boxes. Could each of them have 24 candy bars? 32? 30? 45? 16?

16 Notice that $(24 \div 3) \div 2 = 8 \div 2 = 4$.
Also $24 \div 6 = 24 \div (3 \times 2) = 4$.
(a) Is $36 \div 12 = (36 \div 4) \div 3$?
(b) Is $40 \div 8 = (40 \div 4) \div 2$?

17 Work the following problems without using pencil and paper.
(a) $856 \times (73 \div 73)$ (b) $856 \times (73 - 73)$
(c) $(a \div a) \times 43$ (d) $(a - a) \times 574$
(e) $(16 \times 0) \div 8$ (f) $(16 + 0) \div 8$

18 Determine the number N, given that

(a) $43 \times (N - 7) = 43$ (b) $86 + (N - 20) = 86$

(c) $63 \times (N - 18) = 0$ (d) $N \div 1 = 56$

(e) $84 \times (N - 10) = 84$ (f) $(N + 2) \times (b \div b) = 15$

19 How would you use the fact, $6 \times 9 = 54$, to find the unknown factor in $3 \times n = 54$?

20 Draw a number line for each statement.

(a) $4 \times 6 = 24$ (b) $6 \times 250 = 1{,}500$

(c) $32 \div 8 = 4$ (d) $2{,}800 \div 400 = 7$

TECHNIQUES OF MULTIPLICATION AND DIVISION

We saw in the previous chapter that, in order to multiply two whole numbers, we could construct an array with as many rows as the first number and as many columns as the second, and then count the members of the array. In that chapter we recorded, in our multiplication table, the results of doing this process for every pair of whole numbers less than 10.

For larger whole numbers, such a procedure would be both time-consuming and tedious. It would be boring to work the problem 24 × 38 by laying out 24 rows of pennies on the table with 38 in each row and then counting up all the pennies.

We would welcome an easier method of multiplying whole numbers. Fortunately, thanks to our place-value decimal system of numeration, and what we have learned about the properties of multiplication, there is an easier method.

■ **Question 1** Is there anything in principle which would prevent our enlarging the multiplication table in Chapter 7 to such a size that we could read off the answer to 24 × 38?

AN ALGORITHM FOR MULTIPLICATION

To illustrate this process, we will figure out 24 × 38 twice. First we will use an array of marks on the paper, rather than pennies, and we will count the members of the array in a special way. Then we will do the problem over and will see that if we are willing to rely on the commutative, associative, and distributive principles, we can do without the array. Figure 8.1 shows a 24-by-38 array.

We will count the members of this array in a number of steps.

STEP 1:
Starting at the top and going down, draw a horizontal line across the array just after each tenth row. Starting at the left and going right, draw a vertical line through the array just after each tenth column (see

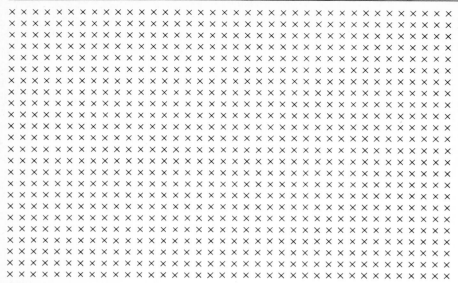

Figure 8.1

Figure 8.2). (As you probably guessed, we chose tenth because the base of our numeration system is ten.) Now the array appears as in Figure 8.2.

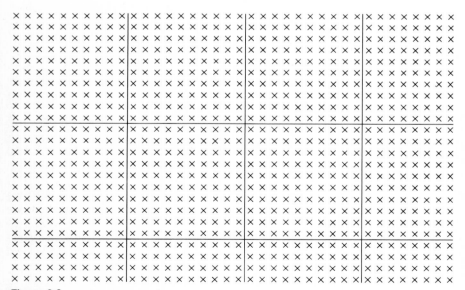

Figure 8.2

STEP 2:
We divide the array into four pieces, as in Figure 8.3.

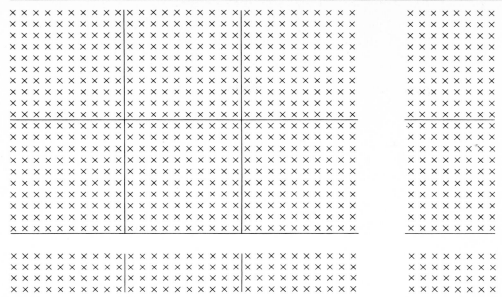

Figure 8.3

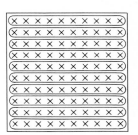

Figure 8.4

STEP 3:

Next we will count the members of the upper left-hand part of the original array. We remember that the horizontal and vertical lines were drawn after every tenth row and column. Therefore, in each of the squares there are 10 marks in each row and there are 10 rows. Group the marks of each row together as we did in Chapter 2 and then group the 10 rows of ten together. Then each square (magnified) appears as in Figure 8.4. We recognize that inside each square there are 10 sets of ten, or 1 hundred. But the squares form a 2-by-3 array. We remember (or look up in the table) that $2 \times 3 = 6$, so there are 600 marks in the top left-hand part of the original array (see Figure 8.5).

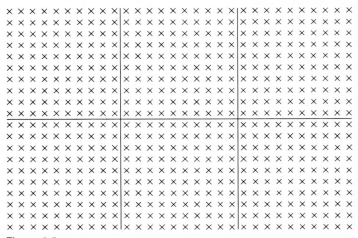

Figure 8.5

STEP 4:
Now we will count the number of marks in the lower left-hand part of the original array. First we group each row into a pile of ten (Figure 8.6). Now we see that the lower left-hand part of the original array is a 4-by-3 array of sets of ten. Since $4 \times 3 = 12$, there are 12 tens, or 120, marks in the lower left-hand part of the original array.

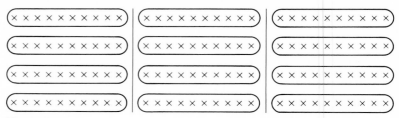

Figure 8.6

STEP 5:
For the upper right-hand part of the original array, we group into columns of ten (Figure 8.7). Now we can see that there is a 2-by-8 array of sets of ten. Since $2 \times 8 = 16$, there are 16 tens, or 160, in the upper right-hand part of the original array.

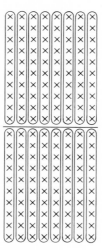

Figure 8.7

Figure 8.8

STEP 6:
Turning now to the lower right-hand part of the original array, we see a 4-by-8 array. Since $4 \times 8 = 32$, this is the number of marks in the lower right-hand part of the original array (Figure 8.8).

STEP 7:
Finally, we just add the number of marks in the four parts of the original array:

$$\begin{array}{r} 600 \\ 120 \\ 160 \\ +\ 32 \\ \hline 912 \end{array}$$

So $24 \times 38 = 912$

■ **Question 2** Suppose we worked 24×37 by the same method. In which of steps 3 to 6 will the arrays be different from those above?

ANOTHER ALGORITHM FOR MULTIPLICATION

Now we will do the same problem over again, but, instead of using arrays, we will use properties of our decimal place-value numeration system and properties of multiplication to justify each step.

STEP 1:

$24 \times 38 = [(2 \times 10) + 4] \times [(3 \times 10) + 8]$

Renaming in our numeration system.

STEP 2:

$[(2 \times 10) + 4] \times [(3 \times 10) + 8] =$
$\qquad (2 \times 10) \times (3 \times 10) + 4 \times (3 \times 10) + (2 \times 10) \times 8 + 4 \times 8$

Distributive principle used twice.

STEP 3:

$(2 \times 10) \times (3 \times 10) = (2 \times 3) \times (10 \times 10) = 6 \times 100 = 600$

Commutative and associative principles, the multiplication table ($2 \times 3 = 6$), and renaming ($10 \times 10 = 100$).

STEP 4:

$4 \times (3 \times 10) = (4 \times 3) \times 10 = 12 \times 10 = 120$

Associativity principle, the multiplication table, and renaming.

STEP 5:

$(2 \times 10) \times 8 = (2 \times 8) \times 10 = 16 \times 10 = 160$

Associative and commutative principles, the multiplication table, and renaming.

STEP 6:

$4 \times 8 = 32$

Multiplication table.

STEP 7:

$24 \times 38 = 600 + 120 + 160 + 32 = 912$

Addition technique.

■ **Question 3** Suppose we worked 4×38 by the same methods. Which of the above steps would we omit?

SHORTER ALGORITHMS Of course, we do not normally go into this much detail when working problems. We could shorten the seven-step procedure above like this:

$$\frac{38}{\times 24} = \frac{(30+8)}{\times(20+4)} = \frac{(30+8)}{\underline{\times 20}} + \frac{(30+8)}{\underline{\times 4}}$$

$$= \frac{30}{\underline{\times 20}} + \frac{8}{\underline{\times 20}} + \frac{30}{\underline{\times\ \ 4}} + \frac{8}{\underline{\times\ \ 4}}$$
$$\quad\ \ 600 \qquad 160 \qquad 120 \qquad 32$$

$$= 912$$

Usually, however, we write a problem like this in vertical form.

$$
\begin{array}{r}
38 \\
\times\ 24 \\
\hline
32 \\
120 \\
160 \\
\underline{600} \\
912
\end{array}
$$

$$
\begin{array}{l}
(4 \times 8) \\
(4 \times 30) \\
(20 \times 8) \\
(20 \times 30)
\end{array}
$$

If we are willing to do some mental arithmetic and to add the "3 tens" from $32 = (3 \times 10) + 2$ to the "2 tens" from $120 = (1 \times 100) + (2 \times 10)$, and similarly to add the "1 hundred" from $160 = (1 \times 100) + (6 \times 10)$ to the "6 hundreds," we can condense the problem above to:

$$
\begin{array}{r}
38 \\
\times\ 24 \\
\hline
152 \\
760 \\
\hline
912
\end{array}
$$

and this in turn can be shortened to:

$$
\begin{array}{r}
38 \\
\times\ 24 \\
\hline
152 \\
76 \\
\hline
912
\end{array}
$$

Of course, any of these variations is as good as any other. The last uses the fewest figures but requires more mental arithmetic. Each individual should be free to put whatever value he wants on this trade-off.

■ **Question 4** Why was it permissible, in the last example above, to omit the 0 at the end of 760?

LARGER NUMBERS To multiply three-digit or larger numbers, the same techniques apply. For example, if

$$n = 234 \times 433$$

we may write

$$n = (200 + 30 + 4) \times 433$$
$$= (200 \times 433) + (30 \times 433) + (4 \times 433)$$
$$= (4 \times 433) + (30 \times 433) + (200 \times 433)$$

```
    433
  × 234
  1,732 ─────┐
 12,990 ──────────┐
 86,600 ──────────────┐
101,322 = 234 × 433
```

Sometimes, particularly if you are interested in estimates rather than exact answers, you can think like this:

$$38 \times 43 = (30 + 8) \times 43$$
$$= (30 \times 43) + (8 \times 43)$$
$$= (30 \times 40) + (30 \times 3) + (8 \times 40) + (8 \times 3)$$
$$= (40 \times 30) + (3 \times 30) + (40 \times 8) + (3 \times 8)$$

A rough estimate of the answer would be 30×40; for a better estimate, add 40×8 to the rough estimate; and for a still better estimate, add 30×3 to the second estimate.

	38 ×43				
(40 × 30)	1,200	1,200	1,200	1,200	1,200
(40 × 8)	320		320	320	320
(3 × 30)	90			90	90
(3 × 8)	24				24
	1,634	1,200	1,520	1,610	1,634
		rough estimate	better estimate	still better	exact answer

Notice that in this case we start multiplying with the digits on the left which represent the larger parts of the given factors.

Notice also the positional system coming into play in each of the intermediate computations. For example,

$$40 \times 30 = 4 \times 10 \times 3 \times 10$$
$$= 4 \times 3 \times 10 \times 10$$
$$= 12 \times 100$$
$$= (10 + 2) \times 100$$
$$= (10 \times 100) + (2 \times 100)$$
$$= (1 \times 1,000) + (2 \times 100)$$

This indicates a 12 in the hundreds position, or a 1 in the thousands position and a 2 in the hundreds position. The pattern of the number of zeros should also be noted in the multiplication process.

■ **Question 5** In this multiplication problem, what does the "32" stand for?

$$
\begin{array}{r}
4,275 \\
\times\,8 \\
\hline
40 \\
56 \\
16 \\
32 \\
\hline
34,200
\end{array}
$$

**AN ALGORITHM FOR
DIVISION**

We saw in the last chapter that division is the process of finding the missing factor in a multiplication problem when we are given the product and the other factor. When we are asked to find $20 \div 4$, we are asked to find a number n such that $4 \times n = 20$ (or $n \times 4 = 20$).

But multiplication means an array, so when we are asked to find $20 \div 4$, we are asked to find out how many columns there are in an array if there are 4 rows and if there are 20 members of the array.

Figure 8.9

One way to go at this is to start out with 20 things (in our case, as usual, marks on the paper) and to try to fit them into a 4-rowed array, and then to count the number of columns. We could do this by filling in one column at a time. In Figure 8.9, we have 20 things, and one set of 4 of them has been selected to fill the first column. After these 4 marks have been put into the first column of the array, the picture appears as in Figure 8.10, with the array on the left and the set of marks on the right.

Figure 8.10

Now we can select another set of 4 and put them into the second column (Figure 8.11).

Figure 8.11

Continuing, pulling out one set of 4 at a time, we have the pictures in Figure 8.12. Now we have used up all the marks, and there are 5 columns in the array, so $20 \div 4 = 5$.

■ **Question 6** If the problem had been $20 \div 5$, how many steps would this method require?

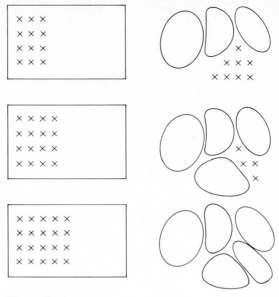

Figure 8.12

ANOTHER ALGORITHM FOR DIVISION

Let's go back and count how many marks are left in the original set after each step. When we started, we had 20 marks, and at the first step we took out one set of 4. Then, by subtraction we can find out how many are left.

$$\begin{array}{r} 20 \\ -\ 4 \\ \hline 16 \end{array}$$

At the next step, we take out 4 more.

$$\begin{array}{r} 16 \\ -\ 4 \\ \hline 12 \end{array}$$

Continuing,

$$\begin{array}{r} 12 \\ -\ 4 \\ \hline 8 \end{array} \qquad \begin{array}{r} 8 \\ -4 \\ \hline 4 \end{array} \qquad \begin{array}{r} 4 \\ -4 \\ \hline 0 \end{array}$$

Notice that we subtracted 4 five times. We could reduce the number of subtractions if we were to move over to the array a larger number of things at a time. For example, we could move over to the array, at the first step, enough marks to fill up two columns (see Figure 8.13). On the right we placed a 2 to the right of 8 to remind us that we removed 2 sets of 4.

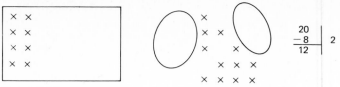

Figure 8.13

Now we could move enough marks over to the array to fill two more columns (Figure 8.14).

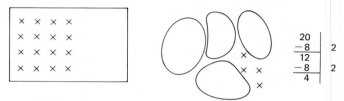

Figure 8.14

One more set of 4 marks takes care of them all (Figure 8.15).

Figure 8.15

The number of sets of 4 that we moved over to the array are indicated to the right of the vertical line, 2 at the first step, 2 more at the second step, and 1 at the last step, a total of 5. So, to compute 20 ÷ 4 we could just add up the numbers to the right of the vertical line.

Of course, if we remembered our multiplication table well enough, we could have done the whole thing in one fell swoop (Figure 8.16).

Figure 8.16

■ **Question 7** If a student started the problem 20 ÷ 4 by writing

$$\begin{array}{r|l} 20 & \\ -12 & 3 \\ \hline 8 & \end{array}$$

what does his array look like at this stage?

Looking back now at this problem, three points stand out. The first is that the process can always be tried, even if it does not work. For example, if we are asked for 22 ÷ 4, we could move five sets of 4 over to the array and fill five columns. However, there would be 2 marks left over, not enough to fill another column; so there is no whole number a such that $4 \times a = 22$. We will come back to this point later.

The second point is that it saves time (and space on the page) to fill more than one column at a time. Closely related to this is the third point: It helps to refer to our multiplication table to see how many columns can be filled at the same time.

SHORTER ALGORITHMS FOR DIVISION

With these points in mind, let us try another example, 42 ÷ 3. We will not draw the arrays or the set of 42 things, but we can easily imagine them if necessary. We first remember that it saves time to fill in more than one column of the array at a time. Also, at this time, we remember that because we use a base ten numeration system, it is convenient to deal in multiples of ten. So let us ask first if we can take out from the set of marks enough to fill 10 columns. This would require 10 sets of 3 or 30. Since 42 is greater than 30, we can do this. Now we have to subtract 30 from 42 to see how many marks are left.

```
3) 42
  −30   10
   12
```

(Notice how we have written the "3" to the left of the "42" to remind ourselves that we are looking for 42 ÷ 3.) Now we remember that $3 \times 4 = 12$. Therefore we can pull out another 4 columns of 3 and put them in the array.

```
3) 42
  −30   10
   12
  −12   4
    0
```

We see that we have used up the whole set of 42 marks, and that the 3-rowed array has 10 + 4 = 14 columns, so 42 ÷ 3 = 14.

Now let us try 75 ÷ 3. We could proceed just as we did in the problem above.

```
3) 75
  −30   10
   45
  −30   10
   15
  −15   5
```

However, we now notice that by remembering that $3 \times 2 = 6$, so that $3 \times 20 = 60$, we could have combined the first two steps.

```
3) 75
  −60   20
   15
  −15   5
```

In either case, we add up the right-hand columns to find that 75 ÷ 3 = 25.

■ **Question 8** Would there be anything wrong in working 96 ÷ 3 this way?

```
3) 96
  -60 | 20
   36
  -30 | 10
    6
   -3 | 1
    3
   -3 | 1
```

Larger problems can be done the same way. We try to fill in more than one column of the array at a time by working with multiples of 10, or multiples of 100, or of 1,000, etc.

For example:

```
6) 768
  -600 | 100
   168
  -120 | 20
    48
    48 | 8
```

Another way of writing this problem is to record the information in the right-hand column up above instead. For example, the first step in the above problem was to fill in 100 columns of the 6-rowed array and to compute the number, 600, of things moved from the pile of 768 over to the array. We could record this as

```
    1
6) 768
   600
```

Note that we write only the digit "1" instead of "100." But we put the digit above the hundreds place of 768.

Next we subtract:

```
    1
6) 768
  -600
   168
```

We see that we can now fill in 20 more columns.

```
   12
6) 768
  -600
   168
  -120
```

Again, we have written just the "2" of the "20," but we placed it over the tens place of 768. Continuing,

```
   128
6) 768
  -600
   168
  -120
    48
    48
```

We could abbreviate this a bit more by not bothering to write those zeros which did not enter into the calculations.

```
    128
6 ) 768
    6
    168
    12
     48
     48
```

Finally, we note that "8" in "168" had a zero subtracted from it, so the same "8" was bound to appear below it in the next step, so we did not really have to write it at all in the second step.

```
    128
6 ) 768
    6
    16
    12
     48
     48
```

■ **Question 9**   A student starts the problem 1,272 ÷ 53 by writing

```
         2
53 ) 1,272
     1 06
```

What does the "106" stand for?

Here is one more example, done in three forms.

```
7 )  16,219              7 ) 16,219       7 ) 16,219
   −14,000    2,000        14,000           14
     2,219                  2,219           22
   − 2,100     300          2,100           21
       119                   119            11
     −  70      10            70             7
        49                    49            49
     −  49       7            49            49
             —————
             2,317
```
Over the right columns: 2,317 and 2,317

DIVIDING BY LARGER NUMBERS

So far, in all our examples we have been dividing by a number less than 10.   Now we have to ask what happens if we divide by a number greater than 10.   As we shall see, the procedure is pretty much the same, except for one thing.   Let us use the problem 996 ÷ 12 as an illustration.

The first thing we remember is that we do not want to fill just one column of the array at a time.   We can easily fill the first 10 columns all at once, since 10 × 12 = 120 is much smaller than 996.   On the other hand, we can see that we cannot fill in 100 columns at once because 100 × 12 = 1,200 is larger than 996.   So we will move a certain number of sets of 10 sets of 12 marks over to the array.

But how many sets of 10 sets of 12 marks can we move?   Here is where things become more complicated than when we are dividing by a number less than 10.   For example, if we had the problem 438 ÷ 6, we could look along the "6" row of our multiplication table.   When we get to the "7" column we find that since 6 × 70 = 420, we could fill in

70 columns all at once, but the "8" column of the multiplication table tells us that there are not enough things in our set to fill 80 columns all at once.

But there is no "12" row in our multiplication table. We have two choices. We could work out the complete "12" row and then go along it until we find the largest number of 10 sets of 12 we could move over to the array.

A second choice would be to construct only part of the "12" row, the part most relevant to our particular problem. We notice that $12 = 1 \times 10 + 2$ and we could certainly move 90 sets of 10 over to our array. But we need sets of 12, and $90 \times 12 = 1,080$ is more than we have. But it is not much more, so let us try 80.

```
12) 996
   -960 | 80
     36
    -36 | 3
```

We find that 80 works fine.

Of course, we could have gone back to the easier, but simpler, method.

```
12) 996
   -120 | 10
    876
   -120 | 10
    756
   -120 | 10
    636
   -120 | 10
    516
   -120 | 10
    396
   -120 | 10
    276
   -120 | 10
    156
   -120 | 10
     36
    -30 | 3
```

What we actually did was to guess that there would be nine steps which could be combined into one. Our guess was wrong, but it suggested that we next guess that the first eight steps could be combined into one.

In general, when we are dividing by numbers larger than 10, we either do it the long way, as above, or we guess. If we guess, our first guess may be wrong, but it should guide us to a better second guess.

■ **Question 10** A teacher taught his students: "In division, always try 5 first. If it is too small, try 6, 7, 8, and 9 one after the other until you find one that works. If 5 is too large, try 4, 3, 2, and 1 until you find one that works." What is wrong with this rule?

Let us try one more example, 1,488 ÷ 48. We can see without much trouble that we can move sets of 10 sets of 48 over to the array, but that we cannot move even one set of 100 sets of 48. So we need to guess how many sets of 10 sets of 48 we can find in our set of 1,488 things.

We note that 48 is close to $50 = 5 \times 10$, and 1,488 is close to $1,500 = 150 \times 10$. There are $1,500 ÷ 5 = 30$ sets of 50 in 1,500, so 30 is a reasonable guess as to how many sets of 48 we can find in 1,488. When we try it we find that 30 is indeed a good guess.

```
48) 1,488
   -1,440  | 30
       48
      -48  | 1
```

We find that $1,488 ÷ 48 = 31$.

Some people concentrate on trying to make their first guess a good one every time. Others would rather do more work but avoid guessing and would do the problem the long but easy way:

```
48) 1,488
   -480   | 10
   1,008
    -480  | 10
     528
    -480  | 10
      48
     -48  | 1
```

Most people compromise. They try to come close with their first guess, but if it is not quite right, they use these calculations to guide them to a better guess.

Either of the problems we have just looked at can be written in a more condensed form, just as in the case where we divided by numbers less than 10. Here is the first one.

```
        83            83
12) 996          12) 996
    960               96
     36               36
     36               36
```

Here is another example, done at two levels of condensation.

```
        462               462
87) 40,194          87) 40,194
    34,800               348
     5,394               539
     5,220               522
       174               174
       174               174
```

■ **Question 11** In the problem 1,162,586 ÷ 486, which would you try first?

2, 20, 200, 2,000, 20,000, . . .

These days, most of the important arithmetic calculations that are made are carried out by some kind of calculating machine. In the past, however, before these machines were invented, arithmetic computations had to be done by human beings. Over the course of history, many tricky shortcuts were invented which made it possible to do multiplication and division problems faster. Most of them involved doing a good deal more mental computation than in even the most condensed forms shown above. These shortcuts are only of historical interest now. Today it is much more important to understand the multiplication and division processes than it is to be able to do computations quickly.

REMAINDERS IN DIVISION

We noted in Chapter 7 that the whole numbers are not closed under the operation of division. Thus, for example, for $37 \div 7$, division as an operation does not yield a whole number. The attempt to set up an array of 37 elements in 7 rows fails, but in the attempt, we were led to a process which gave us a quotient and a remainder—namely 5 and 2.

Let us review what occurs as we attempt to set up an array of 37 elements in 7 rows. First, from the set of 37 elements, we obtain 7 elements and place these 7 in the first column. In the original set of the 37 elements now remain 30 elements. We proceed to obtain another 7 elements from the remaining elements, etc. This process is indicated as follows:

$$
\begin{array}{ll}
37 & \text{elements in the original set} \\
-\ 7 \\
\hline
30 & \text{elements remain after displaying 1 to a row} \\
-\ 7 \\
\hline
23 & \text{elements remain after displaying 2 to a row} \\
-\ 7 \\
\hline
16 & \text{elements remain after displaying 3 to a row} \\
-\ 7 \\
\hline
9 & \text{elements remain after displaying 4 to a row} \\
-\ 7 \\
\hline
2 & \text{elements remain after displaying 5 to a row}
\end{array}
$$

The final step shows that the quotient is 5 and the remainder is 2. Notice that each step may be described as follows:

$$
\begin{array}{ll}
& 37 \\
& -\ 7 \\
37 = (1 \times 7) + 30 & \overline{30} \\
& -\ 7 \\
37 = (2 \times 7) + 23 & \overline{23} \\
& -\ 7 \\
37 = (3 \times 7) + 16 & \overline{16} \\
& -\ 7 \\
37 = (4 \times 7) + 9 & \overline{9} \\
& -\ 7 \\
37 = (5 \times 7) + 2 & \overline{2}
\end{array}
$$

Of course, we could shorten this process, just as we did earlier with the division algorithm.

Thus

$$\begin{array}{r} 5 \\ 7\overline{)37} \\ 35 \\ \hline 2 \end{array}$$

So

$$37 = 5 \times 7 + 2$$

Similarly,

$$\begin{array}{r} 19 \\ 4\overline{)77} \\ 4 \\ \hline 37 \\ 36 \\ \hline 1 \end{array}$$

So

$$77 = 19 \times 4 + 1$$

Many problems have a number relationship which, when put into a number sentence, reveals the need of applying the division process. For example, how many 47-passenger buses will a school have to order to take its 820 pupils on a school excursion? We do not know if they will fit evenly into a certain number of buses, but we rather doubt it, so we set up the number sentence

$820 = (q \times 47) + r$	and we find that
$820 = (10 \times 47) + 350$	so we need more than 10 buses
$350 = (5 \times 47) + 115$	so we need more than $10 + 5 = 15$ buses
$115 = (2 \times 47) + 21$	so we need 17 buses and have 21 children left.

We either order 18 buses and have some empty seats or provide a smaller bus or some private cars for the 21 children remaining after 17 buses have been filled, since we do not want to leave anybody behind. More quickly, of course, we could compute

$$\begin{array}{r} 17 \\ 47\overline{)820} \\ 47 \\ \hline 350 \\ 329 \\ \hline 21 \end{array} \qquad 820 = (17 \times 47) + 21$$

■ **Question 12** What are q and r for $52 \div 52$?

$$\begin{array}{r} 433 \\ \times 234 \\ \hline 1732 \\ 1299 \\ 866 \\ \hline 101,322 \end{array}$$

PROBLEMS

1 In the example to the left, explain why the 866 on the fifth line does not represent 2×433.

2 Fill in the parentheses to explain where each line came from in this multiplication problem.

```
      2,146
  ×    823
         18   (        )
        120   (        )
        300   (        )
      6,000   (        )
        120   (        )
        800   (        )
      2,000   (        )
     40,000   (        )
      4,800   (        )
     32,000   (        )
     80,000   (        )
  1,600,000   (        )
```

3 In finding the product 423 × 624, why can the zeros that are crossed off in the algorithm below be omitted?

```
       624
       423
     1 872
    12 48Ø
   249 6ØØ
   263,952
```

4 One way of working 25 × 17 is

$$25 \times 17 = 25 \times (10 + 7)$$
$$= (25 \times 10) + (25 \times 7)$$
$$= 250 + 175$$
$$= 425$$

Another is

$$25 \times 17 = (2 \times 10 + 5) \times 17$$
$$= (2 \times 10 \times 17) + (5 \times 17)$$
$$= 340 + 85$$
$$= 425$$

Still another is

$$25 \times 17 = (2 \times 10 + 5) \times (10 + 7)$$
$$= (2 \times 10 \times 10) + (2 \times 10 \times 7) + (5 \times 10) + (5 \times 7)$$
$$= 200 + 140 + 50 + 35$$
$$= 425$$

Work this in these three ways:

36 × 82

5 Find the product of 6,439 and 7 first by using 7 as the multiplier and then by using 6,439 as the multiplier.
6 Find the product of 1,457 and 2,004 first by using 2,004 as the multiplier and then by using 1,457 as the multiplier.
7 Multiply in the usual way and then again using the digits of the multiplier in left-to-right order.
   (a) 24 × 207   (b) 155 × 643
8 Show a rough estimate and show successively better estimates for
   (a) 43 × 21   (b) 46 × 404   (c) 518 × 4,682

9 Since $2 \times 5 = 10$, how many zeros follow the 1 in $20,000 \times 50,000$? In which position is the 1 in $20,000 \times 50,000$?

10 What has been done wrong in this problem?

```
    43
   ×32
   120
    90
    80
     6
   296
```

11 We discussed division by numbers less than 10 and by numbers greater than 10. Why did we not discuss division by 10?

12 Is the work in the middle correct? On the left?

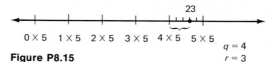

```
                                 3
                                20
                               100            123
   37)4,551            37)4,551          37)4,551
     3,700   100 × 37    3,700             37
       851                 851             85
       740    20 × 37      740             74
       111                 111            111
       111     3 × 37      111            111
```

13 Use the division algorithm to find the quotient and the remainder for each of these.
  (a) $(512 \div 8)$ (b) $(779 \div 18)$
  (c) $(23 \div 14)$ (d) $(50 \div 100)$
  (e) $(6,535 \div 47)$

14 Check each of your answers in Problem 13 by multiplying the quotient by the divisor and adding the remainder to the result.

15 For $37 \div 7$ show the segments of 7s on the number line and locate the point representing 37. Find the quotient q and the remainder r on the number line. See, for example, Figure P8.15 for $23 \div 5$.

```
                          23
 ←———+———+———+———+——|·|·|·|———→
   0×5   1×5   2×5   3×5  4×5  5×5
                                    q = 4
 Figure P8.15                       r = 3
```

16 Do the same for the pair $3 \div 8$.

17 Write in the form $a = (q \times b) + r$ for each of these:
  (a) $23 \div 5$ (b) $3 \div 8$ (c) $37 \div 7$

18 Determine the missing digits in these problems.

```
           34
 (a)  12)***        (b)      36
         3*                 ×1*
         4*                 1*0
         48                  36
          0                 **0
```

19 Use your multiplication table for numbers in base three to compute:
  (a) $2_{three} \times 21_{three}$ (b) $12_{three} \times 202_{three}$

20 Using the table, work these problems.
  (a) $120_{three} \div 12_{three}$ (b) $2222_{three} \div 12_{three}$

MATHEMATICAL SENTENCES

We said in Chapter 1 that whole number arithmetic is a special way of looking at the world which makes it easier for us to answer "how many" questions. Quite often, when we are faced with a problem which asks a "how many" question, we can find the answer to the problem by translating the problem from the English language in which it is written into mathematical language.

NUMBER SENTENCES   In developing the properties of numbers and various operations on numbers, we have been using a rather special language involving

symbols for numbers, such as: 1, 5, 2, 9, 3, . . .
symbols for operations, such as: $+, -, \times, \div$
and symbols showing relations between numbers, such as: $=, \neq, <,$
$>, \leq, \geq$

We have seen that a number may be named by many numerals. For example, $3 + 4$, $9 - 2$, $\frac{28}{4}$, 7, VII, all name the same number, and we may write

$$3 + 4 = 9 - 2 \quad \text{or} \quad 9 - 2 = \tfrac{28}{4} \quad \text{or} \quad 3 + 4 = 7 \ . \ . \ .$$

to show that these are numerals for the same number. In this way, we form mathematical sentences, where the symbol "=" acts as the verb. The numeral to the left acts as the subject, and the numeral to the right acts as a predicate noun.

A statement such as "$7 - 5 = 2$" is in mathematical form, but it can be put into words as in the sentence "When five is subtracted from seven, the result is two." The sentence "The result of adding the number five to the number nine is the number fourteen" can also easily be put in the much shorter form "$9 + 5 = 14$."

A great deal of mathematics is in the form of sentences about numbers or *number sentences* as they are called. Sometimes the sentences make true statements as in both of the above examples;

sometimes the number sentences are false as in the cases "5 + 7 = 11" or "17 − 4 = 3." Whether it is true or false no more disqualifies the statement as a sentence than the statement "George Washington was vice president under Abraham Lincoln" is disqualified as a sentence.

■ **Question 1**   Which of these are *not* number sentences?

$$7 = 7 \qquad 5 - 2 = 1 + 2 \qquad 8 - 3 \neq 4 + 1 \qquad 4 + 5 = 4 + 5$$

$$9 + 2 \neq 2 + 9 \qquad 5 < 5 \qquad 8 - 3 < 3 \times 8 \qquad 1{,}001 \geq 1{,}001$$

OPEN SENTENCES   As we have noted, verbal sentences may be true: "George Washington was the first president of the United States," or false: "Abraham Lincoln was the first president of the United States." We also encounter sentences such as: "He was the first president of the United States." If read out of context, it may not be known to whom "he" referred and it may thus be impossible to determine whether the sentence is true or false. In fact, "□ was the first president of the United States" may be a test question requiring the name of the man for which it would be a true sentence. Such a sentence is called an *open sentence* and is of great usefulness not only in history tests but in many other situations as well. In fact, open number sentences are the basis of a great deal of work in arithmetic. For example, our definition of subtraction really used an open number sentence. "7 − 5 is that number which makes the open sentence 5 + □ = 7 a true statement."

■ **Question 2**   What open sentence corresponds to 8 ÷ 8?

Any number sentence has to have a verb or a verb phrase. The most common ones are: "is equal to," "is not equal to," "is more than," "is less than," "is more than or is equal to," "is less than or is equal to." The symbols which we use for these verbs are listed in the table below to the left, with examples of sentences using them to the right.

Symbol	English Translation	Example
=	is equal to	3 + 4 = 7
≠	is not equal to	5 − 1 ≠ 4
>	is more than	7 − 3 > 1
<	is less than	15 < 10
≥	is more than or equal to	9 ≥ any one-digit number
≤	is less than or equal to	0 ≤ any whole number

None of the examples listed to the right above are open sentences. They make statements about specific numbers which are described or represented by a single numeral such as 7 or by a *mathematical* or number phrase such as 3 + 4. If we want to write an open number sentence, we will use an *open number phrase* such as □ + 7 or 17 − □ where the symbol □ is used to help you remember that the empty space may be filled by some numeral. Because symbols like □ are awkward to type or write, we frequently use a letter such as n or

a or x for the same purpose. Thus, a simple open number phrase may be written as $n + 7$ instead of $\square + 7$ and an open number sentence as $n + 7 = 10$. What whole number or numbers will now make this open sentence a true statement? In this case the answer is easily obtained by trial. It is true that $3 + 7 = 10$, while $0 + 7 \neq 10$, $1 + 7 \neq 10$, $2 + 7 \neq 10$, $4 + 7 \neq 10$, etc., and we see that 3 is the only number which does the trick.

■ **Question 3** What numbers make this open sentence a true statement?

$x + 1 = 0$

What number or numbers will make the open sentence $\square < 5$ a true statement? Again, by trial we find that $0 < 5$, $1 < 5$, $2 < 5$, $3 < 5$, and $4 < 5$ are true statements while $5 < 5$, $6 < 5$, $7 < 5$, etc., are false statements. Thus we see that any member of the set $\{0, 1, 2, 3, 4\}$ makes the statement true. What about the open sentence $n + 6 < 11$? We can translate the sentence into words by saying "the sum of a certain number and 6 is less than 11" and we see the numbers which make this a true statement are again the members of the set $\{0, 1, 2, 3, 4\}$.

SOLVING OPEN SENTENCES

Open number sentences are called equations if the verb in them is "=." Sentences with any of the other verbs listed above are called "inequalities." Those numbers which make the sentences true are called *solutions* of the equations or inequalities. When you have found the entire set of solutions of an open sentence, you can say that you have *solved* the sentence.

We said above that by *solving* an open equation or inequality we mean finding that number, or all those numbers, which make the sentence a true one. At this time, you can do this primarily by trial and error after thinking carefully about what the sentence says. For instance, to solve the equation $n - 4 = 7$ means to find the number which is the result of adding 4 to 7. The answer is, of course, 11. To solve $n + 5 = 32$ is to find that number which added to 5 will give 32, or $n = 32 - 5 = 27$. On the other hand, to solve $n - 4 \leq 7$ means to find all those numbers from which 4 may be subtracted and for which the result will be less than or equal to 7. Is 2 such a number? No, because $2 - 4$ is not a whole number. Is 3? No. But 4, 5, 6, 7, 8, 9, 10, and 11 do make the sentence true. On the other hand, $12 - 4 = 8$, which is more than 7. So 12 and any other larger number make the sentence false. We see that the set of solutions is $\{4, 5, 6, 7, 8, 9, 10, 11\}$.

■ **Question 4** What is the set of solutions of the open sentence

$x + 1 \neq 0$

USE OF MATHEMATICAL SENTENCES

Open number sentences are frequently used to solve problems. To do this, you must be able to describe the numbers in the problems by number phrases and to translate the clues given in the problem into an equation or inequality. To work with number phrases you must be able to translate the phrase into words. The open phrase $\square + 5$ or the equivalent phrase $n + 5$ may be translated as "a number increased by 5." It may, of course, have many different translations, such as:

"a number n added to 5"
or "the sum of a number and 5"
or "5 more than a number n"

However, all of the translations have the same mathematical meaning. Furthermore, all the English translations mean the same as "$n + 5$." With practice, we learn to understand the different ways of expressing a number phrase.

The use of a mathematical sentence to solve a problem may be illustrated as follows:

There are 22 children in a class, and 10 of the children are boys. How many are girls? We can write several different open sentences to express the relationship among the numbers involved. Thus, $10 + n = 22$ or $22 - n = 10$. In each case, we can think "a number added to 10 gives the sum 22." The only number which makes this a true statement is 12. This is the solution of the number sentence, and the answer to the problem is: "There are 12 girls in the class."

■ **Question 5** Rewrite the open sentence $2x + 5 = 17$ in English.

In using number sentences to solve problems, the key to the situation is in recognizing the relationship between the numbers in the problem. This relationship is written as a number sentence. The solution of the number sentence is found and the result used to answer the question posed by the problem. One more example follows.

John put 23 of his marbles in a bag and Jim put 48 of his marbles in the same bag. If Tom takes out 35 marbles, how many are left in the bag? The number relationship can be thought of as "the number of marbles left in the bag plus the number Tom took out equals the number John and Jim put in." This gives the equation $n + 35 = 23 + 48$. Someone else might think of the relationship as "the number of marbles left in the bag is the difference between the number John and Jim put in and the number Tom took out." This yields $n = (23 + 48) - 35$. Of course, both open sentences have the same solution: 36. The answer to the problem is: "There are 36 marbles left in the bag." Notice that $n + 35 = 23 + 48$ if and only if $n = (23 + 48) - 35$.

■ **Question 6** Write an equation for this problem: If 15 children are to be arranged into 3 equal teams for a race, how many children will be on each team?

OPEN VERB PHRASES We can also have open sentences in which, instead of open number phrases, we have open verb phrases. Consider this problem: Jane has a set of marbles. There are a marbles in this set. Sue also has a set of marbles. There are b marbles in her set. Jane can make a

13-by-13 array with her marbles, with none left over. Using all her marbles, Sue can make an 11-by-15 array. Does Jane have more than, or the same number as, or fewer marbles than Sue?

To answer this question, we need to decide, for the open sentence

$$13 \times 13 \ \Box \ 11 \times 15$$

which of the verbs, $<, =, >$, makes the sentence true. A little computation shows that 13×13 is larger than 11×15, so we find that Jane has more marbles than Sue and the verb $>$ makes the sentence true.

SOLUTION SET ON THE NUMBER LINE

Frequently, a picture of a solution set using the number line can be drawn. Consider the following example for the open sentence

$$\Box + 3 = 8$$

This open sentence has the solution 5. The solution set is $\{5\}$. On the number line this solution can be represented as shown in Figure 9.1. Since the only solution for the sentence is 5, a solid "dot" or circle, is marked on the number line to correspond with the point for 5. No other mark is put on the drawing.

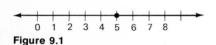

Figure 9.1

■ **Question 7** Show the solution set of $X + 2 = 1$ on the number line.

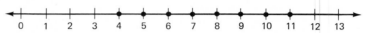

Figure 9.2

The solution set of the inequality $n - 4 \leq 7$ which we solved previously can be represented as in Figure 9.2. Note that on the number line we indicate the solution set by heavy solid dots. The solution set of $n - 4 > 7$ cannot be completely represented because it consists of all numbers greater than 11. We can indicate it, however, as in Figure 9.3 where the heavy dots continue right up to the arrow and the word "incomplete" indicates that all the numbers represented by points still further to the right belong in the solution (see Figure 9.3).

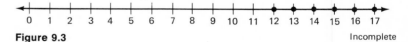

Figure 9.3   Incomplete

OPERATIONS ON THE NUMBER LINES

Number sentences can also be pictured on the number line as shown in Figures 9.4 and 9.5.

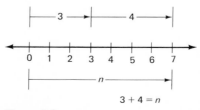

$$3 + 4 = n$$

Figure 9.4

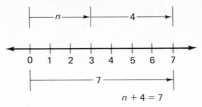

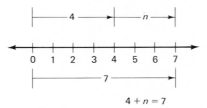

Figure 9.5

Recalling that $n + 4 = 4 + n$, and that $4 + n = 7$ if and only if $n = 7 - 4$, we can observe that

$$n + 4 = 7, \quad 4 + n = 7, \quad \text{and} \quad n = 7 - 4$$

are all statements which say the same thing. We can picture the number sentence $n = 7 - 4$ as in Figure 9.6 and note that the arrow for n agrees with the arrow for n in the picture for $4 + n = 7$ in Figure 9.7.

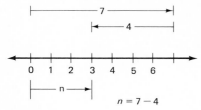

Figure 9.6

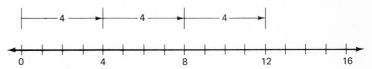

Figure 9.7

Just as addition and subtraction sentences may be shown on the number line, multiplication and division sentences may also be illustrated. For example, to show $3 \times 4 = n$ consider an arrow for 4. Three such arrows laid end to end (tail to head) indicate 3×4 (see Figure 9.8).

Figure 9.8

Figure 9.9 illustrates the sentence for $12 \div 4 = n$. Three 4-arrows fit end to end, showing $n = 3$.

Figure 9.9

■ **Question 8** Show $1 + 2 \times 3 - 4 = n$ on the number line.

PROBLEMS

1 Consider the following sentences:
  (a) The sum of 8 and 7 is 19. (b) $n + 8 = 16$
  (c) $4 + 5 = 10 - 1$ (d) $3 < 2 + 6$
  (e) $5 > 2 + \square$
  Which are false, which are true, and which of them are open sentences?

2 Solve each mathematical sentence.
  (a) $10 + p = 30$ (b) $0 = a + 0$
  (c) $0 - b = 0$ (d) $n - 9 = 20$
  (e) $40 - x = 10$ (f) $15 - t = 12$

3 Tell what operation is used to solve each of these mathematical sentences.
  (a) $5 + 6 = n$ (b) $a = 7 - 4$
  (c) $p + 2 = 43$ (d) $75 = x + 31$
  (e) $5 + b = 6$ (f) $91 - 60 = s$

4 Solve each of these open mathematical sentences:
  (a) $8 \square 6$ (b) $3 + 4 \square 6$
  (c) $(20 + 30) \square (30 + 20)$ (d) $(200 + 800) \square (200 + 700)$
  (e) $(1,200 + 1,000) \square (1,000 + 1,200)$

5 Write a number sentence suggested by each number line in Figure P9.5.

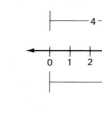

Figure P9.5

(a)

(b)

6 How much of a number line must be shown to picture these mathematical sentences?
  (a) $15 + 18 = n$ (b) $140 - n = 40$ (c) $n = 10 + 20 + 30$

Problems 7 to 20. For each of the following problems:
  (a) Write an open number sentence. (b) Solve it.

7 Jim delivered 35 of his 49 papers. How many more papers does he have to deliver?

8 There are 6 volleyballs in a carton. Our school buys 4 cartons of volleyballs. How many balls do they buy?

9 On Monday Mr. Brown drove 360 miles, on Tuesday 419 miles, on Wednesday 284 miles. How many miles did he drive altogether?

10 Carol paid 99 cents for 3 pairs of anklets. How much did each pair of anklets cost?

11 Louis shares 15 gumdrops equally with 2 of his friends. How many gumdrops does each child receive?

12 Miss Reed has 33 children in her class and 17 children are girls. How many boys are in the class?

13 Ralph delivers 82 papers each day. How many papers does he deliver on Monday through Friday?

14 When Bill went to the school party he counted the children there. After 14 children left he counted again and found 35 were still there. How many had Bill counted the first time?

15 At a popcorn sale, 29 bags were sold in one day. If 12 bags were sold in the morning, how many bags were sold in the afternoon?

16 How many weeks are there in 35 days?

17 Mr. Green drives 12 miles to work each morning and drives 12 miles home each evening. How many miles does he drive in a 5-day work week?

18 Jerry's mother sent him to the store to buy a loaf of bread for 31 cents a loaf, a can of corn for 23 cents, and two candy bars for 5 cents each. Jerry gave the clerk a $1 bill. How much change should Jerry receive from the clerk?

19 There were 300 pupils at the school football game on Tuesday, 250 pupils attended the school football game on Wednesday, and 50 pupils attended both the games. How many pupils attended just one of the games?

20 There are 20 pupils in a class. The class has just two committees to plan a party. The refreshments committee has 7 members. The games committee has 5 members. James, Mary, and Bob are the only ones on both committees. How many pupils in the class are not on any committee?

10. MEASURING LENGTHS AND ANGLES

In the first two chapters of this book we examined the numerical way of looking at the world. In the next two chapters we examined the geometric way. Now we are going to see how these two ways of looking at the world can be connected together. We have, of course, already met the notion of a number line, which gives a geometric picture of the whole numbers. But a much closer connection between arithmetic and geometry develops out of the idea of measurement.

MEASURING LENGTH Let us start by considering how to measure a segment. In Chapter 4 we saw how to compare two line segments $\overline{AB}$ and $\overline{CD}$ in order to say whether $\overline{AB} > \overline{CD}$ or $\overline{AB} \cong \overline{CD}$ or $\overline{AB} < \overline{CD}$.

When $\overline{AB}$ and $\overline{CD}$ can be conveniently represented by a drawing on a piece of paper, this comparison can be carried out, at least approximately, by tracing a copy of $\overline{AB}$ and placing it on top of the drawing of $\overline{CD}$. But, even if $\overline{AB}$ and $\overline{CD}$ were much too long or much too microscopically short to be drawn satisfactorily on a sheet of paper at all, it would still be possible to conceive of $\overline{AB}$ and $\overline{CD}$ as being such that exactly one of the following three statements is true:

1 $\overline{AB} < \overline{CD}$
2 $\overline{AB} \cong \overline{CD}$
3 $\overline{AB} > \overline{CD}$

In mathematics, we think of the endpoints A and B of any given line segment as being *exact locations in space,* although these endpoints can be represented only approximately by penciled dots. Similarly, $\overline{AB}$ is considered as having a certain *exact length,* although this length can be determined only approximately, by "measuring" a drawing representing $\overline{AB}$.

■ **Question 1** If $\overline{AB} < \overline{CD}$ and $\overline{CD} < \overline{EF}$, is it true that $\overline{AB} < \overline{EF}$? How do you know?

Let us describe the process of measurement. The first step is to choose a line segment, say $\overline{RS}$, to serve as a *unit*. This means to select $\overline{RS}$ and agree to consider its measure to be exactly the number 1.

We should recognize that this selection of a unit is an arbitrary choice we make. Different people might well choose different units and historically they have, giving rise to much confusion. For example, at one time the English "foot" was actually the length of the foot of the reigning king and the "yard" the distance from his nose to the end of his outstretched arm. Imagine the confusion when the king died if the next one was of much different stature. Various standard units will be discussed a little later but meanwhile we return to the choice of $\overline{RS}$ as our unit, recognizing the arbitrariness of this choice (see Figure 10.1). Now it is possible to conceive of a line segment, $\overline{CD}$, such that the unit $\overline{RS}$ can be laid off exactly twice along $\overline{CD}$, as suggested in Figure 10.2. Then by agreement the measure of $\overline{CD}$ is the number 2 and the length of $\overline{CD}$ is exactly 2 units. In the same way, line segments of length exactly 3 units, or exactly 4 units, or exactly any larger number of units are conceptually possible, although such line segments can be drawn only approximately.

Figure 10.1

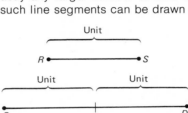

Figure 10.2

■ **Question 2** Would a rubber band be a good unit to use in measuring the length of your desk?

We can also conceive of a line segment $\overline{AB}$ such that the unit $\overline{RS}$ will not "fit into" $\overline{AB}$ a whole number of times at all. In Figure 10.3 $\overline{AB}$ is a line segment such that starting at A the unit $\overline{RS}$ can be laid off 3 times along $\overline{AB}$ reaching Q which is between A and B, although if it were laid off 4 times we would arrive at a point P which is well beyond B. We see that $\overline{AB}$ has length *greater than* 3 units and *less than* 4 units. In this particular case, we can also estimate visually that the length of $\overline{AB}$ is nearer to 3 units than to 4 units, so that *to the nearest unit* the length of $\overline{AB}$ is 3 units. This is the best we can do without considering fractional parts of units, or else shifting to a smaller unit.

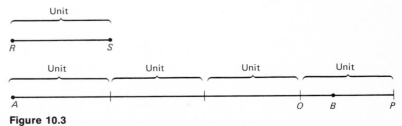

Figure 10.3

To help us in estimating whether the measure of a segment is, say, 3 or 4, we can bisect our unit. In Figure 10.4 $\overline{RS}$ is again shown as a unit with T bisecting $\overline{RS}$ so that $\overline{RT} \cong \overline{TS}$ and $\overline{RS}$ is used to measure another segment $\overline{MN}$.

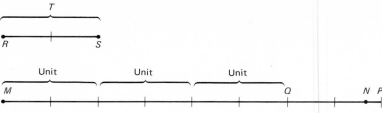

Figure 10.4

In laying off the unit along $\overline{MN}$, label P the endpoint of the first unit that falls on or beyond N, and Q the end of the preceding unit, just as you did for $\overline{AB}$ in Figure 10.3.

We see that $\overline{NP} < \overline{RT}$ and the measure of $\overline{MN}$ is 4. Going back to Figure 10.3, we can check that $\overline{BP} > \overline{RT}$ and so the measure of $\overline{AB}$ is 3.

There is nearly always such a decision to be made about whether or not to count the last unit which extends beyond the endpoint of the segment being measured. The reason for this is that it is rare indeed for the unit to fit an exact number of times from endpoint to endpoint. It is well to realize now that measurement is approximate and subject to error. The "error" is the segment from the end of $\overline{AB}$ to the end of the last unit being counted. In Figure 10.3 the error is $\overline{BQ}$; in Figure 10.4 it is $\overline{NP}$. We note that the error in any measurement is always at most one-half the unit being used.

■ **Question 3** I mailed four identical letters, first class, paying 10 cents postage apiece, for a total of 40 cents. If I had put them all in one envelope, the postage would have been only 30 cents. Why?

As we shall see shortly, the use of different units gives rise to different measures for the same segment. Thus, if we consider in Figure 10.5 a segment congruent to $\overline{MN}$ in Figure 10.4 but use $\overline{KL}$, rather than $\overline{RS}$, as our unit, $\overline{MN}$ has a length of 6 units.

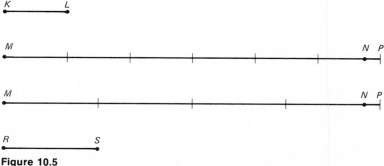

Figure 10.5

STANDARD UNITS If everyone used his own units, we would have difficulty comparing our results or communicating with each other. For these reasons certain units have been agreed upon by large numbers of people and such units are called *standard units.*

Historically there have been many standard units, such as a yard, an inch, or a mile used to measure line segments. Such a variety is a great convenience. An inch is a suitable standard unit for measuring the edge of a sheet of paper, but hardly satisfactory for finding the length of the school corridor. While a yard is a satisfactory standard for measuring the school corridor, it would not be a sensible unit for finding the distance between Chicago and Philadelphia.

Such units of linear measure as inch, foot, yard, and mile are standard units in the British-American system of measures. In the eighteenth century in France, a group of scientists developed the system of measures which is known as the metric system, using a new standard unit.

■ **Question 4** An old farm saying is: "The corn should be knee high by the Fourth of July." Is "knee" a standard unit?

In the metric system, the basic standard unit of length is the *meter,* which is approximately 39 inches or a little more than 1 yard. The metric system is in common use in all countries except those in which English is the main language spoken, and is used by all scientists in the world including those in English-speaking countries. England is moving toward the decimal system, and it is probable that the United States will adopt it within not too many years.

The principal advantage of the metric system over the British-American system lies in the fact that the metric system has been designed for ease of conversion between the various metric units by exploiting the decimal system of numeration. Instead of having 12 inches to the foot, 3 feet to the yard, and 1,760 yards to the mile, the metric system has 10 centimeters to a decimeter, 10 decimeters to a meter, and 1,000 meters to a kilometer. This makes conversions between units very easy.

We have already noted that in the metric system, the *meter* is the unit which corresponds approximately to the yard in the British-American system. The metric unit which corresponds to the inch is the centimeter, which is one-hundredth of a meter. A meter is almost 40 inches, so it takes about $2\frac{1}{2}$ centimeters to make an inch. Figure 10.6 illustrates a scale of inches and a scale of centimeters so you can compare them.

Centimeters

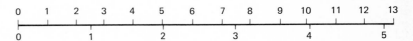

Inches

Figure 10.6

■ **Question 5** About how many centimeters are there in a foot?

So far we have said little about metric units larger than the meter. The most useful of these is the *kilometer,* which is defined to be 1,000 meters. The kilometer is the metric unit which closely corresponds to the British-American mile. It turns out that one kilometer is a little more than six-tenths of a mile. The Olympic race of 1,500 meters is over a course which is just a little less than a mile. We will discuss the metric system in more detail in a later chapter.

We have treated the inch, foot, yard, and mile as "standard" units for linear measure, in contrast to units of arbitrary size, which may be used when communication is not important. Actually, the one standard unit for linear measure even in the United States is the meter, and the correct sizes of other units such as the centimeter, inch, foot, and yard are specified by law with reference to the meter. Various methods for maintaining a model of the standard meter have been used by the Bureau of Standards. For many years the model was a platinum bar, kept under carefully controlled atmospheric conditions. The meter is now defined as being a length which is 1,650,763.73 times the wavelength of orange light from krypton 86. This standard for the meter is preferred because it can be reproduced in any good scientific laboratory and provides a more precise model than the platinum bar.

■ **Question 6** Would you be surprised if you were told that the speed limit in France is 100?

SCALES AND RULERS Once a standard unit such as a yard, meter, or mile is agreed upon, the creation of a scale greatly simplifies measurement.

> A *scale* is a number line with the segment from 0 to 1 congruent to the unit being used.

A scale can be made with a nonstandard unit or with a standard unit.

> A *ruler* is a straightedge on which a scale using a standard unit has been marked.

If we use the inch as the unit in making a ruler, we have a measuring device designed to give us readings to the nearest inch. Most ordinary rulers are marked with $\frac{1}{16}$ inch unit or with 1 millimeter unit.

THE APPROXIMATE NATURE OF MEASURE As we have seen, any measurement of the length of a segment made with a ruler is, at best, approximate. When a segment is to be measured, a scale based on a unit appropriate to the purpose of the measurement is selected. The unit is the segment with endpoints at two consecutive scale divisions of the ruler. The scale is placed on the segment with the zero-point of the scale on one endpoint of the segment. The number which corresponds to the division point of the scale nearest the other endpoint of the segment is the measure of the segment. Thus, *every measurement is made to the nearest unit.* If the inch is the unit of measure for our ruler, then we have a situation

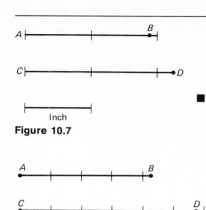

Inch

Figure 10.7

A ——+——+——+—— B

C ——+——+——+——+——+—— D

Centimeter

Figure 10.8

in which two line segments, apparently not the same length, may have the same measure, to the nearest inch. The measure of $\overline{AB}$ to the nearest inch is 2. We write this, m($\overline{AB}$) = 2. The measure of $\overline{CD}$ to the nearest inch is also 2; m($\overline{CD}$) = 2 (see Figure 10.7).

■ **Question 7** What is the width of the room you are in to the nearest mile?

For the same two segments we may get a different measure if we use a different unit segment. It should be clear that if the unit is changed, the scale changes. Thus, if we decide to use the centimeter as our unit, the scale appears as in Figure 10.6, and Figure 10.8 shows that in centimeters m($\overline{AB}$) = 4 and m($\overline{CD}$) = 6. Now the measures do indicate that there is a difference in the lengths of the two segments. Notice that by using a smaller unit (the centimeter) we are able to distinguish between the lengths of two noncongruent segments which in terms of a larger unit (the inch) have the same measure.

If measurements of the same segment are made in terms of different units, the error in the measurements may be different since it is at most one-half the unit being used. Thus, if a segment is measured in inches, the error cannot be more than one-half of an inch, while if it is measured in tenths of an inch, the error cannot be more than one-half of a tenth of an inch. As a result, if greater accuracy is desired in any measurement, a smaller unit should be used.

■ **Question 8** Suppose you are told that the length, to the nearest inch, of the side of a square is 3. What can you say about the perimeter (the distance around the four sides)?

RAYS AND ANGLES

In addition to line segments, there are many other kinds of geometric figures which we can measure. One important set of figures is the set of angles. Before we can define these we need first to make another definition. We remember from Chapter 3 that two points A and B in a plane determine the line segment $\overline{AB}$ and the line $\overleftrightarrow{AB}$.

Besides $\overline{AB}$ and $\overleftrightarrow{AB}$, the two points A and B determine another particular set of points called "the ray $\overrightarrow{AB}$" and for which the symbol is $\overrightarrow{AB}$ (see Figure 10.9).

The ray $\overrightarrow{AB}$ consists of the point A and all those points of the line $\overleftrightarrow{AB}$ on the same side of A as B.

A beam of light emanating from a pinpoint source is an excellent representation of a ray. See also Figure 10.9a.

Since there are only two directions in a line from a fixed point on the line, there can be only two distinct rays on the line with the fixed point as a common endpoint. See Figure 10.9b. However, since any point of the line may serve as the endpoint of a ray on the line, a line contains more rays than can be counted.

A ray may be a less familiar concept than a point or a line, but it is a very useful one, particularly when we come to talk about angles.

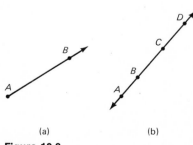

(a)   (b)

Figure 10.9

An *angle* is the union of two rays which have the same endpoint but which are not parts of the same line.

In Figure 10.10a the angle is $\overrightarrow{AC} \cup \overrightarrow{AB}$. Note that in Figure 10.10b the point S *is* a point of the angle since S is in $\overrightarrow{PQ}$ but point W *is not* a point of the angle since it is not in either $\overrightarrow{PQ}$ or $\overrightarrow{PT}$.

■ **Question 9** The vertices of a triangle determine how many rays?

The common endpoint of the two rays is called the *vertex* of the angle. The symbol for an angle is ∠ and the angle is usually named by naming three points of the angle; the first being a point (not the vertex) on one ray; the second being the vertex; and the third being a point (not the vertex) on the other ray. Thus the angle represented in Figure 10.10a is ∠*BAC* or ∠*CAB.* In Figure 10.10b the angle may have various names: ∠*QPR,* or ∠*SPT,* or ∠*RPS,* etc. Note that it is correct to say ∠*QPR* = ∠*TPS* since they are simply different names for the same angle.

Just as a simple closed curve divides the plane into two parts, the interior and the exterior of the curve, so does an angle divide the plane into two parts, which we shall call the interior and exterior of the angle. But which part shall we call the interior? An easy way to decide is as follows: Consider ∠*BAC.* $\overline{CA}$, $\overline{AB}$, and $\overline{BC}$ determine a simple closed curve; in fact, they determine a triangle which has an interior. ∠*BAC* divides the plane into two parts. The interior is that one of the two parts which includes the interior of the triangle. Figure 10.11 will help to make this clear.

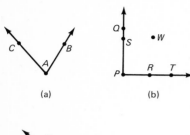

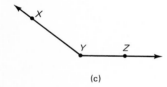

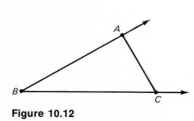

Figure 10.10

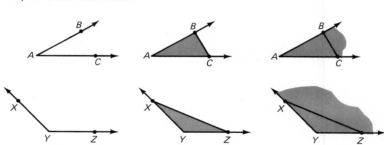

Figure 10.11

Question 10 In Figure 10.10b, is ∠*RPS* = ∠*SPR*?

If we consider two segments $\overline{AB}$ and $\overline{BC}$, we see that they determine rays $\overrightarrow{BA}$ and $\overrightarrow{BC}$, and thus determine an angle. It must be remembered that the angle consists of all the points in both rays and not just those points in $\overline{AB}$ and $\overline{BC}$. This is why we should be careful to observe that while a triangle or a polygon determines its angles, the angles are not part of the triangle. As a set of points, ∠*ABC* consists of the rays $\overrightarrow{BA}$ and $\overrightarrow{BC}$, but △*ABC* contains only the points of the segments $\overline{BA}$ and $\overline{BC}$ (as well as those of $\overline{AC}$). See Figure 10.12.

Figure 10.12

CONGRUENCE OF ANGLES We can compare two angles in much the same way as we compared two segments. Thus if two angles $\angle ABC$ and $\angle PQR$ are given, we can take as a representation of $\angle ABC$ a tracing, say $\angle A'B'C'$, and make the ray $\overrightarrow{B'C'}$ fall on $\overrightarrow{QR}$ with $\overrightarrow{B'A'}$ falling in the same half-plane as $\overrightarrow{QP}$ and with B' falling on Q. See Figure 10.13. Now if $\overrightarrow{B'A'}$ falls on $\overrightarrow{QP}$ we say that $\angle ABC \cong \angle PQR$.

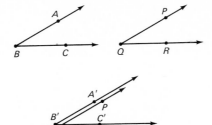

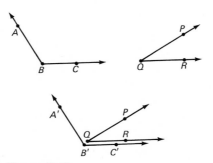

Figure 10.13

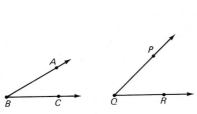

Figure 10.14

But $\overrightarrow{B'A'}$ may fall in the interior of $\angle PQR$ in which case $\angle ABC$ is less than $\angle PQR$ or $\angle ABC < \angle PQR$. See Figure 10.14.

Finally, $\overrightarrow{B'A'}$ may fall in the exterior of $\angle PQR$ and in this case $\angle ABC$ is greater than $\angle PQR$ or $\angle ABC > \angle PQR$. See Figure 10.15.

Figure 10.15

Again, as in comparing segments, we find that these three are the only possibilities. That is: Either $\angle ABC < \angle PQR$ or $\angle ABC \cong \angle PQR$ or $\angle ABC > \angle PQR$.

■ **Question 11** We have seen how segments can be compared and how angles can be compared. Does it make sense to try to compare a segment and an angle?

There is a special angle of great importance called a *right angle*, which we may illustrate as follows. Take any point P on $\overleftrightarrow{AB}$ such that A and B are on opposite sides of P and take any point Q not on $\overleftrightarrow{AB}$. Draw $\overrightarrow{PQ}$. This is illustrated in Figure 10.16. We get two angles, $\angle APQ$ and $\angle BPQ$. If we compare these angles, we may occasionally find that they are congruent. In this case we say that ∠*APQ* and ∠*BPQ* are both *right* angles.

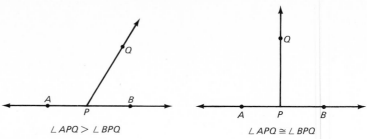

$$\angle APQ > \angle BPQ \qquad\qquad \angle APQ \cong \angle BPQ$$

Figure 10.16

Models of right angles occur in many places in the world as, for instance, at the corner of an ordinary sheet of paper. A model can be made even from an irregular sheet of paper by folding it twice, as shown in Figure 10.17.

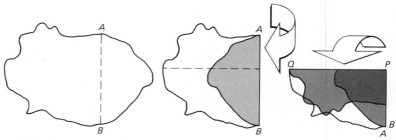

Figure 10.17

■ **Question 12** Are any two right angles congruent?

THE MEASURE OF AN ANGLE

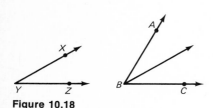

Figure 10.18

Just as we think of every line segment as having a certain exact length, so too we think of every angle as having a certain exact size, even though this size can be determined only approximately by measuring a chalk or pencil drawing of it.

Let us examine a process for measuring angles. As with linear measure, we need to devise a way to assign a number as the measure to each angle. The first step is to select an arbitrary angle to serve as a *unit* and agree that its measure is the number 1. In Figure 10.18 we take ∠*XYZ* as our unit.

Now we can conceive of forming an angle ∠*ABC* by laying off the unit twice about a common vertex *B*, as suggested in the same figure. We say that in terms of the unit ∠*XYZ* the measure of ∠*ABC* is 2. We write:

In terms of the unit ∠*XYZ*,   m(∠*ABC*) = 2

In similar fashion we can conceive of forming angles whose measures are 3 or 4 and so forth until we come to an angle whose interior is nearly a half-plane, as shown in Figure 10.19.

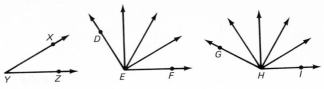

Figure 10.19

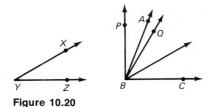

Figure 10.20

We can also conceive of an angle such that our unit ∠*XYZ* will not fit into it a whole number of times. In Figure 10.20 we have an ∠*ABC* in which, if we start at $\overrightarrow{BC}$, the unit ∠*XYZ* can be laid off 2 times about *B* without quite reaching $\overrightarrow{BA}$ (that is, with $\overrightarrow{BQ}$ in the interior of the angle), though if we were to lay it off 3 times, we would arrive at a ray (call it $\overrightarrow{BP}$) which is well beyond $\overrightarrow{BA}$.

What can be said about the size of ∠*ABC*? Surely, it is greater than 2 units and less than 3 units. In Figure 10.20, we can also estimate by eye that the size of ∠*ABC* is nearer to 2 units than to 3 units, so:

To the nearest unit in terms of unit ∠*XYZ*,   m(∠*ABC*) = 2

This is the best that can be done without considering fractional parts of units, or else shifting to a smaller unit. We always assign the measure so that the error is less than one-half the unit angle. But, just as in measuring segments, the measure of a specific given angle is almost always an approximation. A smaller error may be obtained by using a smaller unit but we can never be absolutely accurate. This sort of trouble can never be avoided. It is inherent in the approximations necessary for any measurement process. We just have to learn to live with it both now and later on when we study area and volume and other quantities.

■ **Question 13** Suppose we put two angles side by side, as in Figure 10.21. Will m(∠*APB*) + m(∠*BPC*) = m(∠*APC*)?

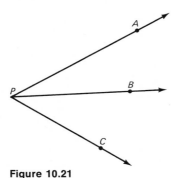

Figure 10.21

THE STANDARD UNIT   Just as, when we considered the measure of segments, we found that a standard unit, a scale, and a ruler were useful, we find the corresponding elements valuable in angle measure. The most common standard unit of angle measure is called a "degree." We write it in symbols as 1°. When we speak of the size of an angle, we may say its size is 45°, but if we wish to indicate its measure, we must keep in mind that a measure is a number and say that its measure, in degrees, is 45. If we lay off 360 of these unit angles using a single point as a common vertex, then these angles, together with their interiors, cover the entire plane. Note that if $\angle ABC$ is a right angle, m($\angle ABC$) = 90 (see Figure 10.22).

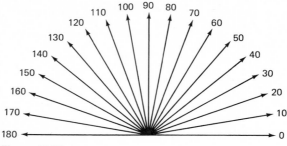

Figure 10.22

A scale like this can be used to measure an angle. Place the angle on the scale with one side of the angle on the ray that corresponds to zero and the vertex of the angle at the intersection of the rays. Then the number which corresponds to that ray along which the other side of the angle falls is the measure of the angle, in degrees. The size or measurement of the angle is that number of degrees.

Since placing an angle on a scale is inconvenient, an instrument called a *protractor* is usually used. Then the scale can be placed on the angle, rather than the angle on the scale.

Consider the drawing of a protractor in Figure 10.23 and think of the rays from point V. In the drawing, two of the rays are shown in broken lines, while on the actual instrument a portion of each ray is shown on the curved part. To measure an angle with the protractor, place the protractor on the angle so that point V is on the vertex of the angle and the ray which corresponds to zero on the protractor lies on one side of the angle. Then the number which corresponds to the protractor ray which is on the other side of the angle is the measure, in degrees, of the angle.

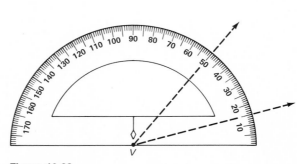

Figure 10.23

■ **Question 14** Draw a large triangle and cut it out. Place it on the protractor in Figure 10.23 to measure each of the three angles. Add up the measurements. Repeat with another, different triangle. Are your two sums about the same?

OTHER UNITS Nowadays scientific and engineering problems demand such accuracy that a common unit of length is an angstrom, which is one hundred millionth of a centimeter. This unit is very small. Astronomers on the other hand for their purposes use very large units such as the "astronomical unit" which is the average distance of the sun from the earth, and the "light-year," which is the distance traveled by light in one year traveling at a rate of about 186,000 miles per second.

But whatever units are used, it should be remembered that measurement is always approximate and answers are expressed to the nearest unit, whatever unit is being used. The decision as to which is the most appropriate unit always has to be made whether we consider the distance to the nearest galaxy of stars to the nearest light-year, the distance from Washington to New York to the nearest mile, the diameter of an automobile engine piston to the nearest thousandth of an inch, or the length of a wave of light to the nearest angstrom.

Measurement is the foundation stone of science and a connecting link between the physical world around us and mathematics.

■ **Question 15** How far does light travel in 1 hour?

PROBLEMS

1 The measures of the sides of a triangle in inch units are 17, 15, and 13.
   (a) What are the measures of the sides if the unit is a foot?
   (b) What is the distance around the triangle in inches? In feet?
   (c) Is there anything curious about your answer?
   (d) How do you explain it?

2 Use A———B $\overline{AB}$ as a unit to measure the segments in Figure P10.2. Is $\overline{CD}$ congruent to $\overline{EF}$? Do your answers contradict each other?

C———D E———F

Figure P10.2

3 A desk is 9 chalk pieces long. What is the largest possible error in the measurement?

4 In which of the following sentences are standard units used?
   (a) He is strong as an ox.
   (b) Put in a pinch of salt.
   (c) We drink a gallon of milk per day.
   (d) I am 5 feet tall.

5 A ray has how many endpoints?

6 If the union of two rays, $\overrightarrow{AB}$ and $\overrightarrow{CD}$, is a line, what will the intersection of $\overrightarrow{AB}$ and $\overrightarrow{CD}$ be?

7 Does a ray separate the plane?

8 (a) What does the statement $\overrightarrow{AB} = \overrightarrow{CB}$ tell us?
   (b) Does $\overrightarrow{BA} = \overrightarrow{BC}$ give the same information?

Figure P10.9

9 Referring to Figure P10.9, rename the sets in simple notation.
(a) Union of $\overline{BC}$, $\overline{CD}$, and $\overline{DE}$.
(b) Intersection of $\overline{AB}$ and $\overline{BC}$.
(c) Intersection of $\overrightarrow{CA}$ and $\overrightarrow{ED}$.
(d) Intersection of $\overrightarrow{CD}$ and $\overrightarrow{DC}$.
(e) Union of $\overrightarrow{EA}$ and $\overrightarrow{BC}$.

10 Represent $\overleftrightarrow{PR}$ and show Q between P and R. Which of the following denote the same ray?
$\overrightarrow{PQ}$, $\overrightarrow{QP}$, $\overrightarrow{QR}$, $\overrightarrow{PR}$, $\overrightarrow{RP}$, $\overrightarrow{RQ}$

11 Name all the angles shown in Figure P10.11.

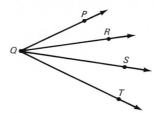

Figure P10.11

12 Can $\overrightarrow{VX}$ and $\overrightarrow{WX}$ be sides of an angle?

13 What is wrong with each of the following?
(a) $\overline{AB} \cong \overline{CD}$ (b) $\overleftrightarrow{AB} = \overline{AB}$ (c) $\angle ABC > \overline{BC}$

14 Referring to Figure P10.14, if $\angle APB \cong \angle BPC$, must these angles be right angles?

15 If $\angle ABC > \angle PQR$ and $\angle PQR > \angle XYZ$, what can be said about $\angle XYZ$ and $\angle ABC$?

16 Which of the pairs of angles in Figure P10.16 are congruent?

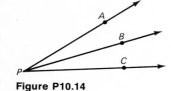
Figure P10.14

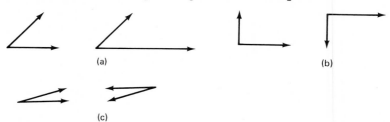

(a)

(b)

(c)

Figure P10.16

17 In reference to Figure P10.17, which angles are right angles? Do not guess, but make a model of a right angle for comparison purposes.

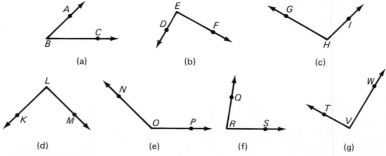

(a)

(b)

(c)

(d)

(e)

(f)

(g)

Figure P10.17

18 Fold a piece of paper twice to get a model of a right angle. How many right angles can you fit together side by side around a point in the plane?

19 Why is it incorrect to say $\overline{AB}$ is a subset of the interior of $\angle MAL$ in Figure P10.19?

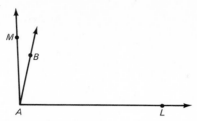

Figure P10.19

20 In Figure P10.20, a protractor is shown placed on a figure with several rays drawn from point A. Find the measure, in degrees, of each of the angles.

(a) $\angle BAK$ (b) $\angle BAC$
(c) $\angle BAE$ (d) $\angle MAF$
(e) $\angle GAM$ (f) $\angle DAE$
(g) $\angle HAK$

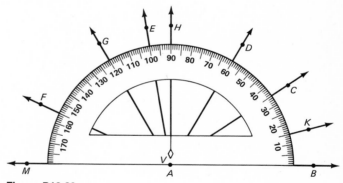

Figure P10.20

11. MEASURING AREAS

We learned, in the last chapter, how to go about measuring lengths and angles. Now we will take up a more complicated problem, measuring the areas of regions in the plane. First, however, it will be convenient to take another look at some of the simpler plane figures and to sort them out into various categories, using some measurement ideas to do so.

CLASSIFICATION OF TRIANGLES

Triangles can be classified by comparing their sides or their angles. *(We really should not speak of an angle of a triangle, since we saw above that the set of points which is the angle is not a subset of the set of points which is a triangle. But since the segments of the triangle do determine three angles, we use the words "angle of a triangle" for the more correct but much longer "angle whose rays are determined by the sides of a triangle.")*

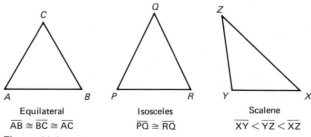

Equilateral	Isosceles	Scalene
$\overline{AB} \cong \overline{BC} \cong \overline{AC}$	$\overline{PQ} \cong \overline{RQ}$	$\overline{XY} < \overline{YZ} < \overline{XZ}$

Figure 11.1

FIRST: Considering the sides of a triangle (see Figure 11.1),

1 If all three sides are congruent, the triangle is *equilateral*.
2 If two sides are congruent, the triangle is *isosceles*.
3 If no two sides are congruent, the triangle is *scalene*.

SECOND: Considering the angles of a triangle,

1 One angle may be a right angle. Such a triangle is called a *right triangle.*

2 All angles may be less than a right angle. In this case the triangle is said to be *acute.*

3 One angle may be greater than a right angle. Such a triangle is called an *obtuse triangle.*

In cases 1 and 3 comparison of angles will show that the other two angles are always less than a right angle (see Figure 11.2).

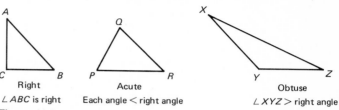

Right	Acute	Obtuse
∠ABC is right	Each angle < right angle	∠XYZ > right angle

Figure 11.2

■ **Question 1** Can an equilateral triangle be obtuse?

It is interesting to note that if you compare the angles of an equilateral triangle you will find that they are all congruent and if you do so for the angles of an isosceles triangle you will find that two of them are congruent. It is also true that if two or three angles of a triangle are congruent then two or three sides will be. We will come back to these ideas in a later chapter.

CLASSIFICATION OF QUADRILATERALS

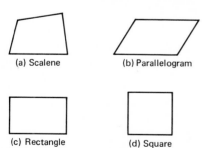

(a) Scalene

(b) Parallelogram

(c) Rectangle

(d) Square

Figure 11.3

Triangles are always convex, but other polygons may not be. However, usually it is convex polygons that we are interested in. There are many interesting types of convex polygons. Those with four sides, the quadrilaterals, may be classified as were triangles by listing special properties of their sides or angles. Such a classification is given in the list below for a few of the best-known cases (see Figure 11.3).

1 The scalene quadrilateral. No two of its sides are congruent.
2 The parallelogram. Its opposite (nonintersecting) sides are segments of parallel lines. They are also always congruent.
3 The rectangle. It is a parallelogram whose angles are all congruent. They are also all right angles.
4 The square. It is a rectangle whose sides are all congruent.

■ **Question 2** Are the horizontal sides of a parallelogram always longer than the other sides?

AREA To measure a plane region is to select a certain unit and to assign in terms of that unit a number which is called the *measure of the area* of the region.

Note that just as in the length of a line segment the area of a region involves both a number and a unit. Thus an *area* may be expressed as 6 square inches. The measure of the area is the number 6. While it is important to have these distinctions clearly in mind when working with lengths, areas, and later, volumes, it becomes too cumbersome to keep mentioning them. The important thing to remember is that we always compute with *numbers,* but we express answers in terms of numbers and the appropriate units.

Let us recall how the subject of linear measurement was approached, since area will be approached in a similar manner. First we encountered the intuitive concept of *comparative length* for line segments: any two line segments can be compared to see whether the first of them is of *smaller length,* or the *same length,* or *greater length* than the second. Corresponding to this we have in the present chapter the idea of *comparative area* for plane regions. (Recall that by definition a plane region is the union of a simple closed curve and its interior.) Even when they are rather complicated in shape, two regions can, in principle at least, be compared to see whether the first of them is of *smaller area,* or the *same area,* or *greater area* than the second.

■ **Question 3** If a segment is 15 inches long, what is the measure of its length?

In the case of line segments, this comparison is conceptually very simple: we think of the two segments to be compared, say $\overline{AB}$ and $\overline{CD}$, as being placed one on top of the other in such a manner that A and C coincide: then either B is between C and D, or B coincides with D, or B is beyond D from C, etc. This conceptual comparison of line segments is also easy to carry out approximately by using physical models (drawings and tracings, etc.) of the line segments involved.

COMPARISON OF REGIONS

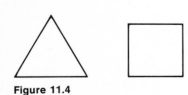

Figure 11.4

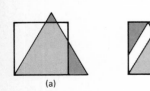

(a)   (b)

Figure 11.5

In the case of plane regions, this comparison is more complicated, both conceptually and in practice. This is because the shapes of the two plane regions to be compared may be such that neither will "fit into" the other. How, for example, do we compare in size (area) the two plane regions pictured in Figure 11.4? If we think of these regions as placed one on top of the other, neither of them will fit into the other. In this particular case, however, we can think of the two pieces of the triangular regions which are shown shaded heavily in Figure 11.5a as snipped off and fitted into the square region in Figure 11.5b. This shows that the triangular region is of smaller area than the square region. As the figures involved become more complicated in shape, this sort of comparison becomes increasingly difficult in practice. We need a better way of estimating the area of a region.

■ **Question 4** A yardstick is 1 inch wide (and 36 inches long, of course). A piece of typing paper is $8\frac{1}{2}$ by 11 inches. When laid on the floor, which covers more area?

In studying linear measure we first found out how to compare two segments. What was the second step in the process? We chose a *unit* of length. That is, we selected a certain arbitrary line segment and agreed to consider its length to be measured, exactly, by the number 1. In terms of this unit we could then conceive of line segments of lengths exactly 2 units, 3 units, 4 units, etc., as being constructed by laying off this unit successively along a line 2 times, 3 times, 4 times, etc. The process of laying off the unit successively along an arbitrary given line segment yielded underestimates and overestimates for the length of the given segment since the segment might have turned out to be greater than 3 units (underestimate) but less than 4 units (overestimate). We selected the closer of these two estimates as the measure of the segment in terms of the selected unit, realizing that any such measure is usually only approximate and subject to error. Since the error was at most one-half of the unit used, by selecting a smaller unit we found we could usually make the measurement more accurate if we wanted to.

We now proceed similarly in the measurement of area. The first step is to choose a unit of area, that is, a region whose area we shall agree is measured exactly by the number 1. Regions of many shapes, as well as many sizes, might be considered. An important thing about a line segment as a unit of length was that enough unit line segments placed end-to-end (so that they touch, but do not overlap) would together *cover*, either exactly or with some excess, any given line segment. Similarly, we need a unit plane region such that enough of them placed so that they touch, but do not overlap, will together *cover*, either exactly or with some excess, any given plane region. Some shapes will not do this. For example, circular regions do not have this property. Thus, in Figure 11.6, if we try to cover a triangular region with small nonoverlapping congruent circular regions, there are always parts of the triangular region left uncovered. On the other hand, we can always completely cover a triangular region, or any region, by using enough nonoverlapping congruent square regions.

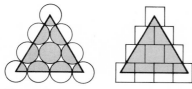

Figure 11.6

■ **Question 5** Can we cover a triangular region with congruent parallelograms?

While a square region is not the only kind of region with this covering property, it has the advantage of being a simply shaped region. The size of the unit of area is determined by choosing it as a square whose side has length equal to one linear unit. It will turn out that the use of such a square region as the unit of area makes it easy to compute the area of a rectangle by forming the product of the numbers measuring the lengths of its sides.

A SCALE TO ESTIMATE AN AREA

Having chosen a unit of length, we then made scales and rulers to help in measuring the length of a given line segment. A corresponding instrument is not usually available for area, but we can easily make one for ourselves. This is a *grid,* which is a regular arrange-

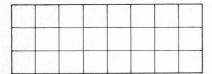

Figure 11.7

ment of nonoverlapping square unit regions. Part of a grid is shown in Figure 11.7. To use such a grid in estimating the area of a given region, we think of it as superimposed on the region. This is illustrated in Figure 11.8. We can verify by counting that 12 of the unit regions pictured are contained entirely in the given region. These are the units heavily shaded in Figure 11.8b. This shows that *the area of this region is at least 12 units.* This is an underestimate. We can also verify by counting that there are 20 additional unit regions lightly shaded in Figure 11.8b, which together cover the rest of the region. Thus the entire region is covered by 12 + 20 or 32 units. This shows that *the area of this region is at most 32 units.* Then 32 is an overestimate of the measure. That is, we now know that the area of the region is somewhere between 12 units and 32 units. Since the difference between the two estimates is 20 units, we see that the accuracy is not very good. The lightly shaded region in Figure 11.8b represents this difference.

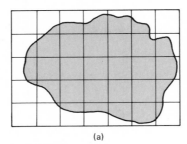
(a)

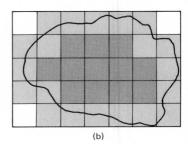
(b)

Figure 11.8

■ **Question 6** How many squares are there in a grid?

In Chapter 10 we saw that more accurate estimates of lengths could be achieved by using a smaller unit. The same is true with area. To illustrate this fact, let us reestimate the area of the same region in Figure 11.8, using this time the unit of area determined by a unit of length just half as long as before (see Figure 11.9).

We can verify by counting that there are 63 of the new unit regions pictured which are contained entirely in the given region. This shows that *the area of the region is at least 63 (new) units.* We can also verify by counting that there are 41 additional unit regions pictured which together cover the rest of the region. Thus, the entire region is covered by 63 + 41 or 104 of the new units. This shows that *the area of this region is at most 104 (new) units.* That is, we now know that the area of the region is somewhere between 63 (new) units and 104 (new) units.

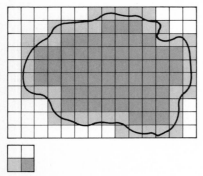

Figure 11.9

Let us compare these new estimates of the area with the old ones. Each old unit contains exactly 4 of the new units, as is clear from Figure 11.9. Each new unit is $\frac{1}{4}$ of the old unit. Thus the new estimates are $\frac{1}{4} \times 63$ or $15\frac{3}{4}$ and $\frac{1}{4} \times 104$ or 26 old units as compared with our former estimates of 12 and 32. The difference is $10\frac{1}{4}$ old units compared to the former 20. Plainly the new estimates based on the smaller unit are the more accurate ones. This may still be quite unsatisfactory. However, in principle it would be possible to estimate the area of this region or even of regions of quite general shape to any desired degree of accuracy by using a grid of sufficiently small units in this way. In practice, the counting involved would quickly become very tedious. Furthermore, where drawings are used to represent the region and grid involved, we would, of course, also be limited by the accuracy of these drawings.

■ **Question 7** If we have a rectangle that is say $2\frac{1}{4}$ inches by $1\frac{1}{2}$ inches, we can cover it exactly by $\frac{1}{4}$-inch unit squares. This means that we can find the area of the rectangle exactly just by counting squares in a grid. Can we ever do this for a triangle?

BASIC IDEAS OF AREA

Let us summarize the discussion to this point. Actually, the emphasis here is not so much on accurate estimates as it is on grasping the following basic sequence of ideas.

1 Area is in some sense a feature of a *region* (and not of its boundary).
2 Regions can be compared in area (smaller, same, greater), and regions of different shapes may have the same area.
3 Like a length, an area should be, in theory, describable or measurable, exactly, by some appropriate number (not necessarily a whole number). In practice this is usually impossible.
4 For this purpose we need to have chosen a unit of area just as we earlier needed a unit of length.
5 The number which measures exactly the area of a region can be estimated approximately, from below and from above, by whole numbers of units.
6 In general, smaller units yield more accurate estimates of an area.

■ **Question 8** Aside from 1 above, are there any of these basic ideas of area which do not have valid corresponding ideas for the measure of angles?

FORMULA FOR AREA OF A RECTANGLE

For some of the more common plane regions such as those whose boundaries are polygons, formulas can be found to compute the measure of the area in terms of the linear measure of appropriate line segments of the figure.

For the rest of this chapter, whenever we want to refer to the area of the plane region bounded by a certain figure we shall, for brevity, refer to it as the area of the figure. Thus, instead of saying the "area of the triangular region bounded by $\triangle ABC$," we shall say the "area of $\triangle ABC$," and so on. Admittedly, this is slightly inaccurate, and if there is any possibility of ambiguity we will go into detail.

If the sides of a rectangle are measured in terms of the same unit, we may find that the lengths of the sides are a and b units where a and b are whole numbers. We then have an a-by-b array of unit squares and we know by the definition of multiplication of whole numbers that the number of squares in the array is $a \times b$. So the rectangle contains $a \times b$ units of area, and we say the measure of the rectangle is $a \times b$. Thus, for this case the measure of area equals the product of the measures of two sides, and we write $A = a \times b$. See Figure 11.10.

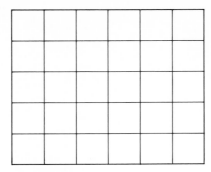

Figure 11.10

Practically, of course, we can measure the sides of a rectangle only to the nearest unit. In Figure 11.11, the length and width of the heavily outlined rectangle may measure 4 and 3 inches respectively to the nearest unit.

By counting unit regions in the superimposed grid we see that the area is between 8 and 15 square inches. If we use the measures of the sides to the nearest inch in the formula

$$A = a \times b$$

we would get $A = 3 \times 4 = 12$ and the area would be 12 square inches. This would be the exact area of the region shaded in Figure 11.11b, which does seem to have about the same size as the given rectangle, and it lies between the underestimate 8 and the overestimate 15 that we got using the grid. It is probably not, of course, the exact area of the original rectangle, but it is the measure of this area to the nearest square inch.

On the basis of these experimental results we make the definition:

The *measure of the area* of a rectangle is the number obtained as the product of the numbers measuring the base and the height.

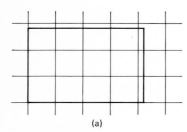

(a)

(b)

Figure 11.11

We usually let A stand for the measure of the area and b and h for the measures of the base and height. For brevity we usually say "the area is the product of the base and height," even though we should say "measure of" each time, and we write the formula for the area of any rectangle as:

$$A = b \times h$$

■ **Question 9** Is the height of a rectangle always less than the base?

We now consider a parallelogram $ABCD$. See Figure 11.12. By cutting off $\triangle BEC$ from parallelogram $ABCD$ and moving it over next to $\overline{AD}$, we can see that the measure of area of parallelogram $ABCD$ is equal to the measure of area of rectangle $ABEF$.

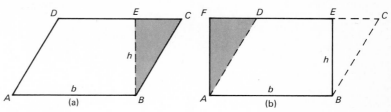

Figure 11.12

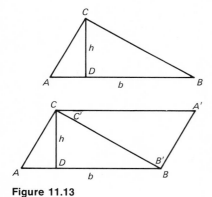

Figure 11.13

Therefore our formula $A = b \times h$ holds for any parallelogram if by h we understand the measure of the height $\overline{BE}$ and not the side $\overline{BC}$ of $ABCD$, i.e., the vertical distance and not the "slant" distance.

The formula for the area of a triangle ABC follows from that of a parallelogram. If a model of $\triangle ABC$ is made and labeled $A'B'C'$, it can be turned over and placed alongside $\triangle ABC$ to form the parallelogram $ABA'C$, whose base and height are $\overline{AB}$ and $\overline{CD}$. $\triangle ABC$ is thus one-half of parallelogram $ABA'C$ and its area must be half that of the parallelogram (see Figure 11.13). In the same fashion any triangle is half of a parallelogram whose base and height are identical with that of the triangle. Therefore, for a triangle,

$A = \frac{1}{2} \times b \times h$

Any other polygon can be divided into triangles by suitably drawn segments, and therefore its area may be found.

■ **Question 10** How many segments do we have to draw to divide a pentagon into triangles?

PROBLEMS

1 Define an "equiangular triangle."
2 Sketch: (a) an obtuse triangle; (b) a triangle which is both obtuse and isosceles; (c) an acute scalene triangle.
3 Make models and compare the regions in Figure P11.3.

(a)

(b)

Figure P11.3

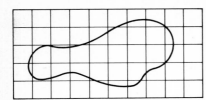

Figure P11.4

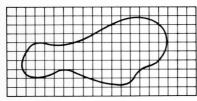

Figure P11.5

4 Consider the region pictured in Figure P11.4 on a grid of units. Fill in the blanks:
(a) There are _____ units contained entirely in the region.
(b) There are _____ units needed to cover the region completely.
(c) The area of the region is at least _____ units and at most _____ units.
(d) The difference is _____ units.

5 Let us choose a new unit of area. A square region has as its side a segment just half as long as before. For every old unit of area we will then have 4 new units of area.

Consider the same region pictured in Figure P11.5 on a grid of new units.
Fill in the blanks:
(a) There are _____ units contained entirely in the region.
(b) There are _____ units needed to cover the region completely.
(c) The area of the region is at least _____ units and at most _____ units.
(d) The difference is _____ units.
(e) Since each new unit is $\frac{1}{4}$ the old unit, this difference in terms of the old units is _____ units.

6 Using unit square S, we find that the area of a rectangle is 24 square units. Each side of square T is one-fifth as long as a side of S. What is the area of the rectangle if T is chosen for the unit square?

7 Find the distance around and the area of a rectangle with a length of 4 meters and a width of 3 meters.

8 The area of a rectangle is 72. Its length is 12. What is its width?

9 What is the distinction between "a 2-foot square" and "2 square feet"?

10 One rectangle whose sides have whole-number lengths has an area of 60. A second rectangle is twice as long and three times as wide as the first. What is the area of the second rectangle?

11 A room measures 5 feet 3 inches by 2 feet 9 inches. What is its area in square inches?

12 If the room mentioned in Problem 11 is measured to the nearest foot, what is its length and width? What is its area in square feet?

13 Compute the areas of these four rectangles.

	Length	Width
(a)	4	6
(b)	8	12
(c)	2	3
(d)	8	3

What is the effect on the area of a rectangle of length a and width b, if a and b are each doubled? What if a and b are each halved? What if a is doubled and b halved?

14 Referring to Figure P11.14, compute the area of the plane figure. (Hint: Partition the figure into rectangles.)

15 The measures of Figure P11.15 are in feet. How many square feet?

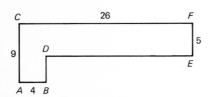

Figure P11.14

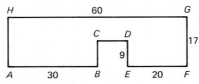

Figure P11.15

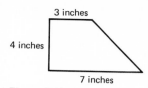

Figure P11.19

16 Two triangles have equal heights. The base of one triangle is twice as long as the base of the other. If the area of the smaller triangle is 42 square inches, what is the area of the other?

17 Two triangles have equal heights. The base of one triangle is 4 inches longer than the base of the second, and its area is 60 square inches more than the area of the second. What is the height of each triangle?

18 Which plane region has the greater area—a region bounded by a square with a side whose length is 3 inches or a region bounded by an equilateral triangle with a side whose length is 4 inches?

A *trapezoid* is a quadrilateral with one pair of opposite sides parallel.

19 Find the area of the trapezoid in Figure P11.19. (Hint: Divide it into a rectangle and a triangle.)

20 Find the area of the trapezoids in Figure P11.20, using the dimensions given. (Hint: Draw a diagonal in each one.)

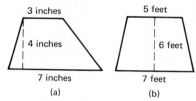

Figure P11.20

12.

RATIONAL NUMBERS

We have seen that whole numbers are useful in describing certain aspects of the world around us. However, there are other important aspects of the world which are too complicated to be described by whole numbers.

Notice that whole numbers refer to single sets. That is to say, if someone specifies a particular set, for example, the set of keys on your key ring, and if someone else mentions a particular whole number, say 7, there is no problem in deciding whether or not your set of keys has the number property 7.

COMPARING PAIRS OF SETS However, there are many cases where we are interested not just in one set, but rather in two sets which we wish to compare. For example, a thief might complain to another thief: "You gave me $100 of the loot from that stickup, but you kept $200 for yourself." The first thief is relating his set of dollars to the second thief's set of dollars.

Here is another example. Little Douglas, at a birthday party, complains: "You cut the cake into eight pieces and I only got one of them." Douglas is comparing his piece of cake to the whole cake.

We are going to describe some arithmetic ideas which can be used to handle comparisons of sets such as in these examples. We will call these ideas "numbers," but to distinguish them from the whole numbers, we will call them *rational numbers*.[1] First, however, we need to see that we usually compare only certain kinds of sets. Suppose that the first thief had complained: "You gave me $100 of the loot from that stickup, but you kept 200 pesos for yourself." We would be surprised at such a remark, because normally we do not compare dollars with pesos in this way.

The concept of "rational number" will apply only when we are com-

---

[1] The word "rational" comes from the idea of ratio, not from the dictionary meaning of "having reason or understanding."

paring a set A with a set B where

1 A has been divided into parts, any two of which are "equivalent,"
2 B has been divided into parts, any two of which are "equivalent," and
3 each part of A is "equivalent" to each part of B.

By *equivalent* we mean *congruent* if A and B are geometric figures, or *having the same number* of elements if A and B are finite sets of objects.

(*We say that two geometric figures which have the same size and shape are* congruent. *We met this notion in Chapter 4, where we discussed congruent segments. In that case, any two segments obviously had the same shape, so to find out whether they were congruent we only had to check to see if they were of the same size. We will investigate congruence of other figures in Chapter 21. In this chapter, it will always be obvious whether or not two given figures are congruent.*)

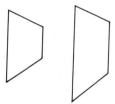

Figure 12.1

■ **Question 1** Why are the two figures in Figure 12.1 not equivalent?

The concept of rational number does apply when A and B look like any of these three pairs of geometric figures. In each case A and B are divided into congruent parts and each part of A is congruent to each part of B (see Figure 12.2).

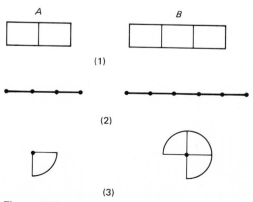

Figure 12.2

The concept of rational number does not apply when A and B look like any of these pairs, either because the parts of A are not congruent to the parts of B, or because the parts of A (or of B) are not congruent to each other (see Figure 12.3).

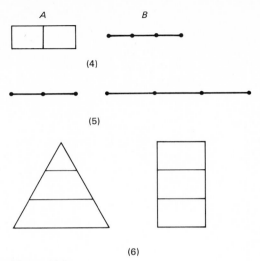

Figure 12.3

The concept of rational number does apply when A and B look like any of these pairs. In each case A and B are divided into parts all of which contain the same number of elements (see Figure 12.4). (The parts are shown by the broken lines.)

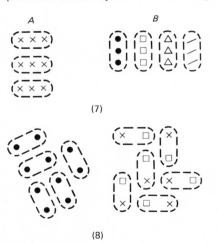

Figure 12.4

The concept of rational number does not apply when A and B look like any of these pairs. In each pair the "same number of members" property fails to hold for some of the parts (see Figure 12.5).

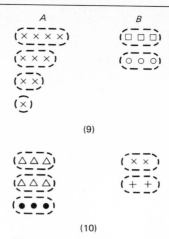

(9)

(10)

Figure 12.5

FRACTIONS AND RATIONAL NUMBERS

To describe the way set A compares with or relates to set B in Example 1, we write the *fraction* $\frac{2}{3}$. We have 2 on the top, because A is divided into 2 parts, and 3 on the bottom because B is divided into 3 parts. The fractions for Examples 2 and 3 are $\frac{3}{5}$ and $\frac{1}{3}$. The *numerator* of a fraction is the number on the top. The *denominator* is the number on the bottom.

Figure 12.6 shows two more pairs of sets that go with the fraction $\frac{2}{3}$. The property shared by all pairs of sets, A and B, for which A is related to B as in Examples 1, 11, and 12 is the *rational* number whose symbol is $\frac{2}{3}$.

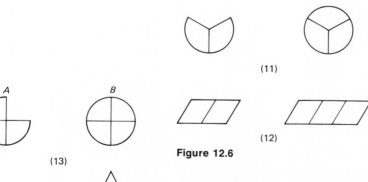

(11)

(12)

Figure 12.6

(13)

(14)

(15)

Figure 12.7

In Figure 12.7, we have three pairs of sets that go with the fraction $\frac{3}{4}$. Example 7 also goes with the fraction $\frac{3}{4}$. The rational number whose symbol is $\frac{3}{4}$ is the property shared by all pairs of sets, A and B, for which A is related to B as in Examples 7, 13, 14, and 15.

Note that Examples 4 and 5 do not have the rational number property $\frac{2}{3}$, because these pairs of sets do not fit the conditions that are required before we can attach any rational number property to them. In each case, the parts of A are *not* equivalent to the parts of B.

■ **Question 2** Draw a pair of sets which have the rational number property $\frac{3}{2}$.

In each of the illustrations above, we compared A with B. We could also compare B with A. The fractions that describe how B is related to A are given in this table:

Example	Fraction that tells how B is related to A	Example	Fraction that tells how B is related to A
1	$\frac{3}{2}$	11	$\frac{3}{2}$
2	$\frac{5}{3}$	12	$\frac{3}{2}$
3	$\frac{3}{1}$	13	$\frac{4}{3}$
7	$\frac{4}{3}$	14	$\frac{4}{3}$
8	$\frac{6}{5}$	15	$\frac{4}{3}$

There is another method of writing the way one set is related to another. In Example 1 above, we could write: "A is $\frac{2}{3}$ *of* B." For these two sets in Example 16 we could write "A is $\frac{5}{2}$ of B" (see Figure 12.8).

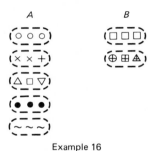

Example 16

Figure 12.8

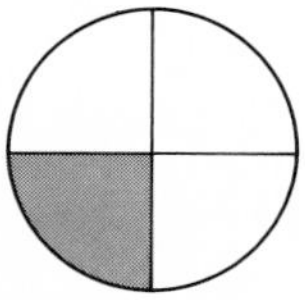

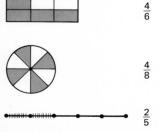

Figure 12.9

$\frac{2}{3}$

$\frac{4}{6}$

$\frac{4}{8}$

$\frac{2}{5}$

Figure 12.10

Finally, in all the examples we have looked at so far, the two sets have been disjoint. This is not necessary. In fact, in many cases A will be a subset of B. For example, if you think you should have $\frac{1}{4}$ of the pie, then the picture you have in mind is shown in Figure 12.9, in which the full circular region, divided into four congruent pieces, is the set B, while set A is the shaded part.

Figure 12.10 shows some more examples of pairs of sets, in each case A being the shaded part and B the whole figure.

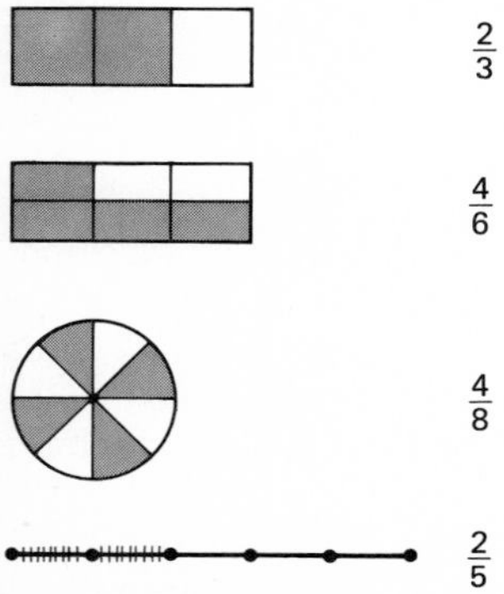

■ **Question 3** If the part outlined by the heavy line in each part of Figure 12.11 is the set B and the shaded part is the set A, what fraction tells how A is related to B in each of these cases?

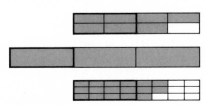

Figure 12.11

We have not given a precise definition of "rational number," but these examples should be enough to give the general idea. However, there is a way of illustrating rational numbers, namely by representing them on a number line, which will help us to better understand not only the meaning of "rational number" but also some of the properties of this kind of number.

RATIONALS AND THE NUMBER LINE

To represent a rational number, such as $\frac{2}{3}$, on the number line, we divide the segment from 0 to 1 into 3 congruent segments. Then, starting at 0, we count out 2 of these segments and mark the right-hand endpoint of the second segment. We attach the label $\frac{2}{3}$ to this point (see Figure 12.12).

Figure 12.12

Following the same procedure, the picture of $\frac{3}{5}$ on the number line is shown in Figure 12.13. For $\frac{5}{3}$, see Figure 12.14.

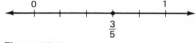

Figure 12.13

Figure 12.14

Notice that, in each of these cases, if we hatch the segments running from 0 to the point representing the fraction and then compare the hatched set to the set of segments from 0 to 1, we have a pair of sets which form an example of the fraction. For example, Figure 12.15 is derived from Figure 12.12 by hatching the segments from 0 to $\frac{2}{3}$. The set A, consisting of these two hatched segments, is related to the set B, consisting of all three segments, according to the rational number $\frac{2}{3}$ (see Figure 12.15).

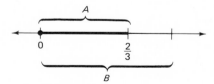

Figure 12.15

There are some very important things about rational numbers which we can illustrate easily by means of the number line.

■ **Question 4** Show $\frac{1}{2}$ and $\frac{2}{4}$ on the same number line.

Figure 12.16 shows a number line on which we have located points corresponding to $\frac{0}{1}$, $\frac{1}{1}$, $\frac{2}{1}$, $\frac{3}{1}$, etc.; another one on which we have located points corresponding to $\frac{0}{2}$, $\frac{1}{2}$, $\frac{2}{2}$, $\frac{3}{2}$, etc.; one on which we have located points corresponding to $\frac{0}{4}$, $\frac{1}{4}$, $\frac{2}{4}$, $\frac{3}{4}$, $\frac{4}{4}$, $\frac{5}{4}$, etc.; and one on which we have located points corresponding to $\frac{0}{8}$, $\frac{1}{8}$, $\frac{2}{8}$, $\frac{3}{8}$, $\frac{4}{8}$, etc.

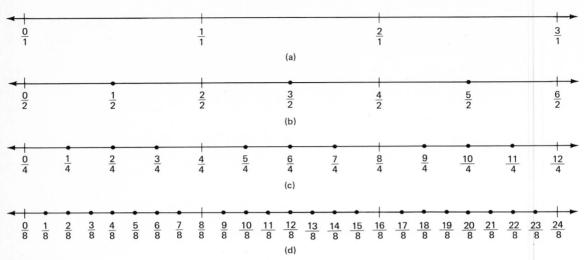

Figure 12.16

As we look at the number lines in Figure 12.16, we see that it seems very natural to think of $\frac{0}{2}$, for example, as being associated with the zero point. For we are really, so to speak, counting off 0 segments. Similarly, it seems natural to locate $\frac{0}{1}$, $\frac{0}{4}$, and $\frac{0}{8}$ as indicated.

Now let us put the four number lines in Figure 12.16 together, as shown in Figure 12.17.

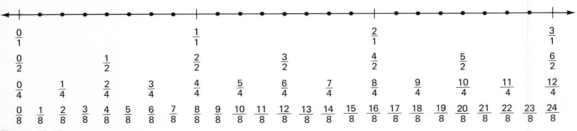

Figure 12.17

EQUIVALENT FRACTIONS   There are a number of things to be noticed in Figure 12.17. First, notice that the three fractions

$\frac{1}{2}$, $\frac{2}{4}$, and $\frac{4}{8}$

all correspond to the same point on the number line. Similarly

$\frac{3}{2}$, $\frac{6}{4}$, and $\frac{12}{8}$

all correspond to the same point on the number line.

■   **Question 5**   Which fraction corresponds to the same point on the number line as $\frac{32}{16}$?

We say that the fractions $\frac{1}{2}$, $\frac{2}{4}$, and $\frac{4}{8}$ are *equivalent*, and similarly $\frac{3}{2}$, $\frac{6}{4}$, and $\frac{12}{8}$ are equivalent. If you locate the point corresponding to $\frac{8}{16}$, you will find that it is the same as the point representing $\frac{1}{2}$, $\frac{2}{4}$, and $\frac{4}{8}$, and so $\frac{8}{16}$ is equivalent to each of these.

Since the representations on the number line of the equivalent fractions $\frac{1}{2}$, $\frac{2}{4}$, and $\frac{4}{8}$ are all the same point, it is reasonable to say that these three fractions all represent the same rational number, and that $\frac{1}{2}$, $\frac{2}{4}$, and $\frac{4}{8}$ are all names (numerals) for this one rational number. The fraction $\frac{8}{16}$ is also a name for this same rational number.

In the same way, all the fractions $\frac{3}{2}$, $\frac{6}{4}$, and $\frac{12}{8}$ are names for another rational number, and $\frac{11}{2}$ and $\frac{22}{4}$ are two different names for still another rational number.

Let us take another look at the equivalent fractions

$$\frac{1}{2}, \ \frac{2}{4}, \ \frac{4}{8}, \ \text{and} \ \frac{8}{16}$$

If we remember some multiplication facts we see that we can rewrite these as

$$\frac{1}{2} \quad \frac{2 \times 1}{2 \times 2} \quad \frac{4 \times 1}{4 \times 2} \quad \frac{8 \times 1}{8 \times 2}$$

This suggests that

if k is a whole number, then

$$\frac{a}{b} \quad \text{and} \quad \frac{k \times a}{k \times b}$$

are equivalent fractions and are names for the same rational number.

The equivalent fractions $\frac{3}{2}$, $\frac{6}{4}$, and $\frac{12}{8}$ are another example of this, since $\frac{6}{4} = (2 \times 3)/(2 \times 2)$ and $\frac{12}{8} = (4 \times 3)/(4 \times 2)$.

■ **Question 6** Show both $\frac{2}{8}$ and $\frac{3}{12}$ on the same number line. Is there a whole number k such that

$$\frac{3}{12} = \frac{(k \times 2)}{(k \times 8)}$$

It is not hard to illustrate this. For example, we can use Figure 12.18 to represent the fraction $\frac{2}{3}$. The large rectangle, divided into 3 congruent rectangles by the vertical lines, is the set B, while the 2 shaded rectangles form the set A.

Now let us divide this large rectangle into 4 parts by means of three equally spaced horizontal line segments (Figure 12.19).

Obviously, the sets A and B, although they have been divided into smaller pieces, have not changed in size, so this figure still represents the fraction $\frac{2}{3}$. However, we can see that A now consists of $8 = 4 \times 2$ small shaded rectangles and B consists of $12 = 4 \times 3$ small rectangles. So this figure represents the fraction $\frac{8}{12}$ or $(4 \times 2)/(4 \times 3)$. This illustrates that

$$\frac{4 \times 2}{4 \times 3} = \frac{2}{3}$$

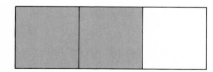

Figure 12.18

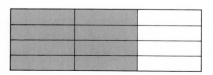

Figure 12.19

Similarly, if we had divided B into smaller pieces by means of 6 equally spaced horizontal line segments, we would have what is shown in Figure 12.20. This figure still represents $\frac{2}{3}$. But it also represents $\frac{14}{21}$. This is an illustration of the fact that

$$\frac{7 \times 2}{7 \times 3} = \frac{2}{3}$$

There is one more thing that we should notice about the figure above. The fraction $\frac{1}{1}$ has for its representation on the number line the same point as the whole number 1 does. Also, $\frac{2}{1}$ and 2 are represented by the same point on the number line, and similarly for $\frac{3}{1}$ and 3, etc.

This suggests that

If k is a whole number, then it is also a rational number represented by the fraction $k/1$.

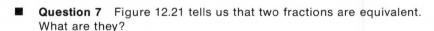

Figure 12.20

■ **Question 7** Figure 12.21 tells us that two fractions are equivalent. What are they?

Figure 12.21

ORDER AMONG RATIONAL NUMBERS

If you think of a whole number, say a, and I think of a whole number, say b, then there are three possibilities:

$a < b$ (a is less than b)

$a = b$ (a is equal to b)

$a > b$ (a is greater than b)

and only one of these possibilities can be true.

A similar statement can be made about two rational numbers, a/b and c/d:

One and only one of these is true

$$\frac{a}{b} < \frac{c}{d} \qquad \left(\frac{a}{b} \text{ is less than } \frac{c}{d}\right)$$

$$\frac{a}{b} = \frac{c}{d} \qquad \left(\frac{a}{b} \text{ is equal to } \frac{c}{d}\right)$$

$$\frac{a}{b} > \frac{c}{d} \qquad \left(\frac{a}{b} \text{ is greater than } \frac{c}{d}\right)$$

If we have two fractions which have the *same* denominator, such as $\frac{4}{11}$ and $\frac{7}{11}$, it is easy to see which of the three order possibilities holds. Draw both on the number line (see Figure 12.22). The fraction $\frac{4}{11}$ tells how the set A of segments from 0 to $\frac{4}{11}$ is related to the set C of 11

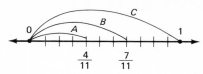

Figure 12.22

segments from 0 to 1. The fraction $\frac{7}{11}$ tells how the set B of segments from 0 to $\frac{7}{11}$ is related to the set C of all 11 segments.

Both A and B are compared with the same set C. Since A is a subset of B, we find it natural to say that $\frac{4}{11}$ is less than $\frac{7}{11}$.

In general, if two fractions have the same denominator, then we need only compare their numerators (which are whole numbers) to find out how the two rational numbers compare. But suppose we have two fractions, such as $\frac{9}{14}$ and $\frac{5}{8}$, with different denominators. In such a case, we can use the fact that a/b and $(k \times a)/(k \times b)$ are equivalent fractions to find two new fractions which do have the same denominator and which are equivalent to the original fractions. The trick is to multiply both the numerator and the denominator of each fraction by the denominator of the other. Thus, for the fractions $\frac{9}{14}$ and $\frac{5}{8}$,

$$\frac{9}{14} \quad \text{and} \quad \frac{8 \times 9}{8 \times 14} = \frac{72}{112} \quad \text{are equivalent}$$

$$\frac{5}{8} \quad \text{and} \quad \frac{14 \times 5}{14 \times 8} = \frac{70}{112} \quad \text{are equivalent}$$

(Notice how the commutative law for multiplication of whole numbers shows up.) Now, since $72 > 70$, $\frac{9}{14} > \frac{5}{8}$.

■ **Question 8** Arrange these in order: $\frac{2}{3}, \frac{3}{4}, \frac{3}{5}, \frac{4}{7}$

Although the arithmetic can become tedious with some problems, this method always works. In a later chapter we will find ways of minimizing the arithmetic.

RATIONAL NUMBERS IN MIXED FORM

Each of us is familiar with the fact that a rational number whose name is $\frac{3}{2}$, for example, also may be named in the "mixed form" sometimes called a "mixed numeral," $1\frac{1}{2}$. (We do not say "mixed number" because $1\frac{1}{2}$ is a numeral, that is, a name for a number, but not a number.) Let us use the number line to examine briefly some of the assumptions underlying our use of the familiar mixed form for naming certain rational numbers (Figure 12.23).

Consider, for instance, the use of $\frac{5}{3}$ and $1\frac{2}{3}$ to name the same rational number. We often state that $\frac{5}{3} = 1\frac{2}{3}$. Behind this statement there is the assumption, among others, that rational numbers can be added: $\frac{5}{3} = \frac{3}{3} + \frac{2}{3} = 1 + \frac{2}{3} = 1\frac{2}{3}$.

In a later chapter we will give systematic consideration to the addition of rational numbers. However, we do wish to point out that this operation is implicit in an interpretation of the mixed form for a rational number.

| $\frac{0}{3}$ | $\frac{1}{3}$ | $\frac{2}{3}$ | $\frac{3}{3}$ | $\frac{4}{3}$ | $\frac{5}{3}$ | $\frac{6}{3}$ | $\frac{7}{3}$ | $\frac{8}{3}$ |

| 0 | $\frac{1}{3}$ | $\frac{2}{3}$ | 1 | $1\frac{1}{3}$ | $1\frac{2}{3}$ | 2 | $2\frac{1}{3}$ | $2\frac{2}{3}$ |

Figure 12.23

RATIONAL NUMBERS AND DIVISION OF WHOLE NUMBERS

We saw in Chapter 7 that not every whole-number division problem has an answer. For example, $12 \div 5$ cannot be done, since there is no whole number which, when multiplied by 5, yields 12.

However, we also remember that the division *process* can always be applied. Pictorially, to apply this process to $12 \div 5$, we start with 12 objects and we try to put them into a 5-rowed array.

We can fill two columns and will have two left over. We now take the remaining two and put them in column three (see Figure 12.24). Now compare the column of two with the column of five. This is an illustration of the fraction $\frac{2}{5}$. The diagram therefore suggests

$$12 \div 5 = 2 + \tfrac{2}{5} \qquad \text{or} \qquad 2\tfrac{2}{5}$$

We will return to this in a later chapter, but notice that it implies that we can always work division problems for whole numbers providing we are willing to accept rational numbers as answers.

Figure 12.24

■ **Question 9** Write the mixed numeral $6\frac{2}{3}$ in the form of a fraction.

A NEW PROPERTY OF NUMBERS

Rational numbers are different in many ways from whole numbers. One such difference is apparent if we recall that for any whole number one can always say what the "next" whole number is and then ask, in a similar vein, what the "next" rational number is after any given rational number. For example, 4 is the next whole number after 3, 1,069 is the next whole number after 1,068, and so on. What is the next rational number after $\frac{1}{2}$? If $\frac{2}{3}$ is suggested as the next one, we can observe that $\frac{1}{2} = \frac{6}{12}$ and $\frac{2}{3} = \frac{8}{12}$, so $\frac{7}{12}$ is surely between $\frac{1}{2}$ and $\frac{2}{3}$. Hence, $\frac{7}{12}$ has a better claim to being next to $\frac{1}{2}$ than does $\frac{2}{3}$. If it is then suggested that $\frac{7}{12}$ be regarded as the next number after $\frac{1}{2}$, we can observe that $\frac{1}{2} = \frac{12}{24}$ and $\frac{7}{12} = \frac{14}{24}$ so $\frac{13}{24}$ is closer to $\frac{1}{2}$ than is $\frac{7}{12}$. To carry this one step further, we can squelch anyone who suggests $\frac{13}{24}$ as being the next number after $\frac{1}{2}$ by pointing out that $\frac{1}{2} = \frac{24}{48}$ and $\frac{13}{24} = \frac{26}{48}$ so that $\frac{25}{48}$ is more nearly "next to" $\frac{1}{2}$ than is $\frac{13}{24}$. It is clear that this process could be carried on indefinitely and, furthermore, would apply no matter what rational number was involved. That is, we can never identify a "next" rational number after any given rational number. A similar argument would show that we cannot identify a number "just before" a given rational number. A number line with a very large unit is shown in Figure 12.25 to illustrate the process we went through in searching for the number "next to" $\frac{1}{2}$.

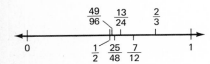

Figure 12.25

Another way of expressing what we have been talking about is to say that between any two rational numbers, there is always a third rational number; in fact, there are more rational numbers there than we could count. Mathematicians sometimes describe this by saying that the set of rational numbers is *dense*. The word is not important to us, but is descriptive of the packing of points representing rational numbers closer and closer together on the number line.

■ **Question 10** Find 100 rational numbers between $\frac{1}{2}$ and $\frac{2}{3}$.

CONCLUSION

So far in this chapter we have but started our study of rational numbers. We will see in later chapters that there are many other aspects of the world around us that can be described in terms of rational numbers. Also, in another chapter, we will see that rational numbers are more like whole numbers than this chapter would suggest. They can be added, subtracted, multiplied, and divided, and the techniques for doing these operations on rational numbers are closely related to the techniques for operating on whole numbers.

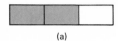

(a)

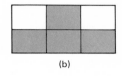

(b)

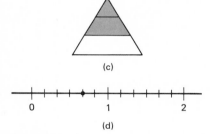

(c)

(d)

Figure P12.1

PROBLEMS

1 Which of the following are models for the same rational number as the ones shown in Figure P12.1?
  (1) Only P12.1a
  (2) Only P12.1c
  (3) Only P12.1a and b
  (4) Only P12.1a, b, and d
  (5) All of these

2 What rational numbers do the models in Figure P12.2 illustrate?

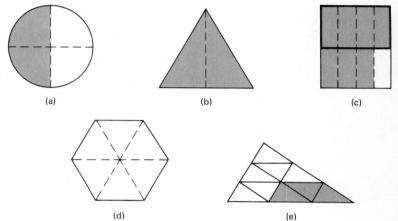

Figure P12.2

3 Most of the figures in Figure P12.3 are models for rational numbers. Some of them are not models because the unit has not been partitioned into congruent parts. For each one that is a proper model, give the rational number which is pictured.

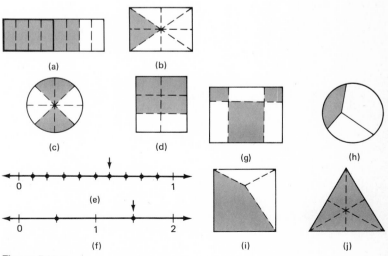

Figure P12.3

4 Which one of the drawings in Figure P12.4 *does not* suggest the same rational number as the other four drawings?

(a)

(b)

(c)

(d)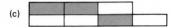

(e)

Figure P12.4

5 Draw models for:
(a) $\frac{2}{3}$ (b) $\frac{4}{6}$ (c) $\frac{7}{7}$
(d) $\frac{0}{6}$ (e) $\frac{3}{4}$ (f) $\frac{1}{7}$

6 Referring to Figure P12.6, what fractional part of the dots is inside the triangle but not inside the rectangle?

7 Locate the point associated with each of the following on a number line.
(a) $\frac{0}{1}$ (b) $\frac{3}{4}$ (c) $\frac{3}{5}$
(d) $\frac{5}{5}$ (e) $\frac{7}{4}$ (f) $\frac{8}{8}$

8 X is the same fraction as shown in Figure P12.8(e). What fraction is X?

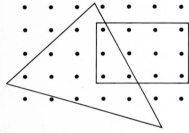

Figure P12.6

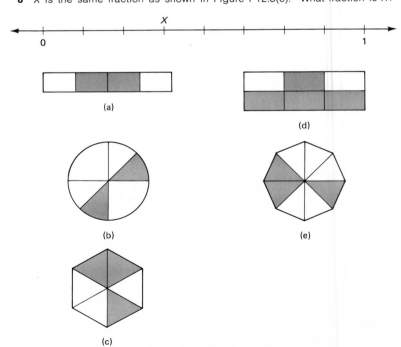

Figure P12.8

9 Interpret on the number line the following:

(a) $\frac{20}{5} = 4$ (b) $\frac{20}{4} = 5$ (c) $\frac{23}{5} = 4\frac{3}{5}$

10 Supply the missing numbers in each of the following:

(a) $\frac{3}{5} = \frac{3 \times \square}{5 \times \square} = \frac{24}{40}$ (b) $\frac{7}{8} = \frac{\square}{32}$ (c) $\frac{\square}{12} = \frac{14}{24}$

11 Determine the missing numbers such that the following are true:

(a) $\frac{5}{7}$ and $\frac{\square}{42}$ are equivalent   (b) $\frac{3}{7}$ and $\frac{9}{\square}$ are equivalent

(c) $\frac{4}{5}$ and $\frac{\square}{30}$ are equivalent   (d) $\frac{2}{15}$ and $\frac{\square}{20}$ are equivalent

We know if k is a whole number, then

$$\frac{a}{b} \quad \text{and} \quad \frac{k \times a}{k \times b}$$

are equivalent fractions and are names for the same rational number. Thus, in Problem 10, for example,

$$\frac{3}{5} = \frac{3 \times 8}{5 \times 8} = \frac{24}{40}$$

We say that $\frac{24}{40}$ is in *"higher terms"* than $\frac{3}{5}$. Also, we say that $\frac{3}{5}$ is in *"lower terms"* than $\frac{24}{40}$.

Similarly, $\frac{6}{12}$ is in "lower terms" than $\frac{12}{24}$, which in turn is in "higher terms" than $\frac{6}{12}$.

Next, we see that $\frac{1}{2} = (6 \times 1)/(6 \times 2) = \frac{6}{12}$, so $\frac{1}{2}$ is in "lower terms" than $\frac{6}{12}$. A little experimentation makes it clear that there is no other fraction which is in "lower terms" than $\frac{1}{2}$. We say that $\frac{1}{2}$ is in *"lowest terms."* Also, going back to $\frac{3}{5}$, there is no other fraction in "lower terms," so $\frac{3}{5}$ is in "lowest terms."

12 For each of the following, give one equivalent fraction in "higher terms" and give three equivalent fractions in "lower terms," including one in lowest terms.

(a) $\frac{24}{36}$ (b) $\frac{30}{60}$

13 Tell which of the following fractions are in "lowest terms."

$\frac{6}{12} \quad \frac{11}{4} \quad \frac{7}{12} \quad \frac{10}{12} \quad \frac{13}{26} \quad \frac{2}{3}$

14 For what pairs of whole numbers a and b are the fractions a/b and b/a equivalent?

15 Make each of the following statements true by writing = or > or < inside the circle.

(a) $\frac{6}{14} \, \bigcirc \, \frac{7}{16}$ (b) $\frac{6}{8} \, \bigcirc \, \frac{9}{12}$

(c) $\frac{30}{63} \, \bigcirc \, \frac{15}{28}$ (d) $\frac{3}{4} \, \bigcirc \, \frac{36}{52}$

(e) $\frac{9}{20} \, \bigcirc \, \frac{45}{100}$ (f) $\frac{143}{13} \, \bigcirc \, \frac{1043}{103}$

16 Order the following sets of fractions from largest to smallest.

(a) $A = \{\frac{0}{2}, \frac{1}{2}, \frac{3}{4}, \frac{5}{16}\}$ (b) $B = \{\frac{200}{300}, \frac{1}{2}, \frac{20}{24}, \frac{9}{12}\}$

(c) $C = \{\frac{9}{9}, \frac{0}{9}, \frac{900}{100}, \frac{50}{150}\}$

17 Express each of these in mixed form.

(a) $\frac{7}{4}$ (b) $\frac{15}{8}$ (c) $\frac{21}{9}$

(d) $\frac{34}{15}$ (e) $\frac{56}{12}$

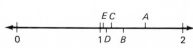

Figure P12.18

18 Name the rational numbers associated with the points A, B, C, D, and E in Figure P12.18, where A is halfway between 1 and 2, B halfway between 1 and A, etc.

19 How many numbers are there between 1 and the number associated with point E?

20 (a) Name five fractions which are less than $\frac{1}{2}$ and greater than $\frac{1}{4}$.

(b) How many answers could be given to part a?

13. PROBABILITY

In this chapter we will take a preliminary look at another particular aspect of the world we live in and the way in which mathematical ideas can help us focus our attention most usefully on a special set of problems. We will see that rational numbers will be very useful in studying this aspect of the world.

ALWAYS, SOMETIMES, NEVER

There are many activities which go on in the world around us for which we can predict the results in advance. If we toss a coin into the air we can be sure, in advance, that it will fall back to the ground. If we go swimming we can be sure, in advance, that we will get wet. If a penny-stamping machine at the U.S. Mint is fed sheets of copper alloy, we can be sure, in advance, that it will not stamp out a series of silver dollars. If we reach into a sack containing 3 red tennis balls and three green ones and take one of them out, we can be sure, in advance, that the ball we take out will be round. And of course the reader can think of any number of similar examples where we can tell in advance what will happen.

However, there are also plenty of cases where we cannot tell in advance what will happen. If we toss the coin into the air again, we cannot tell in advance whether it will land heads or tails. If the penny-stamping machine runs continuously for eight hours, we cannot be sure in advance that none of the pennies it stamps out will be defective. If we plan to check the weather tomorrow at noon, we cannot tell in advance whether or not it will be cloudy.

■ **Question 1** A box contains 5 red marbles and 500 white ones. If you take 6 marbles out, without looking, can you be sure they will all be white? Can you be sure at least 1 will be white?

These examples are enough to remind us that for any particular action and any particular result, we can say either that the result *never* happens, or that it *sometimes* happens, or that it *always* happens. The first and last cases pose no problems for us. We know in advance what the result will be. It is the "sometimes" cases that we need to look at more carefully.

EXPERIMENTS Before we go on, however, let us introduce the standard terminology of this part of mathematics. An activity together with the set of all conceivable results will be called an *experiment*. Thus, the action of tossing a coin in the air together with the set {head, tail} of possible results is an experiment. The action of removing one of the 6 tennis balls from the sack, together with the set {red, green} is another experiment. The action of observing the weather at noon tomorrow together with the set {cloudy, not cloudy} is still another experiment. (Incidentally, the experiment consisting of tossing a coin in the air together with the set {fall back to the ground, float in the air} is also an experiment, but not the same one as the first example above. It is not enough to specify the activity. We must also tell which set of results we are interested in in order to specify an experiment.)

The set of possible results of the action of an experiment is also called the set of *outcomes* of the experiment. Doing the action of an experiment and recording (or at least observing) the actual outcome of the action is called a *trial* of the experiment. We may conduct as many trials of an experiment as we wish. Thus, the coin-tossing experiment has only two possible outcomes, but we may toss the coin as many times as we please.

■ **Question 2** Suppose we toss both a penny and a quarter in the air. What is the set of possible outcomes?

While we cannot predict, in advance, whether a coin will land heads or tails, we know that somehow there is a certain degree of regularity in nature. If we fill a quart jar with pennies and then dump them all on the floor, we are surprised if those landing heads will not just about fill a pint jar. If we repeat the experiment of removing 1 tennis ball from the sack and recording its color (of course putting it back before the next trial of the experiment) 500 times, we would be very surprised if we recorded red only 50 times and green 450 times. If the penny-stamping machine has been turning out fewer than 2 defective coins in each 1,000 for the last two weeks, but today about half are defective, then we would probably conclude that there is something wrong with the machine, that some part was broken or out of adjustment, and hence that today's experiment is not the same as yesterday's.

PROBABILITY *Probability* is the part of mathematics which looks at the kind of regularity which nature displays in the long run. This is a very important part of mathematics since it can provide us with useful information about such things as the kind of weather we can expect tomorrow or next week, the advisability of doing a particular operation on a particular patient, the need for repairs on the penny-stamping machine in the Mint or an engine on a jet plane, etc.

■ **Question 3** If you tossed a coin 100 times and it came up heads each time, what would you expect to happen on the next toss?

Nevertheless, all the basic ideas of probability can be illustrated in much less complicated situations. In what follows, we will use only a few simple experiments, involving coin tossings, drawing cards from a deck, or tennis balls (or marbles) from a sack, etc., to illustrate these basic ideas.

To prepare the background for one of these important ideas about probability, let us compare two simple experiments. In the first one, a black marble and a white one are put in a paper sack. One is pulled out and the color is noted, so the set of outcomes of this experiment is {black, white}. For the second experiment, a black marble and 100 white ones are put in another sack. One is pulled out and the color is noted, so the set of outcomes of this experiment is also {black, white}.

Now suppose you are told that a single trial is to be made of one of these experiments and that if the outcome is white, you are to receive a dollar. Which experiment would you choose? Most people would choose the second experiment, feeling that in this experiment, the outcome "white" is *more probable* than "black."

Now suppose you are told that a single trial will be made of the first experiment. You are to be given one of the colors in advance, and if the outcome is the color you were given, you will receive a dollar. Would you care which color you were given? Most people would not care, feeling that the two possible outcomes are *equally probable.*

■ **Question 4** Suppose you have 50 white and 50 red marbles in a paper bag and you draw one out. Are "white" and "red" equally probable?

In most simple experiments, when two outcomes are compared, it is intuitively clear whether they are equally probable or whether one is more probable than the other. When a coin is tossed, the outcomes "head" and "tail" are equally probable. When a die is thrown, the outcome "one" is less probable than the outcome "greater than one." When a card is drawn from a deck, the outcome "spot card" is more probable than the outcome "face card."

EVENTS Sometimes we are interested in more than the individual outcomes of an experiment. Thus, for example, if we throw a die, we might want to know only if the result is an even number. This means that we are interested in the subset $\{2, 4, 6\}$ of the total set of outcomes $\{1, 2, 3, 4, 5, 6\}$. Or if we draw a card from a deck, we might only want to know if it is a spade. Here we are interested in the subset, consisting of the 13 spades, of the set of all 52 cards.

We will call any subset of the total set of outcomes of an experiment an *event.* We would like to be able to compare events to see if one is more probable than another. Usually this is not as intuitively clear for events as it is for individual outcomes. For example, think of the experiment of throwing two dice. One event is having both the numbers odd. Another event is having the numbers add up to a multiple of 3. Which event is more probable? By appealing to numbers, we can get some help in answering this and similar questions.

■ **Question 5** If you throw two dice, what are the members of the event "the sum of the numbers of spots on the top faces is more than 17"?

Suppose we have an experiment which has a total of N equally probable outcomes. (N would be 2 for tossing a coin, 6 for throwing a die, 52 for drawing a card, etc.) Suppose A is an event, a subset of the total set of outcomes, consisting of a of these outcomes. Then we say that the *probability of A is a out of N*, and we abbreviate this to:

Prob(A) is a out of N

If we throw a die, there are three even number outcomes, so if A is the event "even number," Prob(A) is 3 out of 6. If we draw a card, and if B is the event "ace," then Prob(B) is 4 out of 52.

PROBABILITIES AS FRACTIONS There is another way of specifying the probability of an event. Let us think back to the way we introduced the notion of a rational number in the last chapter. We agreed there that "rational number" made sense only where we were comparing two sets, A and B, each divided into "equivalent" parts. It seems reasonable to say that two outcomes of an experiment, if they are equally probable, are equivalent. Then, instead of saying, as above, that

Prob(A) is a out of N

we could just as well say

$$\text{Prob}(A) = \frac{a}{N}$$

This is the same as saying that the probability of A is the rational number which compares the set A with the set of all possible outcomes.

Thus, if A is the event "even number" for the experiment of tossing a die,

$$\text{Prob}(A) = \tfrac{3}{6} = \tfrac{1}{2}$$

■ **Question 6** What is the probability of the event "less than 3"?

If we have two events, A and B, for the same experiment, and if all outcomes are equally likely, then it would seem reasonable to call A *more probable than B* if the number of outcomes, a, in A is greater than the number of outcomes, b, in B. To answer the question about the two dice, we merely need to count the members of the two events.

Of course, if $a > b$, then

$$\frac{a}{N} > \frac{b}{N}$$

so saying that A is more probable than B is the same thing as saying that

Prob(A) > Prob(B)

This does not seem to say much that is new. But suppose we have two different experiments, for each of which all the outcomes are equally probable, and an event A for the first experiment and an event B for the second. Does it make sense to ask if

Prob(A) > Prob(B)

For example, if A is the event "divisible by 3" for the experiment of tossing a die and B is the event "face card" for drawing a card, does it make sense to ask if

Prob(A) > Prob(B)

It certainly does seem to make sense. To find out the answer, all we have to do is to compare the two rational numbers,

Prob(A) = Prob($\{3, 6\}$) = $\frac{2}{6}$

and

Prob("face card") = $\frac{12}{52}$

A little arithmetic shows that Prob(A) > Prob(B)

■ **Question 7** If you draw a card from a deck, which event is more probable, "spade face card" or "heart spot card"?

COMPOUND EVENTS So far, the only experiments we have looked at have been rather simple. Let us look at a more complicated experiment which consists of a simple experiment done twice in a row. For example, think of the experiment of tossing a coin once and then tossing it again. If we want to calculate the probability of such an event as "at least one head in the two tosses," we need to find the set of all possible outcomes of this compound experiment.

If we toss it once we have exactly two possible outcomes, heads and tails, which we denote by H and T respectively. If we toss the coin twice, we have four possible outcomes from the succession of tosses, all of them equally likely. We can show them in Table 13.1.

Table 13.1

First toss	Second toss
H	H
H	T
T	H
T	T

TREE DIAGRAMS The same information can be given in a "tree diagram," as pictured in Figure 13.1. The tree shows that, for *each* result of the first toss there are two results of the second toss. This is represented in the diagram by *two* arrows from *each* entry in the first column. The possible outcomes for the combination of two tosses can be seen by reading from left to right along the "branches of the tree." They are HH, HT, TH, TT.

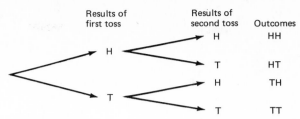

Figure 13.1

It is now an easy matter to see that the probability of the event "at least one head" is $\frac{3}{4}$, since three of the four equally likely outcomes result in one H or more.

■ **Question 8** What is the probability of "no more than one head"?

As another example, consider a spinner with half its area red and half white, like the one shown here. We are assuming that the red region is congruent to the white region. (In the actual construction of a spinner, the red region might have a slightly different area than the white region, but we simplify the mathematical model by assuming congruence of the regions.)

An experiment consists of twirling the pointer with a good push and noting the color the pointer stops on (see Figure 13.2). In this experiment it seems intuitively clear that the two outcomes are equally likely, so

$P(\text{red}) = \frac{1}{2}$ and $P(\text{white}) = \frac{1}{2}$

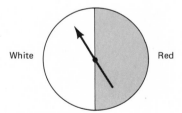

Figure 13.2

We are assuming that the pointer does not land exactly on the boundary between two regions. Indeed, we shall adopt this understanding in all our examples involving spinners. (This agreement represents an ideal situation. Sometimes the spinners in children's games have wide marks for the boundaries, and the stopping of the pointer on such a mark is not unusual. If this happens when you are actually performing an experiment, discard the trial and spin again.)

Now suppose we spin the spinner once and then toss a coin. A tree diagram for this experiment is shown in Figure 13.3.

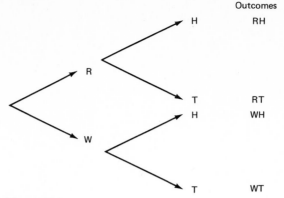

Figure 13.3

Tree diagrams can also be used when a single experiment is repeated more than twice. Figure 13.4 is the diagram for the experiment of tossing a coin four times in a row. Inspection of this diagram makes it easy to compute the probability of such an event as: {at least two consecutive tails}. It is 8 out of 16, or $\frac{1}{2}$.

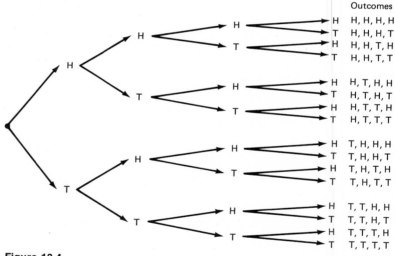

Figure 13.4

■ **Question 9** Suppose we toss a coin three times. What is the probability of "no more than one head"?

MORE THAN TWO OUTCOMES

In these situations, each trial had exactly two outcomes. The use of the tree diagram can often help in analyzing experiments for which the number of outcomes of a single trial is greater than 2.

For example, suppose that we have a box containing three marbles alike except for color: one red, one green, one yellow. If we pick one marble at random, then there are three possible results, which we shall call R, G, and Y, for red, green, and yellow, respectively. If we return the chosen marble to the box and draw again, we again have three possibilities. Each outcome, for the experiment consisting of the pair of drawings, can be described in terms of the color on the first draw and the color on the second draw; for example, RY. The possibilities are shown in the tree diagram in Figure 13.5.

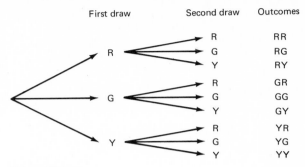

Figure 13.5

Now, let us take a careful look at the experiment of throwing two dice. Experience has shown that it is best to think of this as two separate experiments. To emphasize this, let us say that one die is green and that the other is red. Or, if we wish, we can use just one die, but throw it twice. In either case, one way of displaying the set of all possible outcomes is by means of a tree diagram. However, there is another way which is sometimes more convenient. This is by means of an array. In the array, the rows could stand for the outcome for the green die (or the first throw) and the columns for the outcomes for the red die (or the second throw), as illustrated in Table 13.2.

Table 13.2
Red Die or Second Throw

	1	2	3	4	5	6
1	1,1	1,2	1,3	1,4	1,5	1,6
2	2,1	2,2	2,3	2,4	2,5	2,6
3	3,1	3,2	3,3	3,4	3,5	3,6
4	4,1	4,2	4,3	4,4	4,5	4,6
5	5,1	5,2	5,3	5,4	5,5	5,6
6	6,1	6,2	6,3	6,4	6,5	6,6

Green Die or First Throw (row labels)

■ **Question 10** What is the probability of "6 spots on the top face of each die"?

Once we have the complete set of outcomes of this experiment, of throwing two dice, set out before us, it is easy to settle the problem of comparing the two events such as A: {both numbers odd} and B:

{sum of the numbers a multiple of 3}. Looking through the array, we find that there are 9 elements of the set A, {(1, 1), (1, 3), (1, 5), (3, 1), (3, 3), (3, 5), (5, 1), (5, 3), (5, 5)}. Therefore, Prob(A) is 9 out of 36. But B has 12 elements: {(1, 2), (1, 5), (2, 1), (2, 4), (3, 3), (3, 6), (4, 2), (4, 5), (5, 1), (5, 4), (6, 3), (6, 6)}. Therefore, Prob(B) is 12 out of 36, so B is more probable than A.

We can create a compound experiment by performing one experiment and then following it by a different experiment. The set of outcomes of such a compound experiment can also be displayed in an array (or tree diagram if we wish). For example, think of tossing a coin and then throwing a die. The set of all 12 outcomes can be represented as shown in Table 13.3.

Table 13.3

	1	2	3	4	5	6
H	H,1	H,2	H,3	H,4	H,5	H,6
T	T,1	T,2	T,3	T,4	T,5	T,6

Inspection of this array makes it easy to see, for example, that the event A: {head and even *or* tail and greater than 4}† has 5 elements, so Prob(A) is 5 out of 12.

However, displaying the outcomes of a compound experiment by means of an array is not very convenient if there are more than two simple experiments involved in the compound one. For example, think of the experiment which consists of first tossing a coin, then throwing a die, and finally pulling out a marble from a paper sack containing one red, one green, and one black marble. The array would have to be three-dimensional, which is inconvenient to draw. If we were to add on at the end another simple experiment, then the array would be four-dimensional, which is even harder to draw. Of course, tree diagrams for these more complicated experiments also are awkward, but they can always be drawn, as Figure 13.3 suggests.

■ **Question 11** If we throw a die, toss a coin, and draw a marble out of a sack containing a red, a blue, and a black marble, what is the probability of the event "die shows 5 and coin shows head"?

INDEPENDENT EXPERIMENTS There is one final, and very important, point to be made about tree diagrams. In each of the cases we have examined, the tree diagram for a composite experiment can be obtained by drawing a copy of the tree diagram for the second experiment at the end of each arrow of the tree diagram for the first experiment (and so on if there are more than two experiments). This does *not* always work. In fact, it only works if the two experiments are *independent,* that is, if the outcome of the first has no effect on the second experiment.

---

† In mathematics, the word "or" in a phrase such as "A or B" is always interpreted to mean "either A or B or both."

Go back, for example, to the experiment above of picking one of three marbles out of a box. Suppose the marble is not replaced before the second marble is drawn. Then the second experiment is affected by the outcome of the first trial.

Thus, if the red marble, for example, is drawn on the first trial, then the only possible outcomes for the second trial are G and Y. The tree diagram for this compound experiment is shown in Figure 13.6. This diagram should be compared with Figure 13.5.

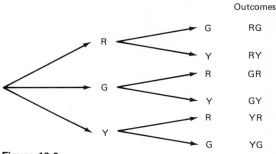

Figure 13.6

■ **Question 12** What is the probability of the event {same color each time} when the marble is replaced after the first drawing? What is the probability when it is not replaced?

SOME EXPERIMENTS FOR FUTURE REFERENCE

By now we have met enough of the basic ideas about probability and have learned enough of the standard terminology to be able to analyze and discuss the results of some simple and some compound experiments. However, it is impossible to obtain a sound intuitive understanding of probability without actually doing and analyzing some experiments.

The rest of this chapter, therefore, describes a few probability experiments. It is essential that the reader carry out each of these exercises completely, recording the results as instructed. The usual problem set at the end of the chapter will include a series of questions designed to help you analyze the results of these experiments.

Be sure to preserve your record of the results of these experiments. We will return to them in a later chapter.

EXPERIMENT I:
Put a penny and a dime in a cup, place the palm of your hand over the top so the coins cannot fall out, shake the cup vigorously, and tip it over so that the coins fall on the table top. Record whether the penny shows a head or a tail, and the same for the dime. For example, if the penny shows a head and the dime a tail, you should record the pair (H, T), giving the penny result first. (Note: *This is really the same experiment as that of tossing a coin twice in a row. The penny corresponds to the first toss and the dime to the second toss.*)

Keep repeating this experiment until you have recorded the results of 100 trials.

EXPERIMENT 2:

Put a penny, a nickle, a dime, and a quarter in a cup. Shake them up and pour them out on the table. Record the outcome head or tail for each coin in a table like this. Repeat until you have gone through 100 trials. Keep the record of the outcomes. (See Table 13.4.)

Table 13.4

Trial	P	N	D	Q
1				
2				
3				
4				

(Note: *This is really the same as the experiment of tossing one coin four times in a row.*)

EXPERIMENT 3:

For this experiment you will need three marbles, one red, one blue, and one green. They should all be the same size and weight, so that they can be told apart only by color. You will also need a paper bag.

Put the marbles in the bag and shake them around. Without looking in the bag, reach in and bring out one marble. Record the color of the marble and put it back in the bag. Then pull out another marble and record its color. There are nine possible outcomes.

Repeat this experiment until you have gone through 100 trials. Keep your record of the outcomes.

EXPERIMENT 4:

This is exactly the same as Experiment 3, except that on each trial the first marble that is taken out is *not* replaced.

Carry out 100 trials of this experiment and record the results.

EXPERIMENT 5:

Toss two differently colored dice and record the outcome as a pair of numbers, as in the array in Figure 13.6. (If you do not have any dice, you can use a pair of different colored six-sided wooden pencils. On each one, mark a dot on one side, two dots on another, and so on up to six dots on the last side. To use these substitute dice, hold each pencil, point up, a few inches above the table and let them drop. Record the number of dots on the top side of each one after it falls.)

Repeat until you have done a total of 100 trials of this experiment. Keep a record of the outcomes.

PROBLEMS

1 A bowl contains red marbles and white marbles. You are told that if you draw one marble it will probably be red. What do you conclude?

2 There are 25 students in a class, of whom 10 are girls and 15 are boys. The teacher has written the name of each pupil on a separate card. If the cards are shuffled and one is drawn, what is the probability that the name written on the card is:

(a) the name of a boy?

(b) your name (assuming you are in the class)?

3 Suppose we have an ordinary die with six faces: 1, 2, 3, 4, 5, 6. Is this game fair or unfair? On a single toss of the die you win if the number on the face which comes up is divisible by 3; you lose if it is not.

4 In the experiment of rolling a die once, what is the probability that the face landing up will have

(a) an even number of dots?

(b) more than one dot?

(c) more than six dots?

(d) at least one dot?

5 Two black marbles and one white marble are in a box. Without looking inside the box, you are to take out one marble. What is the probability that the marble will be black? (Hint: The outcomes "black" and "white" are not equally likely. To get around this, we can pretend that one of the black marbles is labeled with the number 1 and the other with the number 2. Then we can denote the possible outcomes as B_1, B_2, and W. These three outcomes are equally likely, and each has the probability $\frac{1}{3}$. Now, what is the probability of the event $\{B_1, B_2\}$?)

6 Three hats are in a dark closet. Two belong to Mr. Smith and the other to his friend. Mr. Smith reaches in the closet and draws two hats. What is the probability that he will pick his friend's hat and one of his own?

7 If there are three red marbles, four white marbles, and five blue marbles in a box, what is the probability of pulling out:

(a) a red marble?

(b) a white marble?

(c) a red or a blue marble?

8 From the same box, what is the probability of drawing

(a) a green marble?

(b) a red, white, or a blue marble?

9 There are five colored marbles in a container. A friend is blindfolded. He draws a marble, holds it up for you to see, returns it to the container, mixes the marbles, and repeats the process. You record the following colors in order, B (black), R (red), G (green): BGBGRRRBBGBBRBRRGBR. How many marbles of each color are probably in the container?

10 If three honest coins are tossed, what is the probability that three heads will show? Use the tree diagram which shows the eight possibilities for three coins. What is the probability that two heads and one tail will show?

11 (a) You have four coins in a cup, shake them, and pour them on a table. Do you think you will probably get two heads and two tails?

(b) If five coins are tossed at one time, what is the probability that all five will fall heads?

12 An experiment consists of one toss each of an ordinary die and a coin.

(a) Make a tree diagram showing the possible outcomes.

(b) What are $P(3, H)$, $P(6, T)$, $P(\text{even number}, T)$?

13 Two dice are rolled. Compute these probabilities:

(a) The sum of the two numbers is 6.

(b) The sum of the two numbers is greater than 7.

(c) The product of the two numbers is 6.

(d) The two numbers are equal.

(e) One of the two numbers is less than 4.

14 A bowl contains 10 marbles, of which 5 are white, 3 are black, and 2 are red. We will assume that they are identical in size, hence that each marble is equally likely to be picked if you reach into the bowl and take one marble without looking.

  (a) What is the probability that you will pick a white marble in one draw?

  (b) Assuming you pick a white marble the first time and do not replace it, what is the probability that you will pick a black marble the second time?

  (c) Assuming you pick a white marble the first time and a black marble the second time and do not replace them, what is the probability that you will pick a red marble the third time?

15 If, in Problem 14, you do replace the marbles, what would the probabilities be?

Each of the final five problems refers to the five experiments which you did. In each case, three events are listed. For each one you are to:

  (1) Compute the probability of each event, and express it as a fraction.

  (2) Count the number of times the event actually took place in your first 50 trials. Write the fraction whose numerator is this number and whose denominator is 50.

  (3) Count the number of times the event actually took place in all 100 of your trials. Write the fraction with this number as numerator and with 100 as denominator.

  (4) Graph the three fractions on a number line as in Figure P13.15 for each of the three events.

Figure P13.15

16 Experiment 1:

  Events: (a) Two heads

  (b) No more than one head

  (c) T on the first toss

17 Experiment 2:

  Events: (a) More tails than heads

  (b) Two consecutive tails to start with

  (c) Exactly three heads

18 Experiment 3:

  Events: (a) Two reds

  (b) Different colors

  (c) Green on second draw and a color different from red on the first

19 Experiment 4:

  Events: (a) Same as for Experiment 3, part c

  (b) Blue on at least one draw

20 Experiment 5:

  Events: (a) Sum of spots equals 7

  (b) Sum of spots is odd and less than 10

  (c) First toss yields an odd number, the second an even number

14.

FLOW CHARTS

One very important facet of our present society is the existence of large numbers of high-speed electronic computers. Computers directly affect practically everyone in this country. Computers calculate utility bills and bank balances and print out monthly welfare checks. They check income tax returns and predict who will be elected.

In addition, computers do many other things that affect people indirectly. They guide factory operations and actually operate some oil refineries. They are used to help in the planning and supervising of businesses. They are used by the armed forces to keep track of food and supplies. Without computers, landings on the moon would have been impossible.

The average United States citizen has no idea how computers work. As a result, he does not know what to do when his bank statement is incorrect or when his phone bill is wrong. Should he blame the computer or the people who fed information into the computer?

Because computers play such an important part in our society and because so little is known about them by the average citizen, there is no doubt that our educational system will be called on to remedy this knowledge gap, at least for the next generation, and that a part of the school curriculum will soon be devoted to providing an understanding of the uses, capabilities, and limitations of computers.

Most of these curriculum changes will be located in the secondary school program. However, there are some mathematical ideas needed for an understanding of computers which can, and soon will, be included in the elementary school curriculum.

ALGORITHMS Computers are sometimes called "giant brains." Nothing could be further from the truth. Computers can carry out only a few simple operations on symbols. They do no thinking at all. What makes them useful is that it is possible for them to do enormous numbers of numerical calculations very quickly. However, the computer can do nothing until it is provided with a very precise, very detailed set of instructions as to which operations to carry out and the order in which the operations are to be performed.

Such a list of instructions for carrying out a step-by-step process is an *algorithm*. Algorithms are important in almost every branch of mathematics. We have already looked at algorithms for the arithmetic operations on whole numbers. In this chapter we will take a careful look at the nature and structure of some other algorithms.

■ **Question 1** We saw, in Chapter 8, how to divide one whole number by another. Is there more than one algorithm for division?

First, however, let us consider a nonmathematical example of an algorithm. Mrs. Boxer has gone to Michigan for a week to visit her relatives. Mr. Boxer has been charged with seeing that the three children, Jimmy, Emily, and Elsie, have a nourishing breakfast before they leave for school. They can take care of fruit juice, milk, and toast, but Mr. Boxer is to cook their eggs for them.

He is far from being a gourmet chef, so Mrs. Boxer has left a set of directions for him:

1 Place frying pan on burner.
2 Set heat under frying pan to medium-low.
3 Put butter in pan.
4 Break eggs into pan.
5 Baste eggs with the melted butter.
6 Serve eggs.

This list of step-by-step operations is a good example of an algorithm. Of course, some fathers might need to have the steps made more precise while others, with greater culinary experience, could get along with a shorter, less explicit list. But before going into this kind of consideration we rewrite this algorithm in a pictorial form which is called a *flow chart*.

Notice that each box in Figure 14.1 is numbered. This is to make it easier to discuss the various parts of the diagram.

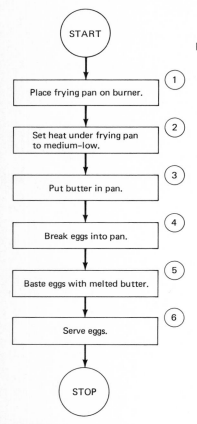

Figure 14.1

ELEMENTS OF A FLOW CHART In this flow chart, as in most, we see

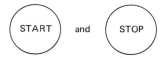

The circular shape reminds us of "Start" and "Stop" buttons for electric motors.

We also observe in our flow chart that each instruction is enclosed in a *frame* or *box*. A little later on we will see that the shape of the frame is used to tell us what kind of instruction appears inside. *Commands* to take some action are written in *rectangular* frames.

In Figure 14.1 all instructions other than START and STOP are of this form, so they have rectangular frames.

To carry out the process shown in a flow chart we go to START, follow the arrow to the first "box" and carry out the instruction given there, then follow the arrow to the next box, and so on.

■ **Question 2** Could this frame be legitimately included in our flow chart?

It is raining today.

After drawing a flow chart it is a good idea to check to see if we can improve it. First we notice that this algorithm assumes that both eggs and butter are on hand. Certainly, if Mr. Boxer were out of one or the other, there would be no point in going through the first two steps of the algorithm. So we have to decide whether both eggs and butter are on hand. To make this decision we introduce a new kind of frame into our flow chart. The frame is oval in shape.

Inside this frame we find a statement on which we make a *decision*.

We have both eggs and butter.

We have two exits from this box, one labeled T (true) and the other labeled F (false). After checking whether the statement is true or false, we leave the box at the corresponding exit and go on to the next box.

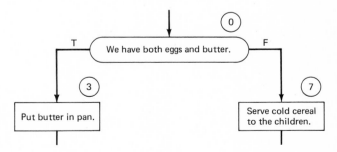

Such an oval box is called a *decision box*. When we put this flow chart fragment into our flow chart of Figure 14.1, we obtain the flow chart of Figure 14.2.

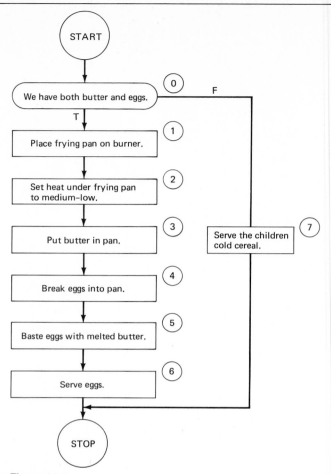

Figure 14.2

■ **Question 3** Could this combination legitimately replace box 7 in Figure 14.2?

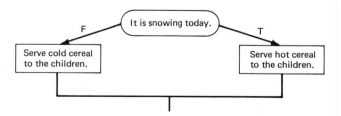

LOOPS There is still another improvement we can make in our flow chart. Box 4 tells Mr. Boxer to break some eggs into the pan, but it does not say how many. Suppose Jimmy wants two eggs each morning while the girls want one each. We could replace box 4 with four boxes.

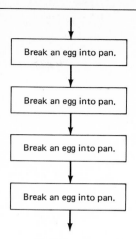

Break an egg into pan.

Break an egg into pan.

Break an egg into pan.

Break an egg into pan.

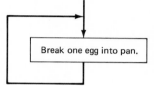

Break one egg into pan.

This is correct but we can simplify this diagram by introducing a *loop*. We see that when we leave this box we are sent right back to repeat the task. The trouble with this idea is that we have no way of getting out of the loop to the next task. We are caught in an endless loop. We can correct this situation by placing a decision box in our flow chart as shown in the figure below.

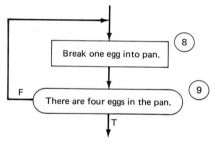

Break one egg into pan. ⑧

F There are four eggs in the pan. ⑨

T

We get our next flow chart by replacing box 4 by boxes 8 and 9.

Another improvement in the flow chart can be made if we remember that the butter in the pan should be melted before the eggs are added. Mrs. Boxer's rule of thumb on this is that the butter should be checked, to see if it has melted, every 30 seconds. So we should insert, between boxes 3 and 4, this loop:

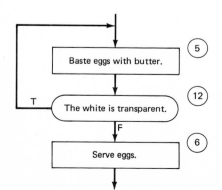

Baste eggs with butter. ⑤

T The white is transparent. ⑫

F

Serve eggs. ⑥

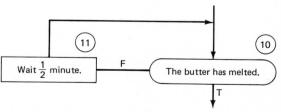

Wait $\frac{1}{2}$ minute. ⑪   F   The butter has melted. ⑩

T

Another refinement is a reminder to Mr. Boxer that the eggs will be done when the white is no longer transparent. This reminder takes the form of another decision box between boxes 5 and 6 (as shown left).

Putting these all together, we get our final flow chart (Figure 14.3).

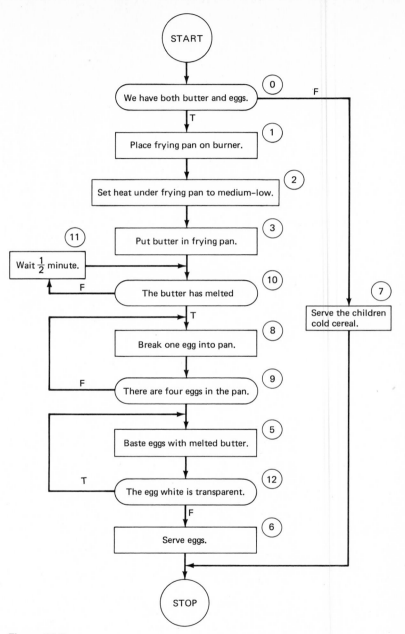

Figure 14.3

■ **Question 4** What would be the effect of replacing box 9 by this one?

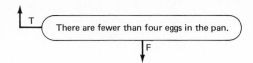

This example has introduced us to two kinds of flow chart boxes, command boxes and decision boxes. In our next example, we will meet a few other kinds of boxes. This example also pointed out that the degree of detail that we need to build into a flow chart depends on the person who is going to use it. One person may need much more detail than the next. The same thing will be true when we come to draw flow charts for computers. The amount of detail will depend on the internal workings of the particular computer we have in mind.

ELEMENTS OF A COMPUTER

Before turning to an example of a mathematical algorithm and a flow chart for it, we need to learn a little bit more about computers. Any computer has a number of components.

One component of a computer is usually called the *arithmetic unit*. We can think of an arithmetic unit as having two "slots" in it into each of which a number can be inserted. The arithmetic unit is able to perform certain arithmetic operations on these two numbers. For example, one kind of arithmetic unit might be able to add the two numbers or, if so instructed, to subtract the second from the first. Another kind of arithmetic unit might be able to do both these operations and in addition be able to multiply the two numbers. A still more powerful unit might be able to do these three operations and also be able to divide the first number by the second.

Two other important components of a computer are the *input* and *output* units. The first of these serves to provide the numbers that the computer is to operate on. The second unit prints out on a strip of paper the results of the specified operations on the numbers that come through the input unit. We will say a little more about these units later on.

Another essential component of a computer is the *storage*. We can think of this as a large warehouse or storehouse containing a quantity of bins, in each of which a single number can be stored for later use. In an actual computer, the number of numbers that can be stored is very large indeed. For the illustrations we will use in this chapter, 26 bins, each labeled with one of the letters $A, B, C, \ldots, Z$ will be more than enough.

The final component of a computer is the *control unit*. We can think of the control unit as consisting of two people, a manager and a messenger. The manager reads the flow chart and, at each step, tells the messenger where to go and what to do when he gets there.

■ **Question 5** Going back to Figure 14.3, who represents the control unit?

COMPUTING A PAYROLL

Let us see how these units could be used to do a numerical job. As an example, each week the payroll for the packing department for the

Watsonville Candy Factory has to be computed. There are 11 workers in this department. The first person, alphabetically, is Sally Boxer. She packs 1-pound boxes of soft-centered chocolates and is paid 17 cents for each box she packs. The next on the list is Billy Chann, who fills 1-pound tins of hard candies and is paid 8 cents per tin. Each of the other workers has a particular task and is paid at a fixed rate, ending with the last on the list, Isabelle Rocker, who fills the large 5-pound boxes of assorted candies and is paid 80 cents per box.

At the end of the week, the accountant, Zaga Svitsky, types two numbers on each of 11 cards. The first number tells the rate at which a particular worker is paid. The second number tells how many containers that worker filled that week. Thus, for example, if Sally Boxer filled 451 boxes, the first card would look like this:

17, 451

(In actual practice, instead of writing the numbers, holes are punched in cards, which happen to have the above shape. The pattern of holes tells the computer what the numbers are.)

If Billy Chann filled 844 tins, his card would look like this:

8, 844

Isabelle Rocker's card might look like this:

80, 79

If the Watsonville Candy Factory did not have a computer, then Zaga Svitsky would have to calculate the payroll by hand. She would start by taking the first card,

17, 451

and multiplying 17 by 451 to find out how much Sally Boxer had earned. Then she would write, on a strip of paper, Sally's rate of pay, the number of containers filled, and the result of the computation (expressed in dollars, rather than cents, for convenience):

17,   451, $76.67

Then she would take the second card

8, 844

do the multiplication, and write another line:

17,   451, $76.67

8,   844, $67.52

and so on to the end.

Now let us suppose that the Watsonville Candy Factory has obtained a computer and that we have been asked to write a flow chart that will compute the payroll just as Zaga Svitsky has done it in the past. The first thing in the flow chart is easy.

MORE FLOW CHART ELEMENTS

The next box in the flow chart is new to us. It is

$$R, Q$$

This box is called an *input* box and is a message to the manager of the arithmetic unit:

> There is a stack of cards in the input unit. Each card has two numbers on it located in the positions labeled "R" and "Q." Send your messenger to the input unit and have him bring back a copy of the numbers on the top card. Have him take the copy to the storage and put the number from the "R" position into the R bin. Have him put the number from the "Q" position into the Q bin.

After carrying out this instruction, the R bin will contain the number 17 and the Q bin will contain the number 451.

The next box in the flow chart will be

$$W \leftarrow R \times Q$$

This is another instruction to the manager of the arithmetic unit:

> Send the messenger to the storage and have him copy the number in the R bin and the number in the Q bin and bring back the copies to the arithmetic unit. Then have him put the first number in one slot in the arithmetic unit and the other number in the other slot. Have the arithmetic unit multiply the two numbers. Put the result in the W bin in the storage.

When this instruction has been carried out, the number 7,667 ($= 17 \times 451$) will be in the W bin.

This is the first, but far from the last, time we meet an instruction box containing a left-pointing arrow. Whenever we do meet such a box, we look at the right-hand side (at the tail of the arrow). We do whatever is called for by the symbols there. (In this case we take copies of the number in the R and Q bins to the arithmetic unit and have them multiplied.) Then we look at the left-hand side of the box (at the head of the arrow) to find out where we are to place the result. In this case, the product is to be stored in the W bin.

The next box in the flow chart will get the numbers in the storage out where Zaga S. can see them:

This is still another instruction, called an *output* box, and it tells the manager:

> Send the messenger to the storage and have him copy the numbers in the R bin, the Q bin, and the W bin. Then have him take these copies to the output unit and have them printed in order.

When this instruction has been carried out, the first line on the output paper will read:

17, 451, 7,667

Of course, we have not finished, since there are still 10 more cards to be processed. We can take care of this by means of a decision box.

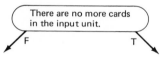

Putting these all together, with a STOP button at the end, our flow chart appears as in Figure 14.4.

■ **Question 6** What is the second number placed in the R bin? The Q bin? The W bin?

There is one thing to which careful attention needs to be paid. When the messenger copies a number in one of the bins in the storage, the number stays in the bin. For this reason, reading the content of a bin is called *nondestructive*. This is important, since we need to use the number in the R bin twice, once for the instruction in the assignment box in the flow chart and once for the output box instruction, and we want the number to be still there when we need it the second time.

However, when we loop back to

and read the second card, we clearly want the messenger to dump out the contents of the R bin and the Q bin so that the new numbers, for

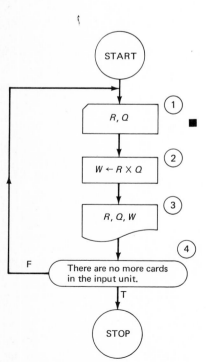

Figure 14.4

Bill Chann, can be put there. For this reason, the placing of numbers in the storage is called *destructive*. Whatever was in the bins before is destroyed when we do this.

COUNTING

Of course, for a payroll for only 11 people, a computer would not make much sense. But for a payroll for many hundreds of workers it would make more sense, especially when the computer can do other useful things at the cost, to us, of merely adding a few boxes to the flow chart.

Suppose, for example, the manager of the Watsonville Candy Factory prefers to have the output lines from the computer numbered in order from 1 to 11, so that the first line would be

1, 17, 451, 7,667

the second would be

2, 8, 844, 6,752

and so on.

■ **Question 7** The Watsonville Candy Factory's computer was used yesterday to compute payrolls. It is to be used today to compute federal and state taxes. Does the storage have to be cleaned out in between?

There is a simple little trick which takes care of this. We merely make use of another bin in the memory, for example, the I bin. We add two new boxes, numbers 5 and 6, to the flow chart and modify one of the original boxes, number 4.

The new flow chart appears as in Figure 14.5.

Box number 5

$$I \leftarrow 0$$

is another case of an instruction box which contains a left-pointing arrow. In this case, what is on the right-hand side of the box is merely the specific number $\underline{0}$, and no arithmetic operation is indicated. This means that we need only equip ourselves with a copy of $\underline{0}$. On the left-hand side of the box, we see "I". This tells us to store the number $\underline{0}$ in the $\underline{I}$ bin.

Box number 6

$$I \leftarrow I + 1$$

is a little trickier. The symbols at the right-hand side of this box tells us to obtain a copy of the number in the I bin and to insert it, together with the number 1, in the arithmetic unit to obtain their sum.

Then the symbol "I" at the head of the arrow tells us to store the sum back in the same I bin. The result of all this is to discard the number in the I bin and to replace it by the number which is larger by one.

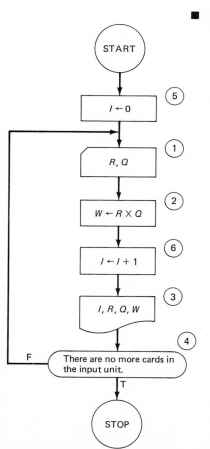

Figure 14.5

The same trick can be used for other purposes. Suppose that the manager wanted to have the total payroll for the packing department printed out after all the individual wages were computed. All we need to do is to use still another memory bin, for example the one labeled "P" (for payroll). To the flow chart we add three new boxes (see Figure 14.6).

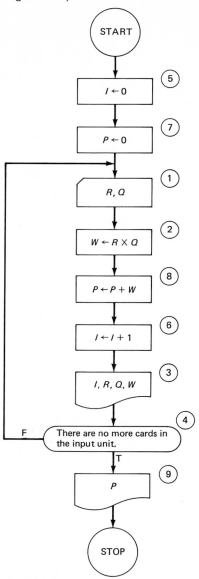

Figure 14.6

Box 8 in the flow chart keeps adding each newly calculated wage to the sum of the ones calculated before. When all have been calculated, the total is in the P box in the storage, and box 9 instructs that this total be printed out.

■ **Question 8** Suppose box 6 in Figure 14.5 were moved to the position just below box 3. What change would be needed in box 5?

In these examples we have been introduced to all the basic ideas about flow charts. We have seen inputs

commands

decisions

and outputs

We have seen how these can be put together in a sequence, running from the beginning

START

to the end

STOP

of an algorithm. In particular, we have seen how the sequence can be looped back on itself.

TRACING AN ALGORITHM

We will look at a few more examples, not to see new aspects of flow charts, but rather to fix and strengthen the ideas we have already met. First, suppose we are asked to compute the sum of the first thousand terms in the sequence which starts with 1 and then takes on every second whole number thereafter. This is the sequence

1, 3, 5, 7, . . .

(You may recall that these are the *odd* numbers.)

Obviously, the particular result of adding the first thousand of these would be of very little interest. However, constructing a flow chart for this job may be instructive. We need three memory bins for such a flow chart, say the N bin, the S bin, and the T bin. The flow chart in Figure 14.7 uses the N bin as a counter, the T bin to store the terms of the sequence, and the S bin to store the results of addition as far as we have gone.

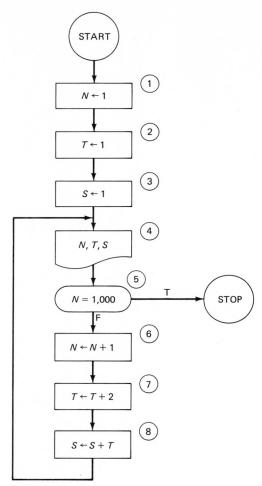

Figure 14.7

It is instructive to trace through what happens step by step in the computer when we use this flow chart. There is a way of recording the results in a table. We use one column of the table for each storage bin, another for any output that is printed, and another for any decisions that are made. (In general, if there were more than one decision box, we would use one column for each.) We also have a column to record which box in the flow chart is operative for any particular step.

For the flow chart in Figure 14.7, the first six rows of the table are shown in Table 14.1.

Table 14.1

STEP	Box	N	T	S	Output	Decision
1	1	1				
2	2	1	1			
3	3	1	1	1		
4	4	1	1	1	1, 1, 1	
5	5	1	1	1		no
6	6	2	1	1		

An explanation is as follows:

STEP 1:
We move to box 1 which commands that the number 1 be put in the N bin.

STEP 2:
We move to box 2, which commands that the number 1 also be put in the T bin.

STEP 3:
We move to box 3, which commands that the S bin also be loaded with another copy of the number 1.

STEP 4:
We move to box 5, and the decision is that N is not equal to 1,000. N, T, and S bins be printed.

STEP 5:
We move to box 5, and the decision is that N is not equal to 1,000.

STEP 6:
We move to box 6, which commands that the number in the N bin be replaced by the number 1 greater.

Notice that once a number has been placed in one of the storage bins, it stays there unchanged for a number of steps. It saves time in filling out such a table if we do not bother to record this unchanged value and only make an entry in a column when a new number is put in the bin. With this convention, the table for the first 15 steps of the algorithm are shown in Table 14.2.

For many purposes, it is sufficient just to keep track of changes in the storage bins. To do this, record the initial number stored in the bin and then, each time a new number is put there, write it below the

Table 14.2

STEP	Box	N	T	S	Output	Decision
1	1	1				
2	2		1			
3	3			1		
4	4				1, 1, 1	
5	5					no
6	6	2				
7	7		3			
8	8			4		
9	4				2, 3, 4	
10	5					no
11	6	3				
12	7		5			
13	8			9		
14	4				3, 5, 9	
15	5					no

Table 14.3

N	T	S
~~1~~	~~1~~	~~1~~
~~2~~	~~3~~	~~4~~
3	5	9

previous number and cross out the previous number. For the above, the result for the first 15 steps would be as shown in Table 14.3.

■ **Question 9** Compare the successive entries in the S column. Compare these with the entries in the diagonal, from the upper left corner to the lower right corner, of the multiplication table on page 87. What do you predict the next entry in the S column will be?

A DIVISION ALGORITHM

As a final example, suppose we have a computer whose arithmetic unit can perform addition, subtraction, and multiplications, but not divisions. Can we write a flow chart that will direct the computer to carry out the division process?

Such a flow chart turns out to be relatively easy to put together. Suppose we go back to the first method for division developed in Chapter 8. We are given the total number of elements in the array and we are given the number of rows. The task is to find how many columns there are in the array. The procedure we started with in Chapter 8 was to remove one column at a time. Arithmetically, this meant repeated subtraction and then a count of the number of subtractions. Let us illustrate with some specific numbers.

Suppose we wish the computer to carry out division by 7. Whatever number N we start with, in order to divide it by 7 we remove sets of 7 at a time from a collection of N elements. Suppose we have done this

Q times. Then

$$\underbrace{N}_{\substack{\text{Total number} \\ \text{of elements}}} = \underbrace{7 \times Q}_{\substack{\text{Number of} \\ \text{elements} \\ \text{removed}}} + \underbrace{R}_{\substack{\text{Number of} \\ \text{elements} \\ \text{remaining}}}$$

Now if we remove one more set of 7, we decrease R by 7 and we increase Q by 1.

These suggest the commands

$$R \leftarrow R - 7$$

and

$$Q \leftarrow Q + 1$$

Since this process is to be repeated over and over we are tempted to write:

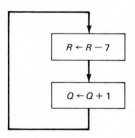

The difficulty with such a loop is that no way is provided for stopping. It just goes on and on forever. How to stop it? The answer, of course, is that we cannot take 7 out of the set unless there are 7 or more there, that is, unless

$$R \geq 7$$

Thus our test for stopping is:

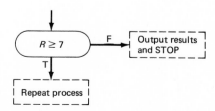

Combining this with the above assignment box we have:

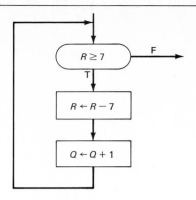

That is the heart of our flow chart. Only minor details remain; namely, to provide output and to give initial values to our variables.

As output we merely give the values of Q and R, thus:

We want to input the value of N:

We start R and Q out with the values they should have before anything is done. These are given by:

Putting all our flow chart fragments together we get the complete flow chart (see Figure 14.8).

The final value of Q, the value that is printed out, is the number of columns in our final array. The final value of R is the number left over at the end. It is one of the numbers

0, 1, 2, 3, 4, 5, 6

The algorithm is an algorithm for *integer division* with a divisor of 7.

■ **Question 10** Trace through this flow chart for the case where $N = 15$.

Of course, the same kind of reasoning would work for any divisor. The divisor does not have to be 7. To make a flow chart for integer division for any divisor, we use a variable, D, to denote the divisor.

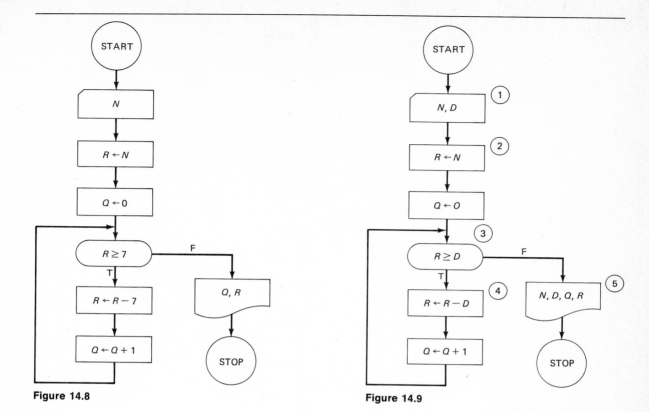

Figure 14.8

Figure 14.9

We call for both the number and the divisor to be input, and we replace each 7 in the preceding flow chart by D. Then we will have the flow chart shown in Figure 14.9.

This time we have called for the output of the values of all the variables N, D, Q, and R in this order, so as to suggest the formula

$$N = \underbrace{D}_{\text{Divisor}} \times \underbrace{Q}_{\text{Quotient}} + \underbrace{R}_{\text{Remainder}}$$

■ **Question 11** Trace through this flow chart starting with $N = 11$ and $D = 2$.

Before concluding this brief introduction to computers, and to the flow charts which guide them, two important reminders are in order. In describing the operations which take place within a computer, we used, for the control unit, the analogy of a manager. This is misleading if it suggests that the control unit has any human abilities such as the ability to think. In reality the control unit is an electronic device, about as complicated as a radio. When certain electronic signals are fed into the control unit, it responds with other electronic signals in a precise, predesigned, and unchanging pattern.

A better analogy would be a typewriter, which responds in one way when one key is struck and in another way when another key is struck. Poems, novels, editorials, and textbooks can be typed out, but it is the human author that gets the credit, not the typewriter. Similarly, computers can carry out very extensive and intricate computations, but the human being who writes the flow chart gets the credit, not the computer.

Our second comment concerns the arithmetic unit. We are all familiar with arithmetic calculators, whether the desk models which are cranked by hand or operated by small electric motors, or the pocket-sized battery-operated electronic models. These correspond very closely to the arithmetic unit of a computer.

Other calculators may be somewhat more complex. Some, for example, have a small number of storage bins. If we want to add three or more numbers it is convenient if the sum of the first two can be stored while the third is being punched in. Still others have a few simple flow charts built in to compute, for example, the average (one-half of the sum) of two numbers, or to compute the reciprocal of a number. Each such built-in flow chart is activated by pushing an appropriate button.

As time goes on, instruments with wider and wider arrays of such accessories become available. However, there is no general agreement as to how complex a calculator must be before it can be called a computer. Usually, though, a computer is considered to have rather large storage facilities, large enough to store not only large quantities of initial data and the results of intermediate computations, but also large enough to hold, in another part of the storage, the flow chart for its current task.

In any case, computers vary tremendously not only in the size of their storage facilities but also in the speed with which they can carry out operations. On the other hand, the basic ideas about flow charts which we have met in this chapter are applicable to all calculators and all computers, no matter how simple or how complex.

PROBLEMS

1 (a) The initial values of A, B, and C are given in the table in Figure P14.1a. The two commands on the right are to be performed in the indicated order. Fill in the values in the table.

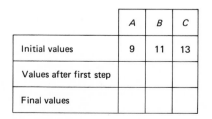
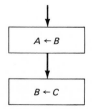

	A	B	C
Initial values	9	11	13
Values after first step			
Final values			

Figure P14.1a

(b)   Use the same instructions as in part a for Figure P14.1b.

	A	B	C
Initial values	9	11	13
Values after first step			
Final values			

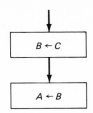

$$B \leftarrow C$$

$$A \leftarrow B$$

Figure P14.1b

2   Trace through the flow chart in Figure P14.2 using the table to the right to list the output values each time through the output box.

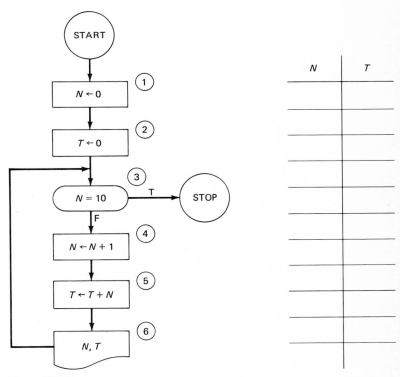

Figure P14.2

N	T

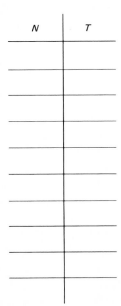

N	T

Figure P14.3

3   Copy the flow chart in Figure P14.2, interchanging boxes 4 and 5.   Trace through this new flow chart using the table in Figure P14.3 to list the output values each time through the output box.

4   Trace through the flow chart in Figure P14.4 and give the output.   Carry your work to the stage where N has the value 7.

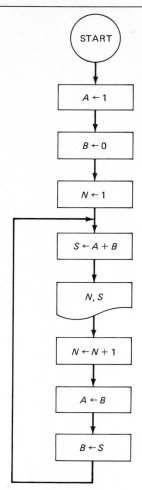

Figure P14.4

K	F
1	1
2	2
3	6
4	24
5	120

Figure P14.5

5 The number "five factorial," written 5!, is defined as

$$1 \times 2 \times 3 \times 4 \times 5$$

and is equal to 120. Similarly, if K is a whole number, then $K!$ is defined as

$$1 \times 2 \times 3 \times \cdots \times K$$

that is, the product of all whole numbers from 1 up to K. We have tabulated a few of the values in Figure P14.5. Fill in two more lines in the table. Now draw a flow chart which computes and prints out $K!$ for each K up to 100.

   A teacher assigned her students a problem of constructing a flow chart as follows: The input consists of the lengths and widths of several rectangles. The purpose is to produce a list with consecutively numbered lines giving the length, the width, and the area of only those rectangles with perimeter greater than 12. The flow charts shown in Figures P14.6a and b were submitted by students as solutions of the problem.

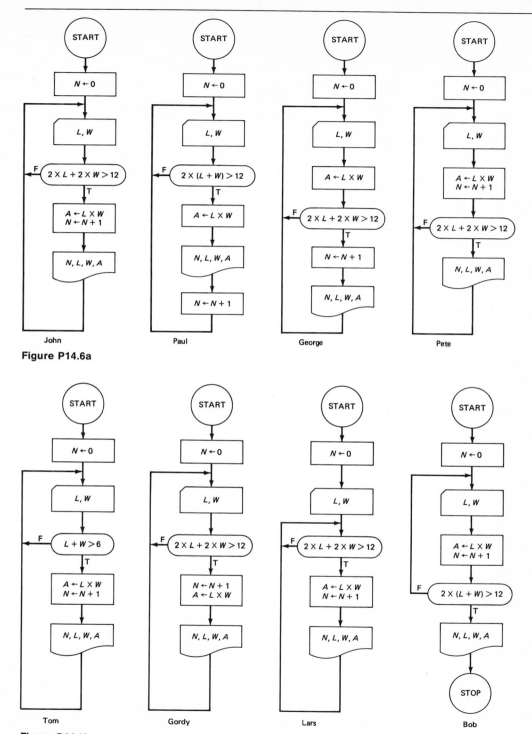

Figure P14.6a

Figure P14.6b

6 Tell which of the solutions are correct and which incorrect.

7 For those which are incorrect, in what way will the answers produced be wrong?

8 For those which are correct, list them in order of efficiency with the one requiring the least amount of computation first.

9 Figure P14.9 is a flow chart for finding the area of certain square regions, given the length of the side L. Problems refer to this flow chart.

   (a) What is the output if the input L is 3?

   (b) What is the output if the input L is 5?

10 The flow chart for $17 \div 7$ is shown in Figure P14.10. Find the numbers in the Q bin and in the R bin at each of the following stages.

   (a) The first time we arrive at the point marked 1.

   (b) The second time we arrive at 2.

   (c) The second time we arrive at 3.

   (d) The third time we arrive at 2.

   (e) The first time we arrive at 4.

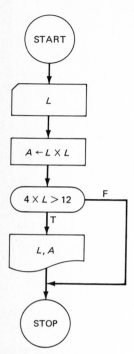

Figure P14.9

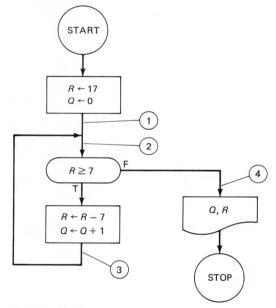

Figure P14.10

11 Trace the flow chart of Figure 14.9 for $39 \div 13$. Write the output in the form: $N = Q \times D + R$.

12 Do the same for $5 \div 24$.   13 Do the same for $125 \div 125$.

14 Do the same for $0 \div 23$.   15 Do the same for $14 \div 1$.

16 Do the same for $7 \div 0$.

17 How many cycles through the flow chart are required to find $45 \div 7$ using the flow chart in Figure 14.8?

18 How many cycles are required to find $3,168 \div 7$ using the same flow chart?

19 The flow chart in Figure P14.19 does division much like our final algorithm in Chapter 8. Trace $856 \div 7$ through it.

20 Check Table P14.20 to see that it is correct.

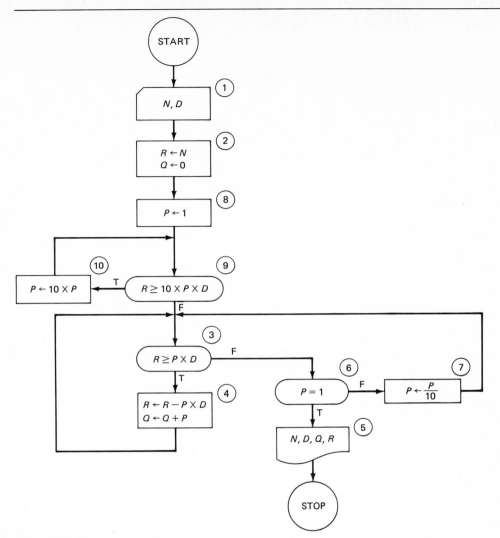

Figure P14.19

Table P14.20 Trace of flow chart for $N = 3{,}168$, $D = 7$

STEP	Flow chart box	Values of variables			Decision
		Q	R	P	
1	1	$N = 3{,}168, D = 7$			
2	2	0	3,168		
3	8			1	
4	9				$3{,}168 \geq 70$ T
5	10			10	
6	9				$3{,}168 \geq 700$ T
7	10			100	
8	9				$3{,}168 \geq 7{,}000$ F
9	3				$3{,}168 \geq 700$ T
10	4	100	2,468		
11	3				$2{,}468 \geq 700$ T
12	4	200	1,768		
13	3				$1{,}768 \geq 700$ T
14	4	300	1,068		
15	3				$1{,}068 \geq 700$ T
16	4	400	368		
17	3				$368 \geq 700$ F
18	6				$100 = 1$ F
19	7			10	
20	3				$368 \geq 70$ T
21	4	410	298		
22	3				$298 \geq 70$ T
23	4	420	228		
24	3				$228 \geq 70$ T
25	4	430	158		
26	3				$158 \geq 70$ T
27	4	440	88		
28	3				$88 \geq 70$ T
29	4	450	18		
30	3				$18 \geq 70$ F
31	6				$10 = 1$ F
32	7			1	
33	3				$18 \geq 7$ T
34	4	451	11		
35	3				$11 \geq 7$ T
36	4	452	4		
37	3				$4 \geq 7$ F
38	6				$1 = 1$ T
39	5	$N = 3{,}168, D = 7$ 452	4		

15.

ARITHMETIC OPERATIONS ON RATIONAL NUMBERS

In Chapters 5 and 7 we learned about the operations of addition, subtraction, multiplication, and division of whole numbers. The techniques for carrying out these operations were treated in Chapters 6 and 8.

In this chapter we will first see what these arithmetic operations on rational numbers mean. The computational techniques will be treated later in the chapter.

The first question to ask is: can we add, subtract, multiply, and divide using rational numbers? If so, how, and do the same properties apply as for addition, subtraction, multiplication and division of whole numbers? In considering these questions we must keep the following things in mind:

1 We have already observed, using the number line model, that such fractions as $\frac{3}{1}$, $\frac{5}{1}$, $\frac{10}{2}$ and the like name points that are named by whole numbers; that such fractions as $\frac{1}{1}$, $\frac{2}{2}$, $\frac{3}{3}$, and, in general, k/k all name the point named by 1; and that $\frac{0}{1}$, $\frac{0}{2}$, $\frac{0}{3}$ and, in general, $0/k$ all name the point named by 0. Since all whole numbers are also rational numbers—though of course not all rational numbers are whole numbers—we will want to make sure that the ordinary properties still apply. For example, "addition" should still be commutative and associative and the special properties of 0 and 1 should still be present.

2 A rational number has many names, hence we want to be sure that the results of an operation do not depend on the particular names we choose to use for the two numbers involved. In other words, we want $\frac{1}{6} + \frac{3}{6}$ to be the same number as $\frac{3}{18} + \frac{6}{12}$.

■ **Question 1** Is there anything incorrect about this sentence?

$\frac{1}{2} + (\frac{2}{3} + \frac{3}{4}) = (\frac{2}{4} + \frac{4}{6}) + \frac{6}{8}$

Incidentally, since we will be talking in the chapters that follow about *rational* numbers, the word "number" is to be taken until further notice to mean "rational number" unless specifically designated otherwise.

ADDITION OF RATIONAL NUMBERS

For two fractions with the same denominator, the matter of addition is easily disposed of and, for small denominators, physical models of the process are easy to draw. As one such model, we might think of a house, a school, and a store along a straight road as shown in Figure 15.1. If it is $\frac{1}{5}$ mile from the house to the school and $\frac{2}{5}$ mile from the school to the store, it is easy to see that it is $\frac{3}{5}$ mile from the house to the store.

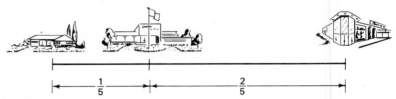

Figure 15.1

Other addition problems, with models that represent them, are shown in Figure 15.2. In each case, both a number line model and a model using regions are given.

From these examples we see that the way to add fractions having the same denominator is simply to add the numerators. Let us, then, make the following definition:

Given two fractions

$$\frac{a}{b} \quad \text{and} \quad \frac{c}{b}$$

with the same denominator $b \neq 0$

$$\frac{a}{b} + \frac{c}{b} = \frac{a+c}{b}$$

The way to deal with fractions that do *not* have the same denominator is simply to use equivalent fractions that *do* have the same denominator.

Back in Chapter 12, when we asked about the order of rational numbers (is $\frac{9}{14}$ greater than $\frac{5}{8}$?), we saw how to find equivalent fractions with the same denominator. By doing this we can add any two fractions. Thus, for example,

$$\frac{9}{14} + \frac{5}{8} = \frac{72}{112} + \frac{70}{112} = \frac{142}{112}$$

■ **Question 2**   Add: $\frac{1}{2} + \frac{1}{3} + \frac{1}{4}$

MIXED NUMERALS

The definition $a/b + c/b = (a+c)/b$ also gives us a way of dealing with so-called "mixed numerals," such as $2\frac{1}{2}$. Such a numeral is read

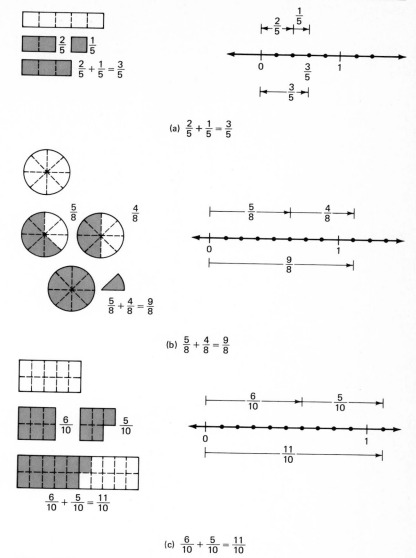

(a) $\frac{2}{5} + \frac{1}{5} = \frac{3}{5}$

(b) $\frac{5}{8} + \frac{4}{8} = \frac{9}{8}$

(c) $\frac{6}{10} + \frac{5}{10} = \frac{11}{10}$

Figure 15.2

"two *and* one-half" and really designates the sum $2 + \frac{1}{2}$. Since $\frac{4}{2}$ is equivalent to the whole number 2, this becomes $\frac{4}{2} + \frac{1}{2}$ or, by our definition, $(4 + 1)/2 = \frac{5}{2}$. Every such mixed numeral can be written as a fraction of the form a/b, and so we could also deal with the equivalence of mixed numeral and fraction names for rational numbers greater than 1. Observe that this also works in the other direction by considering our definition in reverse as $(a + c)/b = a/b + c/b$. For example:

$$\frac{5}{2} = \frac{4+1}{2} = \frac{4}{2} + \frac{1}{2} = 2 + \frac{1}{2} = 2\frac{1}{2}$$

$$\frac{5}{4} = \frac{4+1}{4} = \frac{4}{4} + \frac{1}{4} = 1 + \frac{1}{4} = 1\frac{1}{4}$$

$$\frac{14}{3} = \frac{12+2}{3} = \frac{12}{3} + \frac{2}{3} = 4 + \frac{2}{3} = 4\frac{2}{3}$$

In computing with mixed numerals one can either regroup using the commutative and associative properties (which we discuss later) in order to work with the whole number and fractional parts separately, or one can change the mixed numerals to ordinary fractions and compute using these fractions. The latter procedure is usually easier.

Examples of several addition computations, some in vertical form and some in horizontal form, are given below.

(a) $\frac{7}{8} + \frac{2}{3} = \frac{21}{24} + \frac{16}{24} = \frac{21+16}{24} = \frac{37}{24} = \frac{24+13}{24}$

$$= \frac{24}{24} + \frac{13}{24} = 1 + \frac{13}{24} = 1\frac{13}{24}$$

(b) $4\frac{2}{3} + 1\frac{1}{4} = 4 + \frac{2}{3} + 1 + \frac{1}{4} = 4 + 1 + \frac{8}{12} + \frac{3}{12}$

$$= 5 + \frac{11}{12} = 5\frac{11}{12}$$

(c) $4\frac{2}{3} + 1\frac{1}{4} = \left(\frac{12}{3} + \frac{2}{3}\right) + \left(\frac{4}{4} + \frac{1}{4}\right) = \frac{14}{3} + \frac{5}{4} = \frac{56}{12} + \frac{15}{12} = \frac{71}{12}$

$$= \frac{60+11}{12} = \frac{60}{12} + \frac{11}{12} = 5 + \frac{11}{12} = 5\frac{11}{12}$$

(d) $4\frac{2}{3} = 4\frac{8}{12}$
$\quad\ 1\frac{1}{4} = 1\frac{3}{12}$
$\quad\qquad\ 5\frac{11}{12}$

■ **Question 3**   Can you tell if the solution of the sentence

$6\frac{3}{4} + x = 17\frac{1}{4}$

is larger or smaller than 11 without actually finding the solution?

SUBTRACTION OF RATIONAL NUMBERS   Turning to subtraction of rational numbers we see that if the denominators are the same, a definition similar to that for addition suffices.

Given two fractions

$$\frac{a}{b} \quad \text{and} \quad \frac{c}{b} \quad \text{where} \quad \frac{a}{b} > \frac{c}{b}$$

$$\frac{a}{b} - \frac{c}{b} = \frac{a-c}{b}$$

You will recall from our discussion of "order" that the specification $a/b > c/b$ assures us that for whole numbers a and c, $a > c$, so that the subtraction $a - c$ can be done.

An alternative way of looking at subtraction would be to follow the model given in Chapter 5, namely, "$a - b$ is the whole number n for which $n + b = a$." Translating this using rational numbers, a definition of subtraction would appear as follows:

$$\frac{a}{b} - \frac{c}{b} \quad \text{is that rational number} \quad \frac{n}{b}$$

$$\text{for which} \quad \frac{n}{b} + \frac{c}{b} = \frac{a}{b}$$

In Chapter 5 this was described as the "process of finding a missing addend." We also recognize this as the specification of how we "check" the supposed answer to any subtraction problem.

The definition "$a/b - c/b = (a - c)/b$" gives us an immediate way to find an answer while the second definition, "$a/b - c/b$ is that rational number n/b for which $n/b + c/b = a/b$," is closer to what was done for subtraction of whole numbers. We can see that we can use either definition according to our convenience by noting that the result given by the first definition meets the specification set down by the second definition.

Here is an example:

$$\frac{11}{17} - \frac{4}{17} = \frac{11 - 4}{17} = \frac{7}{17} \qquad \frac{7}{17} + \frac{4}{17} = \frac{7 + 4}{17} = \frac{11}{17}$$

■ **Question 4**   Compute: $(1 - \frac{1}{5}) - \frac{1}{10}$ and $1 - (\frac{1}{5} - \frac{1}{10})$

As to models for subtraction, we can refer back to Figure 15.2 and think in each case of "taking apart" or "undoing" each of the addition problems illustrated there. Such a process would result in two subtraction problems associated with each such addition problem. Figure 15.3 shows the number line models for the two problems associated with Figure 15.2b, $\frac{5}{8} + \frac{4}{8} = \frac{9}{8}$.

$$\frac{9}{8} - \frac{5}{8} = \frac{4}{8}$$

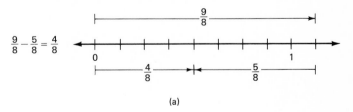

(a)

$$\frac{9}{8} - \frac{4}{8} = \frac{5}{8}$$

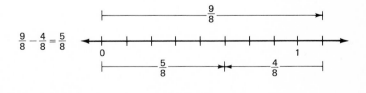

(b)

Figure 15.3

MULTIPLICATION OF RATIONAL NUMBERS

Now let us consider multiplication. With each pair of rational numbers we want to associate a number called the "product." We would like ways of doing this that involve only previously learned operations and concepts. And we would like the properties of such a multiplication to be pretty much the same as those of the now familiar multiplication of whole numbers. Furthermore, in order to be consistent with our efforts so far, we want to find physical models which justify and give content to the procedures we develop.

Now multiplication of rational numbers is really quite a different thing from multiplication of whole numbers, and the models for multiplication of whole numbers do us very little good. "Repeated addition," for example, is all very well for $6 \times 3 = 3 + 3 + 3 + 3 + 3 + 3$ and can even serve for $6 \times \frac{1}{3} = \frac{1}{3} + \frac{1}{3} + \frac{1}{3} + \frac{1}{3} + \frac{1}{3} + \frac{1}{3} = \frac{6}{3}$. But it will not do at all for $\frac{1}{3} \times 6$ because it is surely nonsense to speak of "6 added to itself $\frac{1}{3}$ times." Even if we assume some sort of commutativity to make $\frac{1}{3} \times 6 = 6 \times \frac{1}{3}$ to get us out of this pickle, we are in a hopeless bind if we then ask about $\frac{1}{6} \times \frac{1}{3}$. Nor is the "$n$-by-$m$ array" model of much use, for while it is easy to speak of 6×3 as meaning the number of things in an array with six rows of 3 things each, it is difficult to find a sensible statement to describe a $\frac{1}{6}$-by-$\frac{1}{3}$ "array" of things. Teachers must surely be aware of such difficulties themselves in order to help youngsters understand this operation which has the same name as and similar properties to the corresponding whole number operation but which is quite different conceptually.

MODELS FOR MULTIPLICATION

The principal models available to us as a basis for defining a "multiplication" for rational numbers are closely tied to our intuitive notions of the meaning of the word "of" in such statements as "$\frac{1}{2}$ *of* them," "$\frac{1}{3}$ *of* the $\frac{1}{2}$ gallon of milk," "find what part of a mile $\frac{2}{3}$ *of* $\frac{1}{8}$ of a mile is," and so on. Let us then show some models using an "of" operation, then define multiplication by identifying it with this operation. Admittedly this sounds like a rather slippery procedure, but if we are to provide physical models in terms of concepts now available to us, it is the only procedure we have. The alternative is to give a purely abstract definition using already known operations. But we avoid whenever possible the making of abstract definitions unsupported by models, as we probably should in teaching youngsters.

(a)

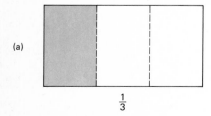

$$\frac{1}{3}$$

Figure 15.4

(b)

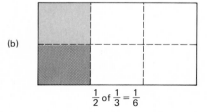

$$\frac{1}{2} \text{ of } \frac{1}{3} = \frac{1}{6}$$

Suppose we want a model for "$\frac{1}{2}$ of $\frac{1}{3}$" in terms, say, of a region. The first step, familiar to us by now, is to represent $\frac{1}{3}$ as one of three congruent parts of the region as shown in Figure 15.4a. Let us now regard the region representing $\frac{1}{3}$ as a region itself and represent $\frac{1}{2}$ of it by marking 1 of 2 congruent regions, as shown in Figure 15.4b. The resulting region, representing $\frac{1}{2}$ of $\frac{1}{3}$, is 1 of 6 congruent parts of the original region, as can be seen from Figure 15.4b. Hence $\frac{1}{2}$ *of* $\frac{1}{3} = \frac{1}{6}$.

Similarly, Figure 15.5 shows $\frac{2}{3}$ *of* $\frac{1}{5}$, which is clearly shown to be 2 of 15 congruent parts of the region we started with. Hence, $\frac{2}{3}$ of $\frac{1}{5} = \frac{2}{15}$.

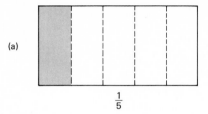

(a)

$\frac{1}{5}$

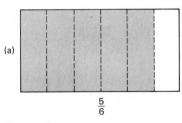

(b)

$\frac{2}{3}$ of $\frac{1}{5} = \frac{2}{15}$

Figure 15.5

■ **Question 5** What "of" statement problem does Figure 15.6 suggest?

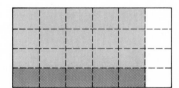

Figure 15.6

Likewise, Figure 15.7 first represents $\frac{5}{6}$ in terms of a region, then shows $\frac{4}{5}$ of the region representing $\frac{5}{6}$, with a resulting region that is $\frac{20}{30}$ of the original region.

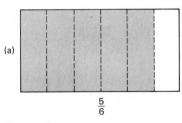

(a)

$\frac{5}{6}$

(b)

$\frac{4}{5}$ of $\frac{5}{6}$

Figure 15.7

Observe that we work such problems just by construction of congruent regions using directions given in terms of counting numbers; then we get our answer by *counting* the number of resulting congruent regions in the original region and how many of these are marked. That is, we need use only quite fundamental notions. We could get closer to "multiplication" by observing that the final result of, say $\frac{4}{5}$ of $\frac{5}{6}$ (Figure 15.7), shows 4 rows, each containing 5 congruent parts, which is a sort of 4-by-5 array. Furthermore, there are

$5 \times 6 = 30$ such congruent parts altogether because there are 5 rows each containing 6 such parts; or a 5-by-6 array. Hence $\frac{4}{5}$ of $\frac{5}{6}$ will be 4×5 parts out of 5×6 parts $= (4 \times 5)/(5 \times 6) = \frac{20}{30}$. Similarly, our model for $\frac{2}{3}$ of $\frac{1}{5}$ (Figure 15-5) ultimately results in a 3-by-5 array of congruent parts and a 2-by-1 array of such parts. That is, $\frac{2}{3}$ of $\frac{1}{5} = (2 \times 1)/(3 \times 5) = \frac{2}{15}$.

■ **Question 6**   Make a 3-by-4 array of marks (as in Chapter 7) and use it to illustrate $\frac{2}{3} \times 12 = 8$.

NUMBER LINE MODELS

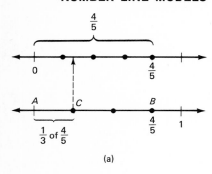

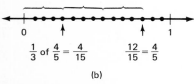

(a)

(b)

Figure 15.8

We can also use the number line as a model of, say $\frac{1}{3}$ of $\frac{4}{5}$, though this model is perhaps harder to follow. As illustrated in Figure 15.8, we first mark $\frac{4}{5}$ on the number line in the usual way. Then, considering the segment $\overline{AB}$, with $m(\overline{AB}) = \frac{4}{5}$, we mark the point C that is $\frac{1}{3}$ of the way along $\overline{AB}$. In terms of our original unit, $m(\overline{AC}) = \frac{1}{3}$ of $\frac{4}{5}$ and we can surely use this to mark a point corresponding to $\frac{1}{3}$ of $\frac{4}{5}$ on the number line. However, our procedure so far has not given us a fraction to name this point, and this is the difficulty with the number line model. As shown in Figure 15.8b, we must resort to equivalent fractions and observe that had we divided the segment into 15 parts and taken 12 of them, the point named $\frac{12}{15}$ would be the same point as that named $\frac{4}{5}$. Now taking a point $\frac{1}{3}$ of the way along on this segment, we arrive at $\frac{4}{15}$. Hence, $\frac{1}{3}$ of $\frac{4}{5} = \frac{4}{15}$. Essentially, we need to rig things via equivalent fractions so that the unit segment is marked in such a way that all our work will come out exactly on one of the original division points from the unit segment. This can always be done by using the product of the two denominators as the number of congruent parts in the original unit segment; you may want to work through enough examples to see why this is so. Figure 15.9 provides two other examples. Clearly, however, the region notion provides a model easier to manipulate than does the number line.

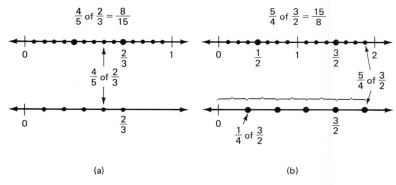

(a)   (b)

Number line models showing $\frac{4}{5}$ of $\frac{2}{3} = \frac{8}{15}$ and $\frac{5}{4}$ of $\frac{3}{2} = \frac{15}{8}$

Figure 15.9

■ **Question 7**   Use the number line in Figure 15.10 to show: $\frac{4}{3} \times \frac{7}{5}$

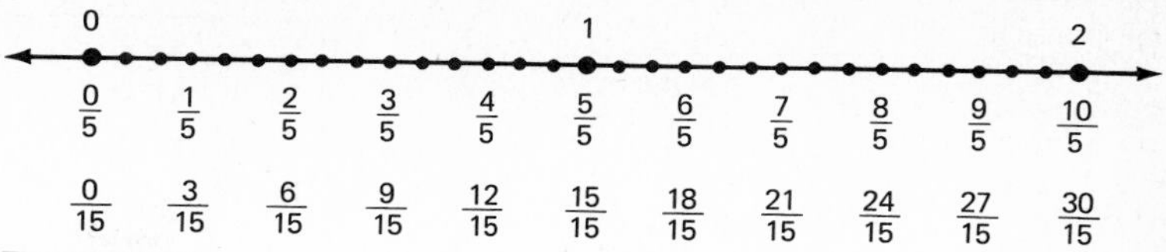

Figure 15.10

THE DEFINITION OF MULTIPLICATION FOR RATIONAL NUMBERS

In order to define multiplication we identify the "of" operation with multiplication; that is, $\frac{1}{2} \times \frac{1}{3} = \frac{1}{2}$ of $\frac{1}{3}$; $\frac{2}{3} \times \frac{4}{5} = \frac{2}{3}$ of $\frac{4}{5}$; etc. The "=" sign means, as usual, that the same number will result. Next, we tabulate the results of all the examples we have considered so far in this chapter and observe that in every case the numerator of the result is the product of the numerators of the fractions, while the denominator of the result is the product of the denominators of the fractions. (You should take a moment to verify this.)

$\frac{1}{2} \times \frac{1}{3} = \frac{1}{2}$ of $\frac{1}{3} = \frac{1}{6}$ $\frac{3}{4} \times \frac{2}{3} = \frac{3}{4}$ of $\frac{2}{3} = \frac{6}{12}$

$\frac{2}{3} \times \frac{1}{5} = \frac{2}{3}$ of $\frac{1}{5} = \frac{2}{15}$ $\frac{2}{3} \times \frac{7}{8} = \frac{2}{3}$ of $\frac{7}{8} = \frac{14}{24}$

$\frac{4}{5} \times \frac{5}{6} = \frac{4}{5}$ of $\frac{5}{6} = \frac{20}{30}$ $\frac{5}{7} \times \frac{8}{9} = \frac{5}{7}$ of $\frac{8}{9} = \frac{40}{63}$

$\frac{2}{4} \times \frac{7}{8} = \frac{2}{4}$ of $\frac{7}{8} = \frac{14}{32}$ $\frac{1}{3} \times \frac{4}{5} = \frac{1}{3}$ of $\frac{4}{5} = \frac{4}{15}$

$\frac{2}{3} \times \frac{3}{4} = \frac{2}{3}$ of $\frac{3}{4} = \frac{6}{12}$

Hence we are led to the definition:

Given two fractions $\dfrac{a}{b}$ and $\dfrac{c}{d}$

$$\frac{a}{b} \times \frac{c}{d} = \frac{a \times c}{b \times d}$$

This definition gives a computational procedure that depends only on multiplication of whole numbers. As with whole numbers, we will call $a/b \times c/d$ the "product" of the "factors" a/b and c/d.

Various refinements of the computational procedure given by the definition, such as "canceling," will be discussed later. For now, let us just observe that whole numbers and mixed numerals can also be handled with this definition by use of fractions equivalent to them, as illustrated below:

$$4 \times \frac{2}{5} = \frac{4}{1} \times \frac{2}{5} = \frac{4 \times 2}{1 \times 5} = \frac{8}{5}$$

$$7 \times 8 = \frac{7}{1} \times \frac{8}{1} = \frac{7 \times 8}{1 \times 1} = \frac{56}{1}$$

$$4\frac{1}{3} \times 2\frac{1}{2} = \frac{13}{3} \times \frac{5}{2} = \frac{13 \times 5}{3 \times 2} = \frac{65}{6}$$

■ **Question 8** What multiplication does the number line in Figure 15.11 suggest?

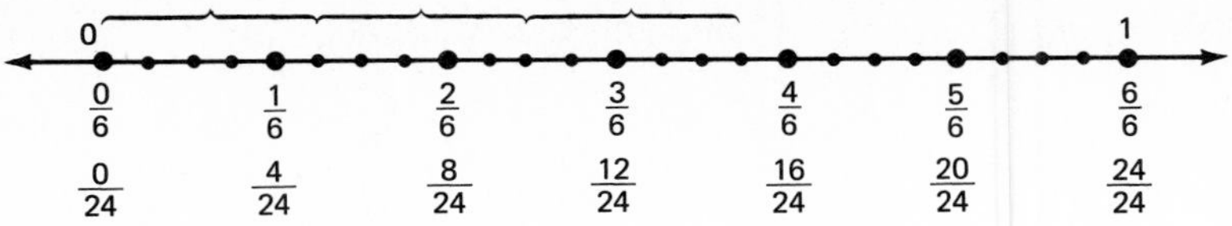

Figure 15.11

**THE MULTIPLICATIVE
INVERSE PROPERTY**

The rational numbers possess an important property not possessed by either the counting numbers or the whole numbers. The product of $\frac{2}{3}$ and $\frac{3}{2}$ is 1. The product of $\frac{5}{2}$ and $\frac{2}{5}$ is 1. The number $\frac{2}{3}$ is called the *multiplicative inverse* or the *reciprocal* of $\frac{3}{2}$. The reciprocal of $\frac{5}{2}$ is $\frac{2}{5}$. The reciprocal of a number is the number by which it must be multiplied to give 1. Every rational number, except 0, has a reciprocal. When a number is named by a fraction, we can easily find its reciprocal by "inverting" the fraction, that is, by interchanging the numerator and the denominator. Thus, the reciprocal of $\frac{5}{8}$ is $\frac{8}{5}$. The reciprocal of $\frac{8}{5}$ is $\frac{5}{8}$. The reciprocal of 2 is $\frac{1}{2}$. (Notice that another name for 2 is $\frac{2}{1}$, which inverted is $\frac{1}{2}$.) In general,

the reciprocal of $\dfrac{a}{b}$ is $\dfrac{b}{a}$

However, it is very important to remember that 0 has no multiplicative inverse. The product of 0 and every number is 0, hence it is *impossible* to find a number such that you can multiply it by 0 and get 1.

■ **Question 9**   What is the multiplicative inverse of 1?

**A PROCEDURE FOR
DIVISION OF RATIONAL
NUMBERS**

We have defined specifically for rational numbers three of the four standard binary operations on numbers. In each case we have observed that rational numbers certainly face us with different situations than whole numbers so that "new" operations must be defined. That is, "addition" for whole numbers is by no means exactly the same operation, conceptually or computationally, as "addition" for rational numbers and these differences are even more marked for "multiplication" of rational numbers.

On the other hand, some concepts do carry over and the definitions of the operations have been formulated in such a way that such standard properties as commutativity, associativity, special properties of 1 and 0, and the like, still apply, with appropriate modifications in stating them, as we shall see.

The pattern we will follow in discussing "division" for rational numbers will, by now, be a familiar one. Except in certain restricted instances the models and concepts discussed for division of whole numbers will not take us very far. To the extent that they do apply, however, they are suggestive of a computational way of proceeding for division of rational numbers. As before, we will want to preserve, as far as possible, the special definitions and properties that apply to division of whole numbers. Multiplication will come into the matter, as you would expect. These considerations lead us to a definition of the operation of division of rational numbers. Our main concern is to

Figure 15.12

make clear the reasons that underlie computational rules for division of rational numbers.

As our first link to previous work, let us consider briefly one model involving multiplication by a rational number. Suppose we want to know how many objects there would be in $\frac{1}{3}$ of a set of 12 objects. This could be done by displaying the 12 objects, dividing the set into three equal parts, then taking one of these, as in Figure 15.12. Hence, $\frac{1}{3}$ of 12 is shown to be 4. Now observe that precisely the same model serves to illustrate $12 \div 3$ if this is interpreted in one of the standard ways, namely, as the number of members in each of three matching subsets of a set with 12 members. The model shows that *division* by 3 and *multiplication* by $\frac{1}{3}$ gives the same result. For reasons that will become apparent, we point out that $\frac{1}{3}$ is the *reciprocal* of 3, as explained above.

■ **Question 10** This array illustrates the whole number multiplication 4×5. What fraction operation might it suggest?

As our second link to previous work let us consider the problem $6 \div \frac{1}{3}$. Now we have nothing at hand yet even to say whether or not this has any meaning (since we do not yet have a "division" operation for rational numbers), but let us suppose that it does and state it as asking the question "How many $\frac{1}{3}$s are there in 6?" which will be recognized as similar to another standard way of interpreting division of whole numbers. Figure 15.13 shows a way of answering this question using the number line. Just by counting the segments of length $\frac{1}{3}$, we see that there are 18 of them in a segment whose length is 6. But we would not need to count, for there are surely 3 such $\frac{1}{3}$ segments in each unit segment, and hence 6×3 of them in 6 units. Thus $6 \div \frac{1}{3}$ gives the same result as 6×3. We observe that 3 is the reciprocal of $\frac{1}{3}$ and note that division by $\frac{1}{3}$ is equivalent to multiplication by the reciprocal of $\frac{1}{3}$.

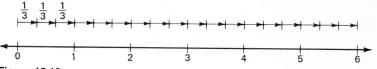

Figure 15.13

To move to a harder example, consider $8 \div \frac{2}{3}$, stated in the form "How many $\frac{2}{3}$s in 8?" Again on the number line, as in Figure 15.14, we observe that it takes 12 segments of length $\frac{2}{3}$ to get a segment of length 8. We are interested in getting a computational procedure to

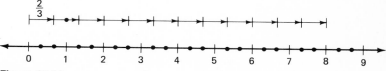

Figure 15.14

replace this number line procedure which would surely become tedious for larger numbers such as $102 \div \frac{2}{3}$. So we observe that each unit contains $1\frac{1}{2}$ of the segments with length $\frac{2}{3}$; hence for our problem, $8 \div \frac{2}{3} = 12$ of the $\frac{2}{3}$ segments in the 8 units as indeed there are. Expressing $1\frac{1}{2}$ as $\frac{3}{2}$, we observe that $8 \div \frac{2}{3}$ gives the same result as $8 \times \frac{3}{2}$; and that $\frac{3}{2}$ is the reciprocal of $\frac{2}{3}$.

■ **Question 11** Show $4 \div \frac{1}{3}$ on a number line.

DIVISION IN TERMS OF MULTIPLICATION

All this is suggestive of a way of making division of rational numbers depend on a previously defined operation: we observe that *division* by a number apparently gives the same result as *multiplication* by the reciprocal of that number. That is,

$$\frac{a}{b} \div \frac{c}{d} \qquad \text{is equivalent to} \qquad \frac{a}{b} \times \frac{d}{c}$$

provided, of course, that none of b, c, or d is zero. This will be recognized immediately in terms of the well-known (and little understood) rule, "invert and multiply."

Hence the "invert and multiply" rule becomes instead a "multiply by the reciprocal of the divisor" rule. But to replace the "invert the divisor and multiply" rule by a "multiply by the reciprocal of the divisor" rule will not, of itself, result in increased understanding or avoid the misapplications that are so familiar and painful to all of us. And the "evidence" we have given so far is pretty flimsy, using only carefully selected examples, all of which have involved at least one whole number. Let us show a couple of harder examples to strengthen this *suggestion* about how to proceed, then turn to a different sort of justification. Figure 15.15a shows $\frac{8}{3} \div \frac{2}{3} = 4$, by showing that the answer to "How many $\frac{2}{3}$s in $\frac{8}{3}$?" is 4. To test our tentative procedure we multiply $\frac{8}{3}$ times $\frac{3}{2}$, the reciprocal of $\frac{2}{3}$, and get $\frac{8}{3} \times \frac{3}{2} = \frac{24}{6} = 4$. Finally, Figure 15.15b shows $\frac{10}{6} \div \frac{2}{3}$ as "How many $\frac{2}{3}$s in $\frac{10}{6}$?" Observe that 2 of the $\frac{2}{3}$s arrows fall short, while 3 of them go beyond. Figure 15.15b suggests that there are $2\frac{1}{2}$ of the $\frac{2}{3}$s segments in a $\frac{10}{6}$ segment. Doing this computation by *multiplying* $\frac{10}{6}$ by the reciprocal of $\frac{2}{3}$, we see

$$\frac{10}{6} \div \frac{2}{3} = \frac{10}{6} \times \frac{3}{2} = \frac{10 \times 3}{6 \times 2} = \frac{5 \times (2 \times 3)}{2 \times (3 \times 2)} = \frac{5}{2} = 2\frac{1}{2}$$

which is the same as our result on the number line (see Figure 15.15).

THE DEFINITION FOR DIVISION OF RATIONAL NUMBERS

We have a possible procedure for division indicated by our previous experience and our models for division of whole numbers and multiplication of rational numbers. Another tack we can take is just to lay aside, for the time being, all such "practical" considerations and formulate a definition for the operation just on the basis that it must be analogous to the formal definition of division of whole numbers. This definition was given in Chapter 7 as

$$a \div b = n \quad \text{if} \quad a = b \times n \qquad \text{where} \qquad b \neq 0$$

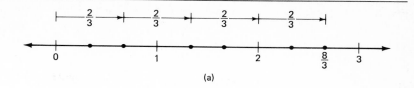

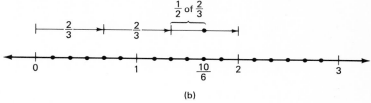

Figure 15.15

We recognize this definition as essentially an instruction to "check" any supposed answer by multiplying quotient times divisor to get back (hopefully) the dividend. The corresponding statement which we can take as a definition of division of rational numbers is

$$\frac{a}{b} \div \frac{c}{d} = \frac{m}{n} \quad \text{if} \quad \frac{m}{n} \times \frac{c}{d} = \frac{a}{b} \qquad \text{where none of } b, c, d, \text{ or } n \text{ can be zero}$$

You may remember that this was described in Chapter 7 as the "missing factor" concept of division. Using this definition one gets an answer for the problem $\frac{8}{3} \div \frac{2}{3} = m/n$ by looking for m/n in the problem $m/n \times \frac{2}{3} = \frac{8}{3}$. This becomes, by our definition of multiplication of rational numbers, $(m \times 2)/(n \times 3) = \frac{8}{3}$ by which it is apparent that m can be 4 and n can be 1. So $\frac{8}{3} \div \frac{2}{3} = \frac{4}{1} = 4$, as we found earlier in this chapter using the number line. Let us take another example, $\frac{4}{9} \div \frac{2}{3} = m/n,$ which we put in the form $m/n \times \frac{2}{3} = \frac{4}{9}$. This can be written $(m \times 2)/(n \times 3) = \frac{4}{9}$, by which it is apparent that m can be 2 and n can be 3, so $\frac{4}{9} \div \frac{2}{3} = \frac{2}{3}$.

But we are faced with a real dilemma if we attempt something like $\frac{2}{3} \div \frac{4}{9} = m/n$ for this becomes $m/n \times \frac{4}{9} = \frac{2}{3}$ or $(m \times 4)/(n \times 9) = \frac{2}{3}$. Now it is pretty hard to see what m and n should be to get the correct result, since there is no whole number times 4 that gives 2 as a product and no counting number times 9 that gives 3. If $\frac{2}{3}$ is replaced by its equivalent, $\frac{12}{18}$, the problem becomes $m/n \times \frac{4}{9} = \frac{12}{18}$ and the answer is now apparent, namely, $m/n = \frac{3}{2}$.

But how would we know what equivalent fraction to use? It certainly is not obvious what one should do to find m/n in, for example, the recasting of $\frac{3}{7} \div \frac{2}{5} = m/n$ in the form $m/n \times \frac{2}{5} = \frac{3}{7}$. There is surely some equivalent of $\frac{3}{7}$ of the form $(m \times 2)/(n \times 5)$ but it is not clear immediately what it would be. (As a matter of fact, $\frac{3}{7} = \frac{30}{70}$ will do the trick, in which case $m/n = \frac{15}{14}$, as the reader should verify.)

It would seem that our formal definition, $a/b \div c/d = m/n$ if $m/n \times c/d = a/b$, does give us a direct tie-in to the definition of division of whole numbers but does not provide us with a very good way of computing answers. Our earlier procedure $(a/b \div c/d = a/b \times d/c)$, on the other hand, rests on some flimsy justification but gives an immediate way to get an answer. The problem just considered, for example, becomes $\frac{3}{7} \div \frac{2}{5} = \frac{3}{7} \times \frac{5}{2} = \frac{15}{14}$, the same result as above.

For the moment let us investigate division from both angles, accepting the formal definition as *the* definition, but using the computational procedure "to divide by a number, multiply by its reciprocal," to get answers. In the course of this investigation we will show that the computational procedure *always* gives quotients that "fit" the formal definition.

■ **Question 12** Compute: $(\frac{1}{2} \div \frac{1}{3}) \div \frac{2}{5}$ and $\frac{1}{2} \div (\frac{1}{3} \div \frac{2}{5})$

AN INVESTIGATION OF DIVISION

In this section we propose to indicate various ways in which a division problem involving whole numbers, rational numbers, or both might appear and indicate the patterns that result from solving such problems. In each case we are interested in a number m/n such that $a/b \div c/d = m/n$. We assume that no denominator can be zero.

Computation using	*Verification using the definition*
$$\frac{a}{b} \div \frac{c}{d} = q = \frac{a}{b} \times \frac{d}{c}$$	$$\frac{a}{b} \div \frac{c}{d} = \frac{m}{n} \quad \text{if} \quad \frac{m}{n} \times \frac{c}{d} = \frac{a}{b}$$

CASE 1:
Two whole numbers:

$$7 \div 3 = \frac{7}{1} \div \frac{3}{1} = \frac{7}{1} \times \frac{1}{3} = \frac{7}{3} \qquad \frac{7}{3} \times \frac{3}{1} = \frac{7 \times 3}{1 \times 3} = \frac{7}{1} = 7$$

In general:

$$a \div c = \frac{a}{1} \times \frac{1}{c} = \frac{a}{c} \qquad c \neq 0 \qquad \frac{a}{c} \times \frac{c}{1} = \frac{a \times c}{1 \times c} = \frac{a}{1} = a$$

(Hence a/c appears as an equivalent way to express $a \div c$.)

CASE 2:
Whole number ÷ rational number or rational number ÷ whole number:

(a) $$7 \div \frac{3}{4} = \frac{7}{1} \times \frac{4}{3}$$

$$= \frac{7 \times 4}{1 \times 3} = \frac{28}{3} \qquad \frac{28}{3} \times \frac{3}{4} = \frac{7 \times 4 \times 3}{3 \times 4} = \frac{7}{1} = 7$$

In general:

$$a \div \frac{c}{d} = \frac{a}{1} \times \frac{d}{c} \qquad \frac{a \times d}{c} \times \frac{c}{d} = \frac{a \times (d \times c)}{(c \times d)}$$

$$= \frac{a \times d}{1 \times c} \qquad = \frac{a}{1} = a$$

(b) $\dfrac{3}{4} \div 7 = \dfrac{3}{4} \div \dfrac{7}{1} = \dfrac{3}{4} \times \dfrac{1}{7}$

$\qquad = \dfrac{3 \times 1}{4 \times 7} = \dfrac{3}{28}$

$\dfrac{3}{28} \times \dfrac{7}{1} = \dfrac{3 \times 7}{28 \times 1}$

$\qquad = \dfrac{3 \times 7}{4 \times 7} = \dfrac{3}{4}$

In general:

$\dfrac{a}{b} \div c = \dfrac{a}{b} \div \dfrac{c}{1} = \dfrac{a}{b} \times \dfrac{1}{c}$

$\qquad = \dfrac{a \times 1}{b \times c} = \dfrac{a}{b \times c}$

$\dfrac{a}{b \times c} \times \dfrac{c}{1} = \dfrac{a \times c}{b \times c} = \dfrac{a}{b}$

CASE 3:

Rational number ÷ rational number:

(a) Same denominator:

$\dfrac{7}{8} \div \dfrac{3}{8} = \dfrac{7}{8} \times \dfrac{8}{3} = \dfrac{7 \times 8}{8 \times 3} = \dfrac{7}{3}$

$\dfrac{7}{3} \times \dfrac{3}{8} = \dfrac{7 \times 3}{3 \times 8} = \dfrac{7}{8}$

In general:

$\dfrac{a}{b} \div \dfrac{c}{b} = \dfrac{a}{b} \times \dfrac{b}{c} = \dfrac{a \times b}{b \times c} = \dfrac{a}{c}$

$\dfrac{a}{c} \times \dfrac{c}{b} = \dfrac{a \times c}{c \times b} = \dfrac{a}{b}$

(The "common denominator case")

(b) Different denominators:

$\dfrac{3}{4} \div \dfrac{5}{7} = \dfrac{3}{4} \times \dfrac{7}{5} = \dfrac{3 \times 7}{4 \times 5} = \dfrac{21}{20}$

$\dfrac{21}{20} \times \dfrac{5}{7} = \dfrac{21 \times 5}{20 \times 7} = \dfrac{(3 \times 7) \times 5}{(4 \times 5) \times 7}$

$\qquad = \dfrac{(7 \times 5) \times 3}{(7 \times 5) \times 4} = \dfrac{3}{4}$

In general:

$\dfrac{a}{b} \div \dfrac{c}{d} = \dfrac{a}{b} \times \dfrac{d}{c} = \dfrac{a \times d}{b \times c}$

$\dfrac{a \times d}{b \times c} \times \dfrac{c}{d} = \dfrac{(a \times d) \times c}{(b \times c) \times d}$

$\qquad = \dfrac{a \times (c \times d)}{b \times (c \times d)} = \dfrac{a}{b}$

The consideration of the various possibilities exhibited above leads us very strongly to believe that the computational procedure given by "$a/b \div c/d = a/c \times d/c$" is equivalent to the formal definition. In fact, the general statement of case 3b suffices as the promised general "proof" that the computational procedure will *always* give a result that can be shown correct by the definition.

Case 1 suggests a new and interesting use of the fraction notation. Observe that the *division problem a ÷ c* becomes the *fraction a/c*. If we permit ourselves to extend this practice to include *any* division problem, we get fractions that no longer resemble those we are used to. For example, $\frac{2}{3} \div \frac{3}{4}$ would become $\frac{2}{3}/\frac{3}{4}$; $14\frac{1}{2} \div 3\frac{1}{5}$ would become $14\frac{1}{2}/3\frac{1}{5}$ and so on. Such a more generalized use of the fraction notation is a very common one, especially in algebra, so that we should not become too attached to the idea that fractions are *always* of the simple form a/b that we have used in this book up to now. Furthermore, while it is true that rational numbers can always be expressed in a fractional form, with a whole number above the line and a counting number below the line; it is *not* true that every fraction with *something* written above the line and something written below the line represents a rational number. Still another common use for the fractional form is with ratios and proportions, as will be discussed in Chapter 18.

■ **Question 13** Solve this equation:

$\frac{3}{4} \times x = \frac{6}{5}$

Observe that our examination of cases reveals that any division involving whole numbers and/or rational numbers can now be performed. This is the really "new" contribution of the division of rational numbers operation. While for whole numbers we have had to caution that division was not "closed" since, for example, $3 \div 7$ does not have a whole-number answer, we have now enlarged our set of numbers and defined the division operation so that all such problems do have answers; in this case $\frac{3}{7}$. Furthermore, such problems as $7 \div 3$ need no longer be stated as having a "quotient" 2 and "remainder" 1, for $7 \div 3$ is now $\frac{7}{3}$ or $2\frac{1}{3}$. It is at this point that it becomes "legal," in the mathematical scheme of things, to use the remainder from a division problem along with the divisor to form a fraction to be included as part of the quotient. Hence, $37 \div 15$ can be regarded as having an exact "quotient," $2\frac{7}{15}$, rather than a "quotient" 2 and "remainder" 7. The corresponding "checks" consist of the multiplication $2\frac{7}{15} \times 15 = 37$ in the first instance rather than the division algorithm $37 = (2 \times 15) + 7$ in the second instance. Observe that "quotient" is used in two senses here, with the second use being that described earlier as a "missing factor" in the requirement of the definition that $a \div b$ is the number n such that $n \times b = a$. Our situation now is that such a missing factor exists for *every* division problem, with the usual exclusion of zero divisors.

As a final way of justifying the statement that *dividing* one rational number by another is equivalent to *multiplying* the dividend by the reciprocal of the divisor, let us write such a division problem in fractional form and assume that we know that such fractional forms have the properties shown in Figure 15.16 no matter what is written in the $\triangle$ above the line or the $\square$ below the line (the usual cautions about zero still apply, of course).

Property A Property B

$$\frac{\triangle}{1} = \triangle \qquad\qquad \frac{\triangle}{\square} = \frac{h \times \triangle}{h \times \square}$$

Figure 15.16

We now outline the solution to such a division problem.

STEP 1:

$$\frac{2}{3} \div \frac{5}{7} = \frac{\frac{2}{3}}{\frac{5}{7}}$$

STEP 2:

$$\frac{\frac{2}{3}}{\frac{5}{7}} = \frac{\frac{2}{3} \times \frac{7}{5}}{\frac{5}{7} \times \frac{7}{5}}$$

STEP 3:

$$\frac{5}{7} \times \frac{7}{5} = 1$$

So

$$\frac{\frac{2}{3} \times \frac{7}{5}}{\frac{5}{7} \times \frac{7}{5}} = \frac{\frac{2}{3} \times \frac{7}{5}}{1} = \frac{2}{3} \times \frac{7}{5}$$

STEP 4:

$$\frac{2}{3} \div \frac{5}{7} = \frac{2}{3} \times \frac{7}{5}$$

STEP 1:
Write the division in fraction form.

STEP 2:
We are interested in disposing of the denominator by making it equivalent to 1 and applying property A. To do this we apply property B using the reciprocal of $\frac{5}{7}$ as the multiplier h.

STEP 3:
Here we have applied the reciprocal property and property A, with $\frac{2}{3} \times \frac{7}{5}$ occupying the $\triangle$ in that property.

STEP 4:
This simply writes the problem we started with and the end result in a single sentence.

Written on one line and in a general form this becomes:

$$\frac{a}{b} \div \frac{c}{d} = \frac{a/b}{c/d} = \frac{a/b \times d/c}{c/d \times d/c} = \frac{a/b \times d/c}{1} = \frac{a}{b} \times \frac{d}{c}$$

This method provides a convincing way of giving meaning to the "invert and multiply" rule (or, in our terms, the "multiply by the reciprocal of the divisor" rule). As we observed earlier in this chapter, the usual properties of fractions of the form a/b (b a counting number, a a whole number) also apply in a wide variety of other uses of the fraction form, including, as in this case, fractions with fraction numerators and fraction denominators.

■ **Question 14** Compute: $1\frac{1}{2} \div 2\frac{2}{3}$

UNFINISHED BUSINESS

You have undoubtedly noticed that our definitions of addition, subtraction, multiplication, and division of rational numbers each require no more than one or two of the arithmetic operations on whole numbers. For addition and subtraction of rational numbers, we need multiplication of whole numbers (in order to get the same denominator) followed by addition of the new numerators, which are still whole numbers. For multiplication and division of rational numbers, we need only multiplication of whole numbers.

Consequently, the techniques for the arithmetic operations on rational numbers seem to require, in addition to our definitions of these operations, no more than the techniques for the arithmetic operations on whole numbers, as they were discussed in Chapters 6 and 8. This is not quite the whole story, however. There are still two computational problems we need to consider. Both of them arise because a rational number can be named by many equivalent fractions. The first has to do with addition and subtraction. Our rule for adding $\frac{1}{4}$ and $\frac{1}{8}$ tells us to write the equivalent fractions

$$\frac{1}{4} = \frac{8 \times 1}{8 \times 4} = \frac{8}{32} \quad \text{and} \quad \frac{1}{8} = \frac{4 \times 1}{4 \times 8} = \frac{4}{32}$$

and then to add:

$$\frac{1}{4} + \frac{1}{8} = \frac{8}{32} + \frac{4}{32} = \frac{12}{32}$$

You must have noticed that it would have been simpler merely to have written a different equivalent fraction for $\frac{1}{4}$.

$$\frac{1}{4} = \frac{2 \times 1}{2 \times 4} = \frac{2}{8}$$

Then the addition would give us

$$\frac{2}{8} + \frac{1}{8} = \frac{3}{8}$$

Of course, this represents the same rational number, since

$$\frac{3}{8} = \frac{4 \times 3}{4 \times 8} = \frac{12}{32}$$

How are we going to handle this problem in more complicated cases? How can we tell when there is a simpler procedure than multiplying the numerator of each fraction by the denominator of the other? The answer will have to be postponed to the end of the next chapter.

The other problem, quite similar, is that of finding a way of deciding whether the result of a computation on rational numbers can be "simplified." For example,

$$\frac{10}{14} \div \frac{9}{6} = \frac{10}{14} \times \frac{6}{9} = \frac{60}{126}$$

Is there an equivalent fraction whose numerator and denominator are smaller? There is.

$$\frac{60}{126} = \frac{6 \times 10}{6 \times 21} = \frac{10}{21}$$

How can we recognize such a situation? Again, this will have to be postponed to the end of the next chapter.

■ **Question 15** We know that $\frac{3}{4}$ and $\frac{4}{3}$ are reciprocals, so $\frac{3}{4} \times \frac{4}{3} = 1$. therefore $\frac{1}{2} \times (\frac{3}{4} \times \frac{4}{3}) = \frac{1}{2} \times 1 = \frac{1}{2}$. Now compute $(\frac{1}{2} \times \frac{3}{4}) \times \frac{4}{3}$.

PROPERTIES OF THE OPERATIONS

On the other hand, since the arithmetic of rational numbers is so intimately connected with the arithmetic of whole numbers, it will come as no surprise to learn that the arithmetic operations on rational numbers have the same properties, such as commutativity, associativity, etc., as the operations on whole numbers. We will not demonstrate all of these, and in fact will merely look at the commutativity of addition.

Suppose we have two rational numbers to be added, expressed as fractions with the same denominator:

$$\frac{a}{b} + \frac{c}{b}$$

Our rule tells us

$$\frac{a}{b} + \frac{c}{b} = \frac{a+c}{b}$$

Since a and c are whole numbers, and since addition of whole numbers is commutative,

$$\frac{a+c}{b} = \frac{c+a}{b}$$

Putting it all together, we have

$$\frac{a}{b} + \frac{c}{b} = \frac{a+c}{b} = \frac{c+a}{b} = \frac{c}{b} + \frac{a}{b}$$

so the order in which we add rational numbers is immaterial. The other properties of the arithmetic operations can be demonstrated equally simply.

There is, however, as we saw, one property for rational numbers which has no counterpart for whole numbers. This is the *reciprocal property.* If a/b is a fraction with neither a nor b equal to zero, then b/a is the *reciprocal* of a/b and

$$\frac{a}{b} \times \frac{b}{a} = 1$$

■ **Question 16** Here is a demonstration that the addition of rational numbers is associative. Fill in the blanks.

$$\left(\frac{a}{b} + \frac{c}{b}\right) + \frac{d}{b} = \frac{a+c}{b} + \frac{}{b} = \frac{(\quad) + d}{b} = \frac{a + (c+d)}{b}$$

$$= \frac{a}{b} + \frac{(\quad)}{b} = \frac{a}{b} + \left(\frac{c}{b} + \frac{d}{b}\right)$$

PROBLEMS

1 Rewrite each of these fractions so that they all have 9 as a denominator:
$\frac{2}{3}, \frac{12}{27}, \frac{14}{63}, \frac{14}{3}, \frac{4}{1}$

2 Illustrate with number-line diagrams each of the following:
  (a) $\frac{1}{4} + \frac{1}{4}$ (b) $\frac{1}{2} + \frac{1}{4}$ (c) $\frac{2}{3} - \frac{1}{6}$

3 (a) What is n, if $(\frac{8}{2} - \frac{3}{2}) - \frac{1}{2} = n$?
  (b) What is n, if $\frac{8}{2} - (\frac{3}{2} - \frac{1}{2}) = n$?
  (c) Does $(\frac{8}{2} - \frac{3}{2}) - \frac{1}{2} = \frac{8}{2} - (\frac{3}{2} - \frac{1}{2})$?
  (d) Does the associative property hold for subtraction?

4 (a) Fill in the blanks in Table 15.4a so that each row, and each column, add up to the same number.

Table 15.4a

$\frac{1}{3}$	$\frac{7}{6}$	2
	$\frac{5}{6}$	$\frac{1}{6}$
$\frac{2}{3}$		$\frac{4}{3}$

(b) Do the same for Table 15.4b.

Table 15.4b

$7\frac{3}{4}$		
	$6\frac{1}{2}$	
$10\frac{1}{4}$	4	$5\frac{1}{4}$

5 Show
    (a) $\frac{1}{2} \times \frac{2}{3}$ (b) $\frac{4}{5} \times \frac{2}{3}$
    first by means of regions and then by means of a number line.

6 (a) When is the reciprocal of a number greater than the number?
    (b) When is the reciprocal of a number less than the number?
    (c) When is the reciprocal of a number equal to the number?

7 Illustrate with a number line the following:
    (a) $2 \div \frac{1}{3}$ (b) $\frac{1}{3} \div 2$ (c) $\frac{1}{3} \div \frac{1}{6}$

8 Fill in each blank below with ">," "<," or "=" to make a true mathematical sentence.
    (a) $2\frac{1}{2} \times 3\frac{1}{4}$ _____ $3\frac{1}{4} \times 2\frac{1}{2}$ (b) $\frac{5}{2} \times \frac{2}{5}$ _____ $\frac{2}{3} \times \frac{2}{3}$
    (c) $6 \div \frac{4}{5}$ _____ $3 \div \frac{2}{5}$ (d) $2\frac{2}{3} \times \frac{1}{4}$ _____ $4 \div \frac{3}{4}$
    (e) $5 \div 3\frac{1}{2}$ _____ $1 \div \frac{2}{3}$

9 Find the rational number quotient in each of the following (*not* a quotient and remainder).
    (a) $633 \div 14$ (b) $1{,}070 \div 51$

10 Carry out the indicated operations and write your answer as a rational number and as an integer plus a rational number.
    (a) $\dfrac{2 + \frac{5}{6}}{2}$ (b) $\dfrac{3\frac{4}{7}}{2\frac{1}{3}}$

11 The heirs to the Stanford Estate receive shares of $\frac{1}{4}$, $\frac{2}{9}$, and $\frac{1}{3}$. The rest is left to charity. What fraction of the estate is left to charity?

12 (a) The quotient of two numbers is $\frac{4}{3}$. The larger number is $\frac{7}{2}$. Find the other.
    (b) The quotient of two numbers is $\frac{4}{3}$. The smaller number is $\frac{7}{2}$. Find the other.

13 If the difference between two numbers is $5\frac{1}{2}$ and the smaller is $2\frac{3}{10}$, what is the larger?

14 Mr. Twiggs changed the price of potatoes in his store from 14 cents a pound to 7 pounds for one dollar.
    (a) Did he raise or lower the price?
    (b) How much was the increase or decrease per pound?

15 The distance from Joe's home to Edward's home is $\frac{3}{4}$ mile. Joe and Edward meet at a point that is $\frac{7}{12}$ of the distance from Joe's home to Edward's home. How far is this point from Joe's home in miles?

For each of the next five problems, write an open number sentence and also indicate which operations are needed to solve it. You need not carry out the operations.

16 If a city block is $\frac{1}{6}$ mile long, how many miles has a man gone when he has walked $5\frac{1}{4}$ blocks south and $8\frac{1}{2}$ blocks east? If he is trying to walk 5 miles during the day, how many miles does he still lack?

17 A piece of wood $1\frac{5}{6}$ inch thick is made by gluing together strips each $\frac{5}{32}$ inch thick. How many strips are there in this piece?

18 A glass is half full of grape juice and another glass twice the size is one-third full. They are then filled with water and the contents mixed. What part of the mixture is grape juice?

19 It is 24 miles from town A to town B. We drive 7 miles and stop for gas. We drive 5 more and stop to fix a flat tire. What fraction of the way is left to go?

20 A jet plane made a flight from San Francisco to New York in $4\frac{1}{4}$ hours. This was $\frac{5}{7}$ of the time it took a turboprop plane. How many hours did the turboprop require for the trip?

16.

FACTORS AND PRIMES

At the end of Chapter 15 we noted two computational questions about fractions. The first was how to find the easiest way to add two fractions which have unlike denominators. The second was how to find out if a particular fraction has a "simpler" form, i.e., if there is an equivalent fraction with smaller numerator and denominator. In this chapter we will develop some mathematical ideas about whole numbers which will lead to answers to these two questions.

FACTORS In Chapter 7, when we looked at the operation of multiplying two whole numbers, we called each of the numbers a *factor* and we called the result the *product*. Thus, 12 is the product of the factors 3 and 4, and 21 is the product of the factors 3 and 7. In Chapter 8 we also found out how to compute the product of any two factors.

Now let us turn the problem around by considering what multiplications a given product could have come from. For example, what multiplications would give 10 as a product? Clearly they would be 1×10 or 2×5. (Of course 10×1 and 5×2 would also give the product 10 but because of the commutative property of multiplication these are not essentially different from those listed.) Similarly, 12 could be obtained as a product from 1×12, 2×6, or 3×4. The number 5 could come only from 1×5 and the number 13 could come only from 1×13.

■ **Question 1** Can any whole number less than 12 be written as a product in three different ways?

Given any whole number b, we could list multiplications that give b as a product by listing the obvious product $b \times 1 = b$ and then asking, as a start, "Does some number times 2 give b as a product?" If the answer is "yes," i.e., if there is an n such that $n \times 2 = b$, we put this in our list of multiplications that give b as product and call both 2 and the number n *factors* of b. We continue then with the question, "Is there a number m such that $m \times 3 = b$?" If so, both 3 and m are factors of b. And so on. For example, 1, 3, 5, and 15 are all factors of 15 because each of these, with some other number, can be used in a multiplication problem which gives 15 as a product.

Of course, if we started with a large number, say 918,273,645 for example, it would take us a long time to find all its factors. If we really needed them we would certainly want to turn the job over to a computer. So let us try to write a flow chart.

■ **Question 2** We saw that the factors of 15 are 1, 3, 5, and 15. Find all the factors of 30 ($30 = 2 \times 15$). How are they related to the factors of 15?

A FLOW CHART FOR FACTORING The first step would be to input into the computer the number, N, whose factors we want, so the first box is

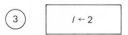

Now we know that the number 1 has the special property that

$$1 \times N = N \quad \text{and} \quad N \times 1 = N$$

We see immediately that both 1 and N are factors of N, so we might as well print them out. Our next box is

Now we have to run through the rest of the numbers after 1 to see which of them are factors of N, so we introduce a counting variable, I, and we start it at the next number after 1. Our next box does this:

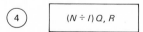

Now we have to test the number 2 to see if it is a factor of N. We recall from Chapter 8 that we can apply the division algorithm to any pair of numbers A and B. The result of the algorithm is a pair of numbers, Q (the quotient) and R (the remainder) such that

$$A = Q \times B + R, \quad \text{with } R < B$$

The next box in our flow chart applies this operation to N and 2, or more generally to N and the integer I we are testing.

This means that the arithmetic unit is given the inputs N and I and outputs numbers Q and R such that

$$N = I \times Q + R$$

(We are assuming that the arithmetic unit in our computer can divide. If it can only add, subtract, and multiply, we can replace box 4 by the entire division program we developed at the end of Chapter 14.)

■ **Question 3** If we started with $N = 10$, what were the first outputs from box 4?

Now we notice that if $R = 0$, then

$$N = I \times Q$$

But this means that I is a factor of N. It also means that Q is a factor of N. So we ask if $R = 0$.

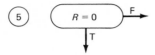

If R is 0, then we exit through the T branch to

which prints out the two new factors of N that we have just found. The next stop would be to go on to the next whole number, so our next box is

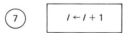

Of course, if $R \neq 0$, then I is not a factor of N (see Problem 14 at the end of the chapter), so we want to go on to test the next whole number. The F branch from box 4 should also lead into box 7.

Now, to test whether this next whole number is a factor of N, we go back to box 4. Our flow chart appears as in Figure 16.1. This flow chart tests each number one after the other, starting with 2, to see if it is a factor of N.

How far do we have to go to be sure we have all the factors of N? When can the process stop? Obviously we can stop when I reaches N. No number larger than N could be a factor of N. Actually, however, we need not test all the whole numbers less than N. In fact, we can stop when

$$I \times I > N$$

To see why this is so, notice that whenever we find one factor I of N, we also find another factor, the Q in

$$N = I \times Q$$

If I is a factor of N, and if $I \times I > N$, then the other factor, Q, must be smaller than I. (If it were larger than I, then $I \times Q$ would be larger than $I \times I$, which is larger than N, and that cannot be, because $I \times Q$ is equal to N.)

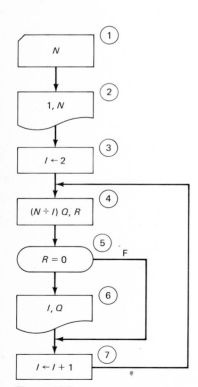

Figure 16.1

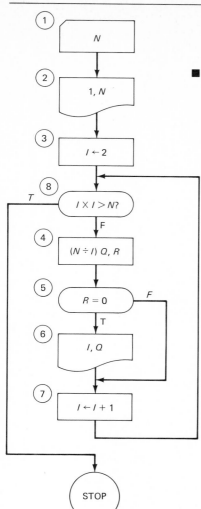

That means that we would have reached the other, smaller factor before we even tested this number I, and we would have printed out both Q (the smaller factor) *and* I.

■ **Question 4** If we start with $N = 30$, what is the largest value of I that appears?

So we see that we can stop the process whenever $I \times I > N$. We can take care of this by inserting a new box between box 7 and box 4.

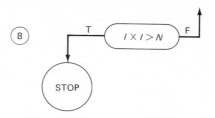

The final flow chart is shown in Figure 16.2. Before going any further, it is important that this procedure be fully understood. Working through a few cases should help. The reader is encouraged to do so for $N = 15$, 16, and 17. You should complete Table 16.1 for the results of using our flow chart on these numbers, in this form.

Figure 16.2

Table 16.1

$N = 15$						$N = 16$						$N = 17$				
I	$I \times I$	Q	R	▭		I	$I \times I$	Q	R	▭		I	$I \times I$	Q	R	▭
1						1						1				
2						2						2				
3						3						3				
4						4						4				
5						5						5				
6						6						6				
						7						7				
						8						8				

PRIME NUMBERS Note that 15 has two factors, 3 and 5, in addition to 1 and itself. Also, 16 has three factors, 2, 4, and 8, in addition to 1 and itself. But 17 has only 1 and itself as factors. It is easy to check that 2, 3, 5, 7, 11, and 13 have the same property of having only two factors. All other whole numbers less than 17, except 0 and 1, have more than two factors. Numbers of this sort are of some importance, so we will put them in a special classification as follows:

Any whole number that has *exactly two different* factors (namely itself and 1) is a *prime number.*

Note that this definition excludes 1 from the set of primes, because 1 does not have two *different* factors, and that it excludes 0 from the set of primes because 0 has more than two factors, in fact infinitely many, because $n \times 0 = 0$ for every number n.

■ **Question 5** Check 19 and 21 to see if they are prime.

Both 0 and 1 are special numbers with special properties and we exclude them from the next set of numbers also. All other numbers are put in a set as follows:

All whole numbers other than 0, 1, and the prime numbers are called *composite* numbers.

The prime numbers have been a subject of mathematical interest for centuries. Over 2,000 years ago the mathematician Eratosthenes invented an easy and straightforward way of sorting out the primes from the list of whole numbers. This method is known as "Eratosthenes' sieve" which describes the fact that it "lets through" only prime numbers. To get the primes less than 49, for example, we would list the numbers from 0 to 49 as shown in Figure 16.3a. Then cross out 0 and 1, since they are not primes; leave 2, since it is a prime, then cross out all other numbers with 2 as a factor since they cannot be primes. This is shown in Figure 16.3a by circling the 2 and crossing out *multiples* of 2 (4, 6, 8, 10, . . .). Leave 3, since it is prime, but cross out all multiples of 3 that remain as shown in Figure 16.3b; that is, cross out 6, 9, 12, . . . if they have not already been crossed out. Of the numbers that remain, leave 5 and cross out its multiples; then leave 7 and cross out its multiples as shown in Figure 16.3c. All the numbers that now remain are prime numbers, as can be verified by attempting to factor them.

~~0~~,	~~1~~,	(2),	3,	~~4~~,	5,	~~6~~,	7,	~~8~~,	9
~~10~~,	11,	~~12~~,	13,	~~14~~,	15,	~~16~~,	17,	~~18~~,	19
~~20~~,	21,	~~22~~,	23,	~~24~~,	25,	~~26~~,	27,	~~28~~,	29
~~30~~,	31,	~~32~~,	33,	~~34~~,	35,	~~36~~,	37,	~~38~~,	39
~~40~~,	41,	~~42~~,	43,	~~44~~,	45,	~~46~~,	47,	~~48~~,	49

(a)

(2)	(3)	5	7	~~9~~
11	13	~~15~~	17	19
~~21~~	23	25	~~27~~	29
31	~~33~~	35	37	~~39~~
41	43	~~45~~	47	49

(b)

(2)	(3)	(5)	(7)	
11	13		17	19
	23	~~25~~		29
31		~~35~~	37	
41	43		47	~~49~~

(c)

Figure 16.3

Hence,

$\{2, 3, 5, 7, 11, 13, 17, 19, 23, 29, 31, 37, 41, 43, 47\}$

is the set of prime numbers between 0 and 49.

The reason why we need to carry out this process only to 7 was pointed out in our discussion of when our program for finding factors could stop.

■ **Question 6** Which numbers between 50 and 60 are prime? (Hint: add another row at the bottom of Figure 16.3a.)

Of course, if we start with a very large number and ask if it is a prime number, it would be very tedious to answer the question by means of Eratosthenes' sieve. A flow chart for a computer would be handy.

Fortunately, our program for finding all the factors of a whole number can easily be modified to print out PRIME if N is a prime number and COMPOSITE otherwise (assuming that neither 0 nor 1 is input). Check the flow chart in Figure 16.4 to see that it does what we want.

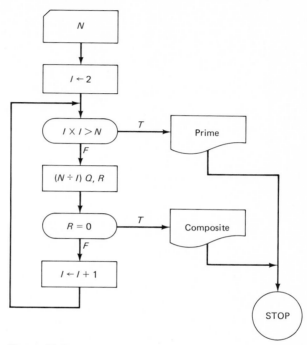

Figure 16.4

PRIME FACTORIZATION

We have seen that any composite number can be written as a product of two factors different from itself. The process of finding these factors is called *factoring,* and the result a *factorization.* Each of these factors may themselves be either prime or composite numbers; if either factor is composite it can be written as a product of two other numbers. If any of these are composite they can be factored. Such a process would end when each factor is a prime number. For example,

$$90 = 3 \times 30 = 3 \times (3 \times 10) = 3 \times 3 \times (2 \times 5) = 3 \times 3 \times 2 \times 5$$

Of course, it is also true that $90 = 9 \times 10$, which is a different factorization. But if the process is continued we get:

$$90 = 9 \times 10 = (3 \times 3) \times (2 \times 5) = 3 \times 3 \times 2 \times 5$$

Likewise:

$$90 = 6 \times 15 = (2 \times 3) \times (3 \times 5) = 2 \times 3 \times 3 \times 5$$

And again:

$$90 = 2 \times 45 = 2 \times (5 \times 9) = 2 \times 5 \times (3 \times 3) = 2 \times 5 \times 3 \times 3$$

Note that in each case the end result, containing only prime numbers, is the same except for the order in which the factors are written.

This property, which can be proved although we will not do so here, is commonly stated as *The Fundamental Theorem of Arithmetic:*

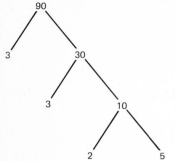

Figure 16.5

Every composite number can be factored as the product of primes in exactly one way, except for the order in which the prime factors appear in the product.

■ **Question 7** What is the prime factorization of 30?

A convenient way of representing the factoring process, as in the examples above, is by means of *factor trees.* The factor tree for the first factorization above is pictured in Figure 16.5. The trees for the other three factorizations are shown in Figure 16.6.

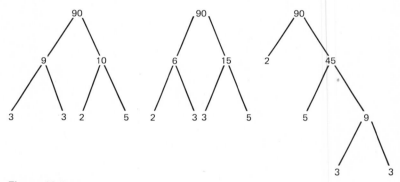

Figure 16.6

Now, how do we get such a prime factorization? The most straightforward way would be simply to take our list of prime numbers and try each one in turn to see if it is a factor of the number in question. And since, for example, the question "Is 2 a factor of 28?" can be answered by seeing whether 2 divides 28 (without a remainder), we see that our test can be carried out by dividing as many times as possible by each prime number in turn. Hence, to factor 28, $28 \div 2 = 14$ so 2 is a factor; $14 \div 2 = 7$, so 2 appears as a factor a second time; and 7 is itself a prime factor. In another form:

```
2| 28
 2 |14
    7
```

Hence the prime factorization of 28 is $28 = 2 \times 2 \times 7$. Now look at the example below and make sure you see why the prime factorization of 1,092 is $1,092 = 2 \times 2 \times 3 \times 7 \times 13$.

```
1        2| 1,092
2          2|546
3          3|273
4          7|91
5            13
```

STEPS 1 AND 2:
2 divides both 1,092 and 546

STEP 3:
2 does not divide 273, so we try 3, which works

STEP 4:
3 does not divide 91; nor does 5; so try 7, which works

STEP 5:
13 is a prime number, so we are finished

It might seem that this process of prime factorization by trying each prime in turn would be lengthy for large numbers. But it turns out that trying the primes through 7, for example, takes care of the factorization of any number up to 49 (7×7), as we have seen; the primes through 13 dispose of numbers through 169 (13×13); and the primes through 50 (which have already been listed) will dispose of all numbers through 2,500 (50×50).

■ **Question 8** Find the prime factorization of 1,001.

Nevertheless, if we have to deal with very large numbers, it would be nice to turn the job over to a computer. The process illustrated above is so straightforward that it will come as no surprise to the reader to be told that a flow chart can easily be constructed. However, it is not necessary to go into it here, so this flow chart is relegated to the problem section at the end of the chapter.

LEAST COMMON MULTIPLE Now we are ready to take up the last two mathematical ideas of this chapter. They will make it possible to handle the two problems stated at the end of Chapter 15. First, if one number is a factor of another, we say that the second is a *multiple* of the first.

For example, "4 is a *factor* of 12" also means that "12 is a *multiple* of 4." That is, the multiples of 4 are all the numbers with 4 as a factor, thus $\{0, 4, 8, 12, 16, 20, 24, 28, . . .\}$ is the set of multiples of 4. Likewise $\{0, 3, 6, 9, 12, 15, 18, 21, 24, 27, . . .\}$ is the set of multiples of 3. The *common multiples* of 4 and 3 are simply those numbers that appear in both sets, that is, in the intersection of the two sets. Hence $\{0, 12, 24, 36, . . .\}$ is the set of common multiples of 4 and 3. Zero is of no interest since it is a common multiple of *any* two numbers. The smallest common multiple, other than zero, we call the *least common multiple;* in the case of 4 and 3 it is 12. We abbreviate the least common multiple as lcm.

■ **Question 9** Write the multiples of 6 and 8 and pick out the least common multiple.

To find the lcm of two numbers, say 60 and 108, we again begin by displaying their prime factorizations:

$$108 = 2 \times 2 \times 3 \times 3 \times 3$$
$$60 = 2 \times 2 \times 3 \times 5$$

Since any multiple of 60 must, of course, contain all the factors of 60 and any multiple of 108 must contain all the factors of 108, any *common* multiple of the two numbers must contain all the factors contained in either 60 or 108. We could get a common multiple by simply taking 60×108 or, using their factors, $2 \times 2 \times 3 \times 5 \times 2 \times 2 \times 3 \times 3 \times 3$, but this would not be the *least* common multiple because we have more 2s and more 3s than we really need as factors for either 60 or 108. Let us use *only* the factors we *need,* namely, two 2s, three 3s, and one 5. Hence $2 \times 2 \times 3 \times 3 \times 3 \times 5$ or 540 will be the lcm of 60 and 108. To examine this result in more detail observe that 108 and 60 have a common block of factors $2 \times 2 \times 3$ so these must be in any common multiple. In addition to these we need the factors 3×3 to make up the 108. The number 5 is not a factor of 108 and the lcm must provide for it in order to be a multiple of 60. Hence, we take the common factors $2 \times 2 \times 3$, throw in 3×3 to get 108, then 5 so as to have all the factors of 60. The result is shown in Figure 16.7. This construction guarantees a number that contains 108 as a factor, hence the number is a multiple of 108, and contains 60 as a factor, hence the number is a multiple of 60. Furthermore, since only the needed factors have been used, this construction gives the smallest such common multiple of 108 and 60.

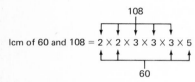

lcm of 60 and 108 = $2 \times 2 \times 3 \times 3 \times 3 \times 5$

Figure 16.7

■ **Question 10** What is the least common multiple of 12 and 30?

To go through another example, let us find the lcm of 18 and 30. Since $18 = 2 \times 3 \times 3$ and $30 = 2 \times 3 \times 5$ our least common multiple must have as factors one 2, two 3s, and one 5 in order to contain the factors of both 18 and 30. Hence the lcm of 18 and 30 will be $2 \times 3 \times 3 \times 5 = 90$. Again, you can easily verify that 90 is a multiple of 18 (namely 5×18) and a multiple of 30 (namely 3×30) and no smaller number is such a common multiple.

Now, as we shall see, this idea of the least common multiple of two numbers is just what we need in order to add two fractions with unlike denominators in as simple a way as possible. The trick is to find the least common multiple of the two denominators and to use it as the denominator of the equivalent fractions. Thus, for example, if our task were to add $\frac{11}{18}$ and $\frac{7}{30}$, then, as we just saw, the least common multiple of 18 and 30 is 90, and in fact

$$5 \times 18 = 90 \qquad \text{and} \qquad 3 \times 30 = 90$$

So we write

$$\frac{11}{18} = \frac{5 \times 11}{5 \times 18} = \frac{55}{90} \quad \text{and} \quad \frac{7}{30} = \frac{3 \times 7}{3 \times 30} = \frac{21}{90}$$

Then

$$\frac{11}{18} + \frac{7}{30} = \frac{55}{90} + \frac{21}{90} = \frac{76}{90}$$

Our original method would lead to

$$\frac{11}{18} = \frac{30 \times 11}{30 \times 18} = \frac{330}{540} \quad \text{and} \quad \frac{7}{30} = \frac{18 \times 7}{18 \times 30} = \frac{126}{540}$$

so

$$\frac{11}{18} + \frac{7}{30} = \frac{330}{540} + \frac{126}{540} = \frac{456}{540}$$

It is obvious that this last is less simple.

■ **Question 11** Find $\frac{7}{12} + \frac{7}{30}$.

GREATEST COMMON FACTOR

Our next idea turns out to be just what we need to "simplify" fractions. This is the idea of the *greatest common factor* of two numbers, which we abbreviate gcf. For example, given the two numbers 12 and 18 the *greatest common factor* is simply the largest number which is a factor of 12 and at the same time a factor of 18. To see what it will be observe that:

The set of factors of 12 = {1, 2, 3, 4, 6, 12}

The set of factors of 18 = {1, 2, 3, 6, 9, 18}

The *common* factors are the factors that appear in *both* sets, namely 1, 2, 3, 6; and of these the *greatest* common factor is, of course, 6. Two more examples are:

EXAMPLE 1:
Set of factors of 20 is {1, 2, 4, 5, 10, 20}

Set of factors of 30 is {1, 2, 3, 5, 6, 10, 15, 30}

Set of common factors of 20 and 30 is {1, 2, 5, 10}

Greatest common factor of 20 and 30 is 10

EXAMPLE 2:
Set of factors of 18 is {1, 2, 3, 6, 9, 18}

Set of factors of 27 is {1, 3, 9, 27}

Set of common factors of 18 and 27 is {1, 3, 9}

gcf of 18 and 27 is 9

Let us now take a harder example and see how we can find the gcf. Suppose we want the greatest common factor of 180 and 420. First note that

$180 = 2 \times 2 \times 3 \times 3 \times 5$

$420 = 2 \times 2 \times 3 \times 5 \times 7$

The primes that appear in both factorizations are 2 (twice), 3 (once), and 5. So we could write:

$$180 = (2 \times 2 \times 3 \times 5) \times 3$$
$$420 = (2 \times 2 \times 3 \times 5) \times 7$$

Clearly $2 \times 2 \times 3 \times 5 = 60$ will be a common factor of both numbers. Furthermore, it will be the largest such common factor, since any larger common factor would have to be 60 times some number which is a common factor of both 180 and 420, and we have already used up all the common factors. To take still another example, let us find the gcf of 72 and 54. Examine the solution given below and see that you understand it:

$$54 = 2 \times 3 \times 3 \times 3 \quad \text{and} \quad 72 = 2 \times 2 \times 2 \times 3 \times 3$$

If we group the factors as follows:

$$54 = (2 \times 3 \times 3) \times 3$$
$$72 = (2 \times 3 \times 3) \times 2 \times 2$$

we see that $2 \times 3 \times 3$ is the "common block" of prime factors from each. Hence, $2 \times 3 \times 3 = 18$ must be a common factor and since we have used up all the common prime factors, 18 must be the *greatest* such common factor. Hence, the gcf of 54 and 72 is 18.

It should be noted that for two *prime* numbers, say 13 and 29, the gcf of the two numbers would be 1, for 1 is the *only* common factor of two different prime numbers and so is certainly the greatest common factor.

■ **Question 12** What is the greatest common factor of 14 and 15?

Now we can apply this idea to our last problem in computation with fractions. Let us consider, for example, the fraction $\frac{180}{420}$. How can we find another fraction, equivalent to this one, with as small a denominator (and numerator) as possible? The answer is simple. Find the greatest common factor of the numerator and the denominator. We did this above. It was 60, and

$$180 = 60 \times 3$$
$$420 = 60 \times 7$$

We can write

$$\frac{180}{420} = \frac{60 \times 3}{60 \times 7} = \frac{3}{7}$$

by our rule for equivalent fractions in Chapter 12. Clearly, we cannot "simplify" or *"reduce"* $\frac{3}{7}$ any further; 60 has used up all the prime factors that are common to 180 and 420.

In the same way, we can see that

$$\frac{20}{30} = \frac{10 \times 2}{10 \times 3} = \frac{2}{3} \qquad \frac{18}{27} = \frac{9 \times 2}{9 \times 3} = \frac{2}{3} \qquad \text{and} \qquad \frac{54}{72} = \frac{18 \times 3}{18 \times 4} = \frac{3}{4}$$

You may remember using a process called "canceling" to simplify products of fractions. Here is an example:

$$\frac{\overset{7}{\cancel{28}}}{\underset{5}{\cancel{15}}} \times \frac{\overset{3}{\cancel{9}}}{\underset{5}{\cancel{20}}} = \frac{21}{25}$$

If you do not remember exactly how this process goes or why it works, here is an explanation.

The student who worked the problem first noticed that both 9 and 15 have 3 for a factor: $9 = 3 \times 3$ and $15 = 3 \times 5$. So the common factor, 3, was "canceled" by crossing out the 9 and the 15 and writing in their places the remaining factors, 3 and 5. Next, the student noticed that 28 and 20 have the common factor 4: $28 = 4 \times 7$ and $20 = 4 \times 5$. So this common factor also was "canceled" by striking out the 28 and the 20 and by writing in their places the remaining factors, 7 and 5. Finally, the numbers that remained after the "canceling" were multiplied together.

Such a procedure is perfectly correct. It is merely a shortcut, as we can see by doing the same problem by the methods we worked out above.

$$\frac{28}{15} \times \frac{9}{20} = \frac{4 \times 7}{3 \times 5} \times \frac{3 \times 3}{4 \times 5} = \frac{4 \times 7 \times 3 \times 3}{3 \times 5 \times 4 \times 5}$$

$$= \frac{(3 \times 4) \times (3 \times 7)}{(3 \times 4) \times (5 \times 5)} = \frac{3 \times 7}{5 \times 5} = \frac{21}{25}$$

What we have done is to factor the given numbers and then to use the fact that a/b and $(k \times a)/(k \times b)$ are equivalent fractions. In the example above, k was 3×4.

CONCLUSION Now we have finished our discussion of the arithmetic operations on rational numbers. In Chapter 15, we saw how to add, subtract, multiply, and divide fractions. In this chapter, we found how to make addition of fractions as easy as possible and we have seen how to simplify the result of any calculations with fractions as far as possible. The ideas we have developed in this chapter about factors, primes, and prime factorization were needed to make it possible to find least common multiples and greatest common factors, which in turn made it possible to simplify arithmetic operations on fractions.

■ **Question 13** Compute $\frac{2}{3} \times \frac{7}{5}$ and write it in as simple a form as possible.

There is one final remark to make. In our very first chapter we stated that mathematics consists of a number of special ways of looking at the world we live in. Yet in this chapter we have dealt not with sets of concrete objects or with the shapes of physical objects. Instead, we have concentrated our attention on the set of whole numbers, each of which is a purely mathematical idea.

However, we claim that we have not really departed from the spirit of the initial statement in Chapter 1. In fact, we claim that by now we have devoted so much attention to, and have had so much experience with, the whole numbers that they have become real to us and form a real extension of the real world we started with. To have mathematical ideas about the whole numbers is no different from having mathematical ideas about concrete objects.

PROBLEMS

1 Find all the factors of each of these numbers.
   (a) 12   (b) 35   (c) 36
   (d) 74   (e) 146   (f) 252

2 Using the Eratosthenes' sieve method, find all prime numbers larger than 50 and less than 100.

3 Which of these numbers are prime?
   (a) 233   (b) 269   (c) 323

4 Find the prime factorization of each of the following, using the factor tree method.
   (a) 105   (b) 75   (c) 320   (d) 3,600

5 Find the prime factorization of each of the following, trying each of the primes 2, 3, 5, . . . in order, as in the text.
   (a) 100   (b) 162   (c) 198   (d) 276

6 Find the lcm of each of these pairs.
   (a) 24 and 36   (b) 60 and 72
   (c) 40 and 48   (d) 48 and 144

7 Find the gcf of these pairs.
   (a) 8 and 12   (b) 24 and 36
   (c) 60 and 72   (d) 40 and 72

8 Find the lcm of each of these sets.
   (a) 2, 6, 15   (b) 40, 48, 72

9 Find the gcf of each of these sets.
   (a) 40, 48, 72   (b) 44, 92, 124

10 Combine these fractions and simplify the result, if possible.
   (a) $\frac{5}{12} + \frac{3}{20}$ (b) $\frac{2}{3} - \frac{1}{6}$
   (c) $\frac{13}{30} + \frac{7}{20}$ (d) $\frac{5}{4} + \frac{2}{3} + \frac{5}{6}$
   (e) $\frac{1}{2} - \frac{1}{6} + \frac{1}{8}$ (f) $\frac{1}{2} + \frac{1}{3} + \frac{1}{4} + \frac{1}{6}$

11 Simplify each of these fractions.
   (a) $\frac{75}{105}$ (b) $\frac{264}{480}$ (c) $\frac{105}{168}$ (d) $\frac{320}{3600}$

12 Some bells are set so that their time interval for striking is different. Assume that at the beginning both of the bells strike at the same time.
   (a) One bell strikes every 3 minutes and a second strikes every 5 minutes. If both bells strike together at 12:00 noon, when will they again strike together?
   (b) One bell strikes every 6 minutes and a second bell every 15 minutes. If both strike at 12:00 noon, when will they again strike together?
   (c) One bell strikes every 4 minutes, a second every 5 minutes, and a third every 8 minutes. If all three strike together at 12:00 noon, when will they again strike together?

13 (a) If 25 is input into the flow chart (Figure 16.2) for finding all factors of a number, then at one step, the output will be

   5, 5

How can the flow chart be modified so that such a "repeated" factor will be printed only once?

(b) The flow chart in Figure P16.13b is supposed to provide us with the prime factorization of the input, N. Try it with $N = 60$.

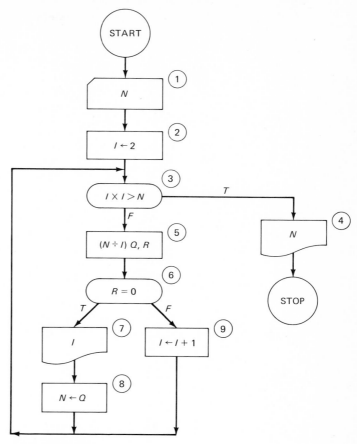

Figure P16.13b

(c) In the process illustrated in the text, the divisors tried out were the successive primes. Box 9 in the flow chart indicates that we are to run through all whole numbers successively, not just through the primes. Is there any danger of having a factor printed out which is not a prime? Run 42 through the flow chart. Is 6 ever printed out? Why not?

14 In the text it is stated that if

$$N = I \times Q + R, R < I$$

and if $R \neq 0$, then I cannot be a factor of N. Suppose $N = 219$ and $I = 27$. Then, by using the division algorithm, we find

$$N = 27 \times 8 + 3$$

(a) Multiply 27 by each of 7, 6, and 5 (the first three numbers less than 8).

How do the results compare with $N = 219$? Does it seem likely that the product of 27 and any number smaller than 8 could be 219?

(b) Multiply 27 by each of 9, 10, and 11 (the first three numbers larger than 8). How do the results compare with $N = 219$? Does it seem likely that the product of 27 and any number larger than 8 could be 219?

(c) Since no number less than 8 and no number larger than 8, when multiplied by 27, gives 219, 27 could be a factor of 219 only if $27 \times 8 = 219$. Does it?

(d) Carry out the same procedure with $N = 455$ and $I = 23$.

Note: *The next few problems are aimed at developing some shortcuts in the process developed in the text for finding the prime factorization of a number. In that process, the number is divided by the primes in order, starting with 2, to find out which primes are factors.*

Suppose, for example, we had a large number N and wanted to find out if 3 is a factor of N. The standard way is to use the division algorithm. Then, if the remainder is zero, we know 3 is a factor. Would it not be helpful if there were some way of just looking at the digits in N and discovering whether or not the remainder will be zero? If we could do that, we would not have to go through the whole division algorithm.

It turns out that there are some easy ways of doing this, and the following problems spell these out for 2, 3, 5, 7, and 11.

The rule for 2 is easy: If 2 is a factor of the last (from left to right) digit of a number N, then 2 is a factor of N. Let us illustrate this with a few examples.

$N = 48$ 2 is a factor of 8 ($8 = 4 \times 2$), so according to the rule, 2 is a factor of 48

To see that the rule is correct, we remember that

$48 = 4 \times 10 + 8$

Now we remember that

$10 = 5 \times 2$

We just saw that

$8 = 4 \times 2$

so

$$48 = 4 \times 10 + 8$$
$$= 4 \times (5 \times 2) + 4 \times 2$$
$$= (4 \times 5) \times 2 + 4 \times 2$$
$$= ((4 \times 5) + 4) \times 2$$

(Note: We have used both the associative and distributive properties.) The last equation shows that 48 is the product of two whole numbers, one of which is 2, and so 2 is indeed a factor of 48.

The rule for 3 is also easy to state: If the sum of the digits of a number N has 3 for a factor, then 3 is also a factor of N.

Let us illustrate why this rule is correct by investigating $N = 531$. We can write

$$531 = 5 \times 100 + 3 \times 10 + 1$$
$$= 5 \times (99 + 1) + 3 \times (9 + 1) + 1$$
$$= 5 \times 99 + 5 + 3 \times 9 + 3 + 1$$
$$= 5 \times 99 + 3 \times 9 + (5 + 3 + 1)$$

now we note that

$$99 = 33 \times 3 \quad \text{and} \quad 9 = 3 \times 3$$

So

$$
\begin{aligned}
531 &= 5 \times (33 \times 3) + 3 \times (3 \times 3) + (5 + 3 + 1) \\
&= (5 \times 33) \times 3 + (3 \times 3) \times 3 + (5 + 3 + 1) \\
&= [(5 \times 33) + (3 \times 3)] \times 3 + (5 + 3 + 1)
\end{aligned}
$$

This shows that 531 (and in fact any number) can be written as the product of a whole number and 3 plus the sum of the digits of the number. If the latter has 3 as a factor, then, by using the distributive property again, we find that the original number can be written as the product of a whole number and 3, so 3 is a factor.

The rule for 5 is also easy to state: If the last digit of N is either 0 or 5, then 5 is a factor of N.

Again we will illustrate with a particular case. Let $N = 635$. We can write

$$635 = 63 \times 10 + 5$$

But

$$10 = 2 \times 5$$

so

$$635 = 63 \times (2 \times 5) + 5$$
$$635 = (63 \times 2) \times 5 + 5$$

By the distributive property

$$635 = [(63 \times 2) + 1] \times 5$$

so 5 is a factor of 635.

15 Test each of these numbers to see which have 2 or 3 as factors.

(a) 111	(b) 132	(c) 303	
(d) 330	(e) 222	(f) 1,007	
(g) 5,236	(h) 25,236	(i) 25,256	
(j) 86,107	(k) 91,962	(l) 968,190	

16 Which of the above have 5 as a factor?

17 Why is it true that any prime number larger than 3 is either one less or one more than the product of some whole number and 6?

The rule for 7 is more complicated, and we will illustrate the rule with a particular case, $N = 511$.

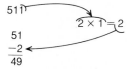

```
511
 51
 −2
 49
```

now, 49 has 7 for a factor. The rule says that 511 also has 7 for a factor. Here is another illustration:

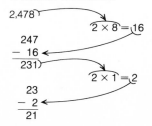

```
2,478
 247
 − 16
 231
  23
 − 2
  21
```

We know that 21 has 7 for a factor. The rule says that 2,478 also has 7 for a factor.

Our final rule, for 11, somewhat resembles the rule for 3: Add the first, third, fifth, etc., digits of the number N. Then add the second, fourth, sixth, etc., digits. Subtract the smaller of the two sums from the larger. If the result has 11 for a factor, so does N.

To illustrate, consider $N = 4,351,677$. Then

$$4 + 5 + 6 + 7 = 22$$

$$3 + 1 + 7 = 11$$

$$22 - 11 = 11$$

Since 11 has itself for a factor, 11 is also a factor of 4,351,677.

18 Test each of these numbers to see which has 7 as a factor.

(a) 407   (b) 569   (c) 671
(d) 861   (e) 903   (f) 4,257
(g) 5,291   (h) 25,525   (i) 84,455

19 Which of the above have 11 as a factor?

20 A number is *even* if it has 2 as a factor. It is *odd* if it is one more than an even number.

(a) What is the remainder after an odd number has been divided by 2?
(b) The sum of two even numbers is _____.
(c) The sum of two odd numbers is _____.
(d) The sum of an even and an odd number is _____.
(e) The product of two even numbers is _____.
(f) The product of two odd numbers is _____.
(g) The product of an even and an odd number is _____.
(h) If a is even and $a + b$ is odd, then b is _____.
(i) If ab is odd and a is odd, then b is _____.

17. DECIMALS

There is one particular set of rational numbers which needs special consideration. This is the set of *decimals,* rational numbers which can be named by fractions whose denominators are 10 or 100 or 1,000 or any other multiple of 10 by itself. (Such products as $100 = 10 \times 10$, $1,000 = 10 \times 10 \times 10$, etc., are called *powers* of ten.)

Decimal fractions force themselves on the attention of youngsters very early because of their use in our monetary system. More important is the fact that decimal notation is used in virtually all technical, scientific, and business computing.

Still more important is the fact that there is a "shorthand" way of writing fractions and mixed numerals, when the denominators are powers of ten, which makes arithmetic computation a good deal simpler than it would be if we had to use the procedures outlined in Chapters 15 and 16.

DECIMAL NOTATION We are all familiar, of course, with decimal notation and we recall that .1 is just another name for $\frac{1}{10}$, .27 for $\frac{27}{100}$, 4.314 for $4\frac{314}{1000}$, and so on. For such decimals all "denominators" are powers of 10, and since the particular denominator in question is revealed by the way the number is written, we just omit writing it. What needs to be settled is why the way a number is written reveals what "denominator" is involved.

We begin by recalling from Chapter 2 the expanded notation for a whole number using the base ten and the idea of place value. Thus:

$$3,842 = (3 \times 1,000) + (8 \times 100) + (4 \times 10) + (2 \times 1)$$

In our base ten place-value system each digit represents a certain value according to its place in the numeral. In the above example, the 3 is in the thousands place, the 8 is in the hundreds place, and so on.

The whole idea of the place-value notation (with base ten) is that the value of each place immediately to the *left* of a given place is 10 *times* the value of the given place. But then the value of a place immediately to the *right* must be *one-tenth* of the value of the given place. To make our place-value system serve for naming rational numbers as well as whole numbers we simply extend this idea of place value by saying that there are places to the *right* of the ones place and that the value attached to each place will, as before, be one-tenth that of the value of the place immediately to its left. When writing whole numbers the *last* place of the whole number was the ones place, but in writing decimals for fractions we have to fix where the ones place is with a dot (.), placed after the ones place. We will call this dot a *"decimal point"* reserving the word "decimal" for the actual numeral. Hence, the place value of the first place to the right of the ones place is $\frac{1}{10}$ of 1 equals, $\frac{1}{10}$, that of the second, $\frac{1}{10}$ of $\frac{1}{10}$ equals $\frac{1}{100}$, that of the third, $\frac{1}{10}$ of $\frac{1}{100}$ equals $\frac{1}{1000}$, and so on. Figure 17.1 illustrates this and shows the names of each of five positions to the right and to the left of the decimal point. Each of the positions to the right we will call a *decimal place.* A term such as "four-place decimal" is meant to designate a numeral with four places to the right of the decimal point.

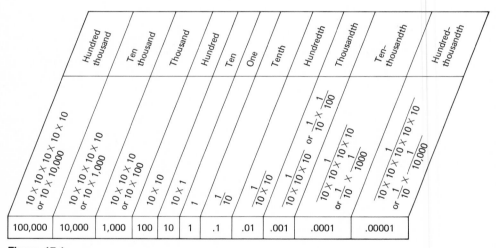

Figure 17.1

■ **Question 1**   Write $1,000 + \frac{1}{1000}$ as a decimal.

Hence, the numeral 435.268 expanded according to place value would be:

$$(4 \times 100) + (3 \times 10) + (5 \times 1) + (2 \times \tfrac{1}{10}) + (6 \times \tfrac{1}{100}) + (8 \times \tfrac{1}{1000})$$

Such a numeral would be read as "four hundred thirty-five *and* two hundred sixty-eight thousandths." Observe that the "and" serves to designate the decimal point. (Careless use of "and" in reading numerals is quite common and sometimes leads to confusion, as indicated by Problem 3 below.) Observe also that just as we say "four hundred thirty-five" instead of "four hundreds, three tens, and five ones" we say "two hundred sixty-eight thousandths" rather than "two-tenths, six-hundredths, and eight-thousandths." The place value of the final digit tells whether we should say "tenths," "hundredths," or what have you. Finally, observe the symmetry about the ones place (*not* around the decimal point):

Tens 1 place to the left of 1 and ten*ths* 1 place to the right of 1; hundreds 2 places to the left of 1 and hundred*ths* 2 places to the right of 1; and so on.

■ **Question 2** Which of the following name the same number as .0702?

1 $702 \times \frac{1}{100000}$

2 $\frac{7}{100} + \frac{2}{10000}$

3 702 ten thousandths

You will recall that in our models of rational numbers the denominator of a fraction denotes how many congruent parts a unit segment or region is divided into while the numerator tells how many of these parts are to be considered. If we regard the "denominator" for a decimal as being implicit in the situation though not explicitly written, so that, for example, .5 means 5 of 10 parts and .38 means 38 of 100 parts, it is easy to justify one very handy property of decimals. This property is exemplified by the fact that .3 = .30 = .300 = .3000 and .37 = .370 = .3700 = .370000 and is simply that you can put as many zeros on the end of a decimal as you like and still have equivalent decimals. Writing these examples in terms of their fraction equivalents, we get $\frac{3}{10} = \frac{30}{100} = \frac{300}{1000} = \frac{3000}{10000}$ and $\frac{37}{100} = \frac{370}{1000} = \frac{3700}{10000} = \frac{370000}{1000000}$ which is clearly just a matter of multiplying both numerator and denominator by the same number. On the other hand, we can take off as many zeros from the end of a decimal as we wish, as can be seen by rewriting the above as .3000 = .300 = .30 = .3. For decimal numbers, then, equivalence is immediately evident. Furthermore, changing two decimals to a "common denominator" is no problem at all; one simply tacks on the required number of zeros. For example, we get a common denominator for .47 (that is, $\frac{47}{100}$) and .5387 ($\frac{5387}{10000}$) by simply tacking two zeros onto .47 to make it .4700 ($\frac{4700}{10000}$).

To tell when one of a pair of decimal numbers is "greater than" or "less than" the other number is also no problem. One simply gives them the same "denominator," in the sense explained above, and compares them directly. For example, .0387 is clearly less than .2 since we can compare .0387 ($\frac{387}{10000}$) with .2000 ($\frac{2000}{10000}$).

■ **Question 3** Arrange these in increasing order:

9.9, $\frac{9}{10}$, $\frac{101}{10}$, $\frac{99}{1000}$, 99

OPERATIONS WITH DECIMALS Each time we meet a new set of numbers, or as in this case a new way of writing familiar numbers, we need to develop ways of dealing with equivalence, less than or greater than relations, and ways of doing standard operations. Equivalence and order for decimals have just been dealt with. To begin a discussion of operations, let us remind ourselves that any such discussions should provide both conceptual models for the process at hand and efficient computational procedures. The conceptual aspect of the operations using decimals can be very quickly disposed of by remarking that since they are only different ways of writing rational numbers, exactly the same models that were used for fractions suffice to give meaning to the operations with decimals. That is to say that for each such decimal used in an operation there is an exactly equivalent fraction of the form a/b, where b is some power of 10, so that using these equivalent fractions, the models and concepts previously discussed will apply. For that matter, any operation could be done merely by changing the decimals to fractions and using the computational procedures already discussed. It is convenient, however, to have ways of dealing *directly* with decimals via the usual operations.

We cannot dispose of these computational procedures very easily because though we have fairly simple rules of thumb to tell us *how* to get answers, these rules are seldom well understood and consequently are sometimes incorrectly applied. This is especially true for the operation of division. In the remainder of this section we will consider each operation in turn by stating the procedure for getting an answer, then explaining why it is that this procedure works, as justified by the various properties and procedures that we have listed for operations with whole numbers and rational numbers and for equivalent decimals. In most cases the procedure will amount to first a computation with whole numbers, then some rule to place the decimal point in the answer that results.

ADDITION OF TWO OR MORE DECIMALS

PROCEDURE:

(a) Add enough zeros to each decimal so that all of them have the same number of decimal places.

(b) Forgetting about the decimal point, add them *as if* they were whole numbers.

(c) Place the decimal point so that the resulting sum has the same number of decimal places as each of the numerals in Procedure a.

EXAMPLE:

(1) $34.8 + .008 + 73.74 + 147 = 34.800 + .008 + 73.740 + 147.000$
$$= 255.548$$

(2) 34.800
        .008
      73.740
      147.000
      255.548

JUSTIFICATION:

Instruction a is simply an instruction to write the decimal parts of each numeral with a "common denominator," in the sense explained earlier. Instruction b has the effect of adding "numerators" of the decimal parts of the mixed numerals, "carrying" the excess from these numerators over into the whole number part of the addition and adding the whole number parts.

Instruction c simply says that the "denominator" for the sum is the same as the common denominator of the addends.

Close examination of this problem computed using the fraction equivalents of the decimal numerals should make each of these points clear. This computation is exhibited in some detail below. Notice the "carry" from the fractional to the whole number part in the next to last step.

$$34.8 + .008 + 73.74 + 147 = 34\frac{8}{10} + \frac{8}{1,000} + 73\frac{74}{100} + 147$$

$$= 34\frac{800}{1,000} + \frac{8}{1,000} + 73\frac{740}{1,000} + 147$$

$$= 34 + 73 + 147 + \frac{800}{1,000} + \frac{8}{1,000} + \frac{740}{1,000}$$

$$= 34 + 73 + 147 + \frac{800 + 8 + 740}{1,000}$$

$$= 34 + 73 + 147 + \frac{1,548}{1,000}$$

$$= 34 + 73 + 147 + \frac{1,000}{1,000} + \frac{548}{1,000}$$

$$= 34 + 73 + 147 + 1 + \frac{548}{1,000}$$

$$= 255 + \frac{548}{1,000} = 255.548$$

■ **Question 4** Which of these name the same number as .0702?

1 $\frac{70}{1000} + \frac{2}{10000}$

2 $(7 \times \frac{1}{10}) + (2 \times \frac{1}{10000})$

3 $.06 + .0101 + .0001$

SUBTRACTION The procedure for subtraction of decimals is exactly analogous to that of addition. The example worked out below should make this clear:

Using Decimals

$$7.58 = 7.580$$
$$\underline{5.689 = 5.689}$$
$$1.891$$

Using Fractions

$$7.580 = 7\frac{580}{1,000} = \quad 6 + \frac{1,000}{1,000} + \frac{580}{1,000} = 6 + \frac{1,580}{1,000}$$

$$5.689 = 5\frac{689}{1,000} = 5 \quad + \quad \frac{689}{1,000} = 5 + \frac{689}{1,000}$$

$$1 + \frac{891}{1,000} = 1\frac{891}{1,000}$$

Observe that the regrouping necessary when using "fractions" is taken care of by the ordinary subtraction of whole numbers in the problem using decimals.

MULTIPLICATION

PROCEDURE:
(a) Pretend for the moment that the decimal points do not exist and do an ordinary multiplication *as if* only whole numbers were involved.
(b) Place the decimal point in the resulting product by counting the number of decimal places in each factor, adding these two counts, and putting the decimal point so that the product has this many decimal places. If there are not enough digits in the product to accommodate the required number of decimal places, one must supply zeros between the decimal point and the first digit of the whole number product. This process is illustrated below.

Using Decimals

(a) $.5 \times .73$

$5 \times 73 = 365$

Three decimal places are required, so

$.5 \times .73 = .365$

(b) $2.1 \times .032$

$$\begin{array}{r} 32 \\ \times\ 21 \\ \hline 32 \\ 64\ \\ \hline 672 \end{array}$$

Four decimal places are required, so

$2.1 \times .032 = .0672$

Using Fractions

$$\frac{5}{10} \times \frac{73}{100} = \frac{5 \times 73}{10 \times 100} = \frac{365}{1,000}$$

$$2\frac{1}{10} \times \frac{32}{1,000} = \frac{21}{10} \times \frac{32}{1,000} = \frac{21 \times 32}{10 \times 1,000}$$

$$= \frac{672}{10,000}$$

JUSTIFICATION:
Examination of the problems using fractions shows that in each such problem we always end up getting a numerator for the product by multiplying exactly the whole numbers that one gets by pretending that the decimal points do not exist. This justifies the first part of our procedure. Likewise, in the fraction problem, we end up multiplying, for example, tenths times hundredths to get thousandths (so a one-place decimal times a two-place decimal requires a three-place decimal as the product); tenths times thousandths to get ten-thousandths (so a one-place decimal times a three-place decimal requires a four-place decimal as the product); and so on. To consider more possibilities, hundredths (two decimal places) times hundredths (two decimal places) would give ten-thousandths (four decimal places), hundredths times thousandths (three decimal places) gives hundred-thousandths (five decimal places) and so on.

Perhaps the best way to convince oneself of the validity of the procedures used in multiplying decimals, or alternatively to "discover" what rules might work, is to work a number of such problems using the fraction equivalents of the decimals and the definition $a/b \times c/d = (a \times c)/(b \times d)$. In this case observe that one does *not* work with the numbers in "simplest form" but must retain the denominators as powers of ten.

■ **Question 5** Change the fractions into decimals and multiply:

$\frac{101}{10} \times \frac{2}{100} \times .010203$

DIVISION

For many people division of decimals is by far the most mysterious and troublesome of all the operations using decimals as far as justifying the rules and procedures used in the various algorithms for getting quotients goes. Again we do the operation *as if* only whole numbers were involved and rely on well-known procedures to place the decimal point. But these procedures are tricky. Furthermore, new possibilities are open to us. For example, if the division does not "come out even," that is, if there is a remainder left, we can now add more zeros after the last decimal place in our divisor and go merrily on our way until either it *does* "come out even" or we stop for some other reason, e.g., boredom, instructions given us, or the conditions of the problem. If it still does not come out even, what do we now do with the "remainder"? Shall we "round off" or let it go? The discussion that follows will not deal with all the possible ramifications of these problems but will, hopefully, make clear why the principal maneuvers we use are sensible and justifiable.

FIRST METHOD OF DIVISION:

From Chapter 8 we know how to do division problems and justify our results at least to the extent of getting a quotient and a remainder for any problem involving whole numbers. Chapter 15 points out briefly that *any* division problem using whole numbers will give a single rational number as quotient without any remainder. For example, $17 \div 4$ becomes $4\frac{1}{4}$ rather than a quotient of 4 and a remainder of 1 and the corresponding "check" becomes $17 = 4 \times 4\frac{1}{4}$ rather than $17 = 4 \times 4 + 1$. Hence, if we can convert our division of decimals problem into a division of whole numbers problem, it can certainly be handled using procedures already discussed. Of course, our quotient may involve fractions rather than decimals, but there is a way to change fractions to decimals which will be discussed near the end of this chapter so even this need not disturb us.

It remains, then, to show how a division of decimals problem can be changed to a division involving *only* whole numbers. The most efficient way to handle this is to use the fraction notation $a \div b = a/b$ to designate the division, carrying on the presumption introduced in Chapter 15 that such fractions behave in pretty much the same way as fractions involving only whole numbers and counting numbers. Now the examples below should make clear what our procedure is. The first two examples detail what is going on by the use of fractions equivalent to our decimals, but the rest proceed directly without this step.

(a) $2.5 \div .2 = \dfrac{2.5}{.2} = \dfrac{\frac{25}{10}}{\frac{2}{10}} = \dfrac{\frac{25}{10} \times \frac{10}{1}}{\frac{2}{10} \times \frac{10}{1}} = \dfrac{25}{2}$

$= 25 \div 2 = 12\dfrac{1}{2}$ or 12.5

(b) $2.5 \div .02 = \dfrac{2.5}{.02} = \dfrac{\frac{25}{10}}{\frac{2}{100}} = \dfrac{\frac{25}{10} \times 100}{\frac{2}{100} \times 100} = \dfrac{250}{2}$

$= 250 \div 2 = 125$

(c) $.9 \div .7 = \dfrac{.9}{.7} = \dfrac{.9 \times 10}{.7 \times 10} = \dfrac{9}{7} = 9 \div 7 = 1\dfrac{2}{7}$

(d) $53.75 \div .5 = \dfrac{53.75}{.5} = \dfrac{53.75 \times 100}{.5 \times 100} = \dfrac{5,375}{50}$

$= 5,375 \div 50 = 107\dfrac{25}{50}$

$= 107\dfrac{1}{2}$ or 107.5

$$.5\overline{)53.75} \longrightarrow 50\overline{)5,375} \longrightarrow 107\dfrac{25}{50}$$

with the long division showing:
$$\begin{array}{r} 107 \\ 50\,)\overline{5,375} \\ \underline{50} \\ 375 \\ \underline{250} \\ 25 \end{array}$$

(e) $1.072 \div .4 = \dfrac{1.072}{.4} = \dfrac{1.072 \times 1,000}{.4 \times 1,000} = \dfrac{1,072}{400}$

$= 2\dfrac{272}{400} = 2\dfrac{68}{100}$ or 2.68

$$.4\overline{)1.072} \longrightarrow 400\overline{)1,072} \longrightarrow 2\dfrac{272}{400}$$

with the long division showing:
$$\begin{array}{r} 2 \\ 400\,)\overline{1,072} \\ \underline{800} \\ 272 \end{array}$$

Observe that the procedure is to multiply both dividend and divisor by a large enough power of ten so that *both* are whole numbers, then divide in the way usual for whole numbers. Observe also that in Examples a, d, and e the answer could easily be changed to an equivalent decimal answer. How the $1\frac{2}{7}$ in Example c could be changed to a decimal will be the subject of the last section of this chapter.

■ **Question 6** What power of ten is used in this division?

$.0102 \div .00102$

SECOND METHOD FOR DIVISION: The procedure just described is not the usual one, as you know. On the other hand we would probably, in teaching youngsters, arrive at our usual procedure for handling all problems efficiently only as the end result of a number of simpler special cases and less efficient but more easily explained procedures. In the end, however, we would typically go about a division of decimals problem as follows:

PROCEDURE:

(a) Move the decimal point in both divisor and dividend the same number of places so that the divisor (but not necessarily the dividend) is a whole number.

(b) Do the problem as if it were division of whole numbers, i.e., ignore the decimal point.

(c) Place the decimal point in the quotient in such a way that the quotient has exactly the same number of decimal places as the *revised* dividend obtained by step a above.

(d) If there is still a remainder when all the digits in the dividend have been used up, one can, if he likes, add more zeros in the dividend and continue the division process. The example at the right illustrates this. Since adding zeros increases the number of decimal places in the dividend, and since we insist that the quotient have exactly as many decimal places as the dividend, this automatically increases the number of decimal places in the quotient. In our example at right, the process just described gets an exact quotient very soon, but this does not always happen and one must decide when to stop and what to do with the last remainder. These last questions will be dealt with later.

Example 1

$3.36 \div .8$

$$8\overline{)3.36}$$

$$\begin{array}{r} 4.2 \\ 8\overline{)33.6} \\ \underline{32} \\ 1\,6 \\ \underline{1\,6} \end{array}$$

$$\begin{array}{r} 4.2 \quad \text{tenths} \\ 8\overline{)33.6} \quad \text{tenths} \end{array}$$

Example 2

$3.38 \div .8$

$$\begin{array}{r} 4\,225 \\ 8\overline{)33.800} \\ \underline{32} \\ 1\,8 \\ \underline{1\,6} \\ 20 \\ \underline{16} \\ 40 \\ \underline{40} \end{array}$$

$$\begin{array}{r} 4.225 \quad \text{thousandths} \\ 8\overline{)33.800} \quad \text{thousandths} \end{array}$$

■ **Question 7** Suppose we know the answer to $2.3 \div 3.7$. How could we find the answer to $2.3 \div .037$?

JUSTIFICATION:

Step a is most easily justified by the arguments used earlier. We write the division as a fraction, then multiply dividend and divisor by the power of ten which will make the divisor a whole number. In the present example,

$$3.36 \div .8 = \frac{3.36}{.8} = \frac{3.36 \times 10}{.8 \times 10} = \frac{33.6}{8} = 33.6 \div 8$$

This procedure has the effect of moving the decimal point the same number of places in both numbers.

We are clearly justified in adding as many zeros as we please after the final decimal place, as in step d, for this is just a matter of using equivalent decimals, as was discussed early in this chapter.

The real problem, that of justifying the placement of the decimal point, is handled by remembering that each division of dividend by divisor must, by definition, give a quotient such that the multiplication of the quotient times the divisor must give the dividend as product. Since we have made the divisor a whole number, so that it has no decimal places, the number of decimal places in the quotient must be the same as in the dividend so that the whole number divisor times the decimal quotient will give exactly the same result as the decimal dividend. In the present case, starting with the revised problem with whole-number divisor, $33.6 \div 8 = n$, means that $n \times 8 = 33.6$ and since only a whole number times a decimal to tenths gives tenths in the product, n must be a decimal expressed to tenths.

FRACTIONS TO DECIMALS

As we pointed out earlier, all the operations with decimals could have been done by changing the decimals to their fraction equivalents and using procedures already considered in some detail. In other words, we have just been doing things that raise no fundamentally new issues but are only alternates to known ways of proceeding. But, if we consider how to change a fraction name to an equivalent decimal name, it turns out that some really new issues are raised.

To go from a fraction name to a decimal name, we again identify a/c as $a \div c$ and perform a division. Some such conversions are shown below.

(a) $\dfrac{1}{2} = .5$　　　(b) $\dfrac{1}{25} = .04$　　　(c) $\dfrac{175}{10} = 17.5$

since

```
   .5
2)1.0
   1 0
```

since

```
    .04
25)1.00
   1 00
```

since

```
     17.5
10)175.0
   10
    75
    70
     5 0
     5 0
```

In the cases shown here the division process terminates. For many fractions, however, the division process does not terminate. When we try it for $\frac{1}{3}$ or $\frac{3}{11}$ for example, we get $\frac{1}{3} = .333 \cdots$ and $\frac{3}{11} = .2727 \cdots$ as shown in these examples.

$$\frac{1}{3} = .333 \cdots \quad \text{and} \quad \frac{3}{11} = .2727 \cdots$$

since

```
      .333 · · ·                          .2727 · · ·
   3) 1.000 · · ·        and         11) 3.0000 · · ·
      9                                  2 2
      ──                                 ──
      10                                 80
       9                                 77
      ──                                 ──
      10                                 30
       9                                 22
      ──       and so on                 ──
       1                                 80
                                         77
                                         ──
                                          3    and so on
```

■ **Question 8** Try to find a decimal expression for $\frac{1}{7}$. How far do you have to go before the results begin to repeat?

For the present, we will just agree to stop the process after a reasonable number of steps, recognizing that our resulting decimal is not quite equal to the original fraction but nevertheless is quite close.

To clarify this last remark, let us observe how we can use the result

$$\tfrac{3}{11} = .2727 \cdot \cdot \cdot$$

to locate $\frac{3}{11}$ on the number line. We first take that part of the number line between 0 and 1 and divide it into 10 congruent segments (Figure 17.2).

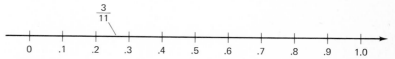

```
     3
    ──
    11
```

```
 ┼────┼────┼────┼────┼────┼────┼────┼────┼────┼────
 0   .1   .2   .3   .4   .5   .6   .7   .8   .9  1.0
```

Figure 17.2

Now $\frac{3}{11} = .2727 \cdot \cdot \cdot$, and we can rewrite this as

$$\tfrac{3}{11} = .2 + .07 + .002 + .0007 + (\underline{\hphantom{XXXX}})$$

From this we see that $\frac{3}{11}$ is certainly greater than .2. On the other hand, .2 is less than .3, .27 is less than .30, .272 is less than .300, .2727 is less than .3000, and so on, so $\frac{3}{11}$ is less than .3. Therefore $\frac{3}{11}$ lies between .2 and .3, as is indicated by the arrow.

Now let us magnify the part of the number line between .2 and .3 and divide it into 10 congruent segments (Figure 17.3).

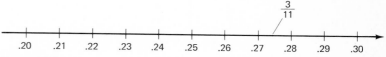

```
                                            3
                                           ──
                                           11
```

```
 ┼────┼────┼────┼────┼────┼────┼────┼────┼────┼────
.20   .21  .22  .23  .24  .25  .26  .27  .28  .29  .30
```

Figure 17.3

Since $\frac{3}{11} = .27 + .002 + .0007 + (\underline{\hphantom{XXXX}})$, we can see that $\frac{3}{11}$ lies between .27 and .28.

One more step shows us that $\frac{3}{11}$ lies between .272 and .273, and we can continue in this way as long as we please, each time pinning $\frac{3}{11}$ down into a smaller and smaller segment (Figure 17.4).

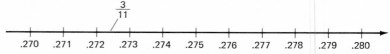

Figure 17.4

Thus, for example, if we use .272 as an approximation, we are less than $\frac{1}{1000}$ from the exact location of $\frac{3}{11}$. If we use .2727, we are less than $\frac{1}{10000}$ away, etc.

■ **Question 9** How much does .142857 differ from $\frac{1}{7}$?

In general, then, to represent a fraction as a decimal we apply the division process which was originally worked out for whole numbers. If the process eventually ends, we have what we want. If it does not, there is nothing we can do except to realize that the more decimal places we calculate, the closer the resulting decimal is to the fraction we started with.

PROBLEMS

1 Write the decimal numeral for
(a) $(9 \times 1{,}000) + (8 \times 100) + (7 \times 10) + (6 \times 1) +$
$(5 \times \frac{1}{10}) + (4 \times \frac{1}{100}) + (3 \times \frac{1}{1000})$
(b) $(7 \times 100) + (2 \times 1) + (9 \times \frac{1}{10}) + (7 \times \frac{1}{1000})$
2 Write in expanded form, shown in Problem 1:
(a) 927.872　　(b) 40.09　　(c) 21.0204
3 Regarding the "and" in each of the following as marking the decimal point, write the following as decimal numerals. Then write one or more numerals that one might get from a careless use of "and."
(a) Four hundred and sixty-one thousandths
(b) Two thousand three hundred and forty ten-thousandths
(c) One hundred and sixteen millionths
4 Change the following decimal fractions to common fractions and reduce to lowest terms:
(a) 0.024　　(b) 1.3625　　(c) 45.0400
5 (a) Put in the proper symbol $<, =, >$ to make a true statement:
(1) 0.47 _____ .0838　　(2) $2\frac{1}{4}$ _____ 2.2
(3) 1.7 _____ 1.70000　　(4) 4.5 _____ 4.49
(b) Arrange from least to greatest:

.25, 2.25, 1, .02, 1.02, 2.002, 2.2
6 (a) On the number line in Figure P17.6a, show:
(1) $.2 + .5$　　(2) $.8 - .5$　　(3) $.1 + .3 + .5$

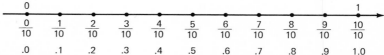

Figure P17.6a

(b) On the number line in Figure P17.6b show:

(1) .02 + .06 − .03 (2) 3 × .03 (3) .1 − .01

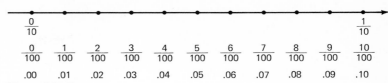

$\frac{0}{100}$	$\frac{1}{100}$	$\frac{2}{100}$	$\frac{3}{100}$	$\frac{4}{100}$	$\frac{5}{100}$	$\frac{6}{100}$	$\frac{7}{100}$	$\frac{8}{100}$	$\frac{9}{100}$	$\frac{10}{100}$
.00	.01	.02	.03	.04	.05	.06	.07	.08	.09	.10

Figure P17.6b

7 Do each of the following problems first using decimals and second using fraction equivalents of the decimals. Do not reduce the answers.

(a) 8.9 + 3 + 5.375 (b) Subtract 12.57 from 40

8 Make each of the following a true statement by supplying the missing decimal point and missing zeros in factor or product as required.

(a) .7 × .9 = 63 (b) 7 × .02 = 14
(c) .006 × .0004 = 24 (d) 1.704 × 2 = 3,408
(e) 3.1 × 400 = 124

9 Rewrite each of the following numerals as a fraction with a whole number as denominator.

(a) $\frac{73.6}{.25}$ (b) $\frac{.097}{3.26}$ (c) $\frac{685}{8.2}$
(d) $\frac{350}{.007}$ (e) $\frac{.649}{.36}$

10 If

$$15\overline{)2{,}310}\quad\frac{154}{}$$

find decimal numerals for the following without actually doing the division.

(a) 231.0 ÷ 15 (b) 23.10 ÷ 15
(c) 2.310 ÷ 15 (d) 23.10 ÷ .15
(e) .2310 ÷ 1.5 (f) .02310 ÷ 15

11 Compute the following:

(a) 9.6 + .08 + 13.452 (b) .004 × .6
(c) .0693 ÷ .36 (d) 6.93 ÷ .00036
(e) 5.143 − 2.7 (f) 100 − .005
(g) (.024 − .02) × (.004)

12 Although it does not always work, there is a quick way of locating the decimal point in multiplication and division problems, namely by *estimating*. Thus, for example, 21 × 483 = 10,141. Now, is .21 × 4.83 equal to 1.0141 or 10.141 or 101.41, or what?

Notice that .21 is approximately equal to $\frac{2}{10}$ and 4.83 is approximately equal to 5. So we can estimate .21 × 4.83 to be about $\frac{2}{10}$ × 5 = 1. So the correct answer is 1.0141.

In each of the following problems select the number in parentheses at the right which appears to be the correct answer.

(a) (.04) × (1.2) = _____ (4.8), (48), (.048), (.48)
(b) (14.6) × (.2) = _____ (29.2), (2.92), (.292), (292)
(c) (.0015) × (1.5) = _____ (.00225), (.0225), (.225)
(d) 4.354 ÷ 7 = _____ (.0622), (.622), (6.22), (62.2)
(e) 8.56 ÷ .008 = _____ (1.07), (10.7), (1070), (107)
(f) .10938 ÷ .06 = _____ (.01823), (.1823), (1.823), (18.23)

13 (a) How much would it cost to accept 0.3 of a $100 gift instead of $\frac{1}{3}$ of $100?

(b) How many places in a decimal approximation of $\frac{1}{3}$ would you use in order to have an error of less than 1 one-millionth?

14 Can $\frac{1}{2}$ be written as a decimal? Yes, because $\frac{1}{2} = \frac{5}{10} = .5$. Can $\frac{1}{5}$ be written as a decimal? Yes, because $\frac{1}{5} = \frac{2}{10} = .2$. Can $\frac{1}{3}$ be written as a decimal? No. Any fraction equivalent to $\frac{1}{3}$ is obtained by multiplying both the numerator and the denominator by the same whole number. There is a factor of 3 in the denominator to begin with, so there would be the same factor of 3 in the denominator of any fraction equivalent to $\frac{1}{3}$. But to write $\frac{1}{3}$ as a decimal we have to express $\frac{1}{3}$ as a fraction whose denominator is a power of 10. But the only prime factors of 10, and hence of any power of 10, are 2 and 5. So 3 is never a factor of any power of 10, and $\frac{1}{3}$ cannot be written as a decimal.

Can $\frac{1}{4}$ be written as a decimal? Yes, because $\frac{1}{4} = \frac{25}{100} = .25$. We did this by noting that $4 = 2 \times 2$. To get a power of 10, we need a factor of 5 to go with each 2. So we multiplied both numerator and denominator by $5 \times 5 = 25$.

Which of the following *cannot* be written as a decimal? Express each of the others as a decimal.

(a) $\frac{13}{8}$ (b) $\frac{9}{10}$ (c) $\frac{5}{6}$

(d) $\frac{73}{25}$ (e) $\frac{3}{16}$

15 Rewrite each of these base three expressions in terms of fractions in base ten.

(a) $.021_{three}$ (b) 22.01_{three}

(c) 101.011_{three} (d) 2.1102_{three}

16 In a decimal symbol, the ones place is always just to the left of the decimal point. This is the usual agreement. Suppose, however, that we had adopted the convention of writing a caret above the ones place. For example, we could write 21.17 as $2\hat{1}17$.

(a) Read the decimal $5326\hat{4}3$.

(b) Give the decimal representation for the sum

$$5(1{,}000) + 2(100) + 8(10) + 7 + 8(\tfrac{1}{10}) + 2(\tfrac{1}{100}) + 5(\tfrac{1}{1000})$$

in the new notation.

(c) Write $\frac{1}{10}, \frac{1}{10000}, \frac{1}{100}$ as decimals in the new notation.

(d) Write $1{,}000 + \frac{1}{1000}$ as a decimal in the new notation.

17 (a) Figure P17.17 is a flow chart. Record the output after going through the printout box the fourth time.

     (1) $N = 2$ $D = 3$ (2) $N = 1$ $D = 5$

     (3) $N = 1$ $D = 50$ (4) $N = 5$ $D = 2$

     (5) $N = 50$ $D = 2$ (6) $N = 3$ $D = 7$

(b) Compute the decimal expression, to three figures to the right of the decimal point, for the fractions

     (1) $\frac{2}{3}$ (2) $\frac{1}{5}$

     (3) $\frac{1}{50}$ (4) $\frac{5}{2}$

     (5) $\frac{50}{2}$ (6) $\frac{3}{7}$

(c) What is the first numeral printed in each case in part a?

(d) For part 2 of Problem 17a, the last three numerals printed were all zero and were not needed. What would you add to the flow chart to STOP when this begins to happen?

18 Four men enter a hardware store, and the first wants to buy 10.5 feet of copper wire, the second wants 15.5 feet, the third wants 8.5 feet, and the fourth wants 16.5 feet. The storekeeper has 50 feet of wire in his store. Can he give each man what he wants?

19 There are 16 ounces in 1 pound. Which is heavier, 7 ounces or .45 pounds?

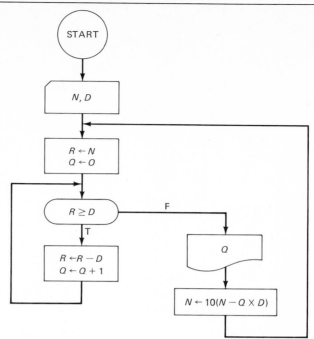

Figure P17.17

For each of these problems, write an appropriate mathematical sentence.

20 (a) The monthly salary of Mr. Kirk increased from $860.50 to $1,634.95 over a period of 9 years. On the average, how much did his monthly salary increase per year?

   (b) A jar of 100 vitamin tablets costs $4.98 and a bottle of 250 of the same tablets costs 8.94. How much more is the cost per tablet if you buy the vitamins by the jar instead of by the bottle?

18.

RATIOS, RATES, AND PERCENT

This is the first of three chapters in which a number of ways are shown of using fractions to describe and study certain aspects of the real world.

In Chapter 12, we agreed to use fractions, or rational numbers, to compare two sets A and B only when A consisted of "equivalent" parts, B consisted of "equivalent" parts, and each part of A was "equivalent" to each part of B. The two possible meanings of "equivalent" given there were that of *congruence*, when A and B were geometric figures, and *having the same number* of elements, when A and B were finite sets of objects, not necessarily congruent.

RATIOS Most of the illustrations in that chapter were of the first kind, where A and B were geometric figures. However, the other case turns up quite frequently. In fact, this case is used so often that a special terminology and symbolism has become associated with it. The concept of *ratio*, which we will develop in this chapter, will give us still one more way of using fractions or rational numbers to indicate how certain physical situations are alike.

Consider the following problem. Eddie can buy 2 candy bars for 6 cents, while Doug can get 6 of the same candy bars for 20 cents. They wonder who is getting the better "buy." It is assumed that there is no special discount for large purchases. Since Eddie knows that he must present 6 cents for every 2 candy bars, he can visualize his candy-purchasing ability as pictured in Figure 18.1. Every 2 candy bars must correspond to 6 pennies.

The last frame clearly indicates that Eddie is doing better than Doug is doing under these arrangements, for he is paying 18 cents for 6 candy bars while Doug is paying 20 cents for 6 candy bars.

Exactly how did we reach this conclusion? At first we asked ourselves what sort of purchase would be like the purchase of 2 candy bars for 6 cents. The situations represented above are a partial

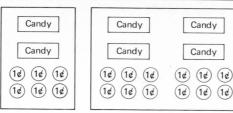 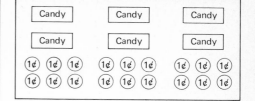

Figure 18.1

Candy bars	2	4	6	8	10
Pennies	6	12	18		

Figure 18.2

answer. To sharpen our understanding of how these situations are alike, let us summarize the essentials of each situation in a table (Figure 18.2). How can we make further entries in our table? If we are able to visualize or draw a picture of the situation, we can make the corresponding table entry with ease.

■ **Question 1** How much does Eddie have to pay for 100 candy bars?

We notice that an essential aspect of each situation we have described can be represented by using a pair of numerals: (2, 6) for the first frame, (4, 12) for the second, and (6, 18) for the third. These pairs can be used to represent a property common to all these situations. Using the pair (2, 6) we introduce the symbol 2:6 (read 2 to 6). In terms of the above model, this can be interpreted as telling us that there are 2 candy bars for every 6 pennies. This same correspondence could have been described using the pair (4, 12) and the associated symbol 4:12. For the above model this would tell us that there are 4 candy bars for every set of 12 pennies. Clearly, 4:12 and 2:6 are different symbols which we can use to indicate the same kind of correspondence, and we write 2:6 = 4:12. Once more, as in the case of numerals for rational numbers, we have an unlimited choice of symbols to represent the same property. The common property is called a *ratio*. In the preceding example, the ratio of candy bars to pennies is said to be 2 to 6, or 4 to 12, or 6 to 18.

RATIOS AS FRACTIONS In fact, a ratio is nothing but a special case of a fraction. It is the case where the "equivalent" parts of A(the set of pennies) are the subsets consisting of the individual pennies. These subsets are obviously equivalent since the number of elements in each of these subsets is the same, namely 1. Similarly, the "equivalent" parts of B are the subsets consisting of the individual candy bars.

We can therefore write the ratio of candy bars to pennies as $\frac{2}{6}$ or $\frac{4}{12}$ instead of 2:6 or 4:12. We will, however, continue to use the latter notation part of the time, since it is used regularly in certain situations and we therefore need to remember what it means. And we usually read a ratio as "1 to 2" rather than "one-half."

■ **Question 2** What is Eddie's ratio of pennies to candy bars?

Now that we see that a ratio is a fraction, we can also see how to decide, without drawing any pictures, whether Eddie or Doug is get-

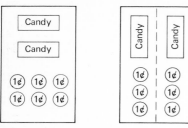

Figure 18.3

ting the better buy. Eddie's ratio of pennies to candy bars is $\frac{2}{6}$. Doug's ratio is $\frac{6}{20}$. We learned how to compare fractions in Chapter 12. Using what we learned, we find that

$$\frac{2}{6} > \frac{6}{20}$$

This means that if we arrange a common denominator, so both will be spending the same number of pennies, Eddie's numerator, the number of candy bars he can buy, will be greater than Doug's.

Can we tell how much Eddie will have to pay for 1 candy bar? If we take another look at the first frame of Figure 18.1, we see that 2 candy bars cost 6 cents. By rearranging this frame as in Figure 18.3, it becomes apparent that 1 candy bar should cost 3 cents.

How much candy can Eddie buy for 1 cent? In trying to answer this question, we find ourselves incapable of describing the situation by a suitable pair of numbers, unless we consider the candy bars to be divisible. In fact, the candy bars are divisible although the store owner is not likely to sell us part of a candy bar. If he would, we would expect to get $\frac{1}{3}$ of a candy bar for a penny. Hence, $\frac{1}{3} : 1$ is another name for the ratio we have been studying. However, if the candy-store owner will not cut up the candy bar, this particular pair of numbers does not describe a situation that will actually occur at the candy store.

The property described by 2:6 is exhibited in a wide variety of situations and is not restricted to sets of candy bars and pennies. Consider each of the following:

1 There are 2 texts for every 6 students.
2 There are 2 boys for every 6 girls in class.
3 The motor scooter does 2 miles in 6 minutes.

After a brief consideration, you should conclude that the table and the associated pictures which we developed for our example of candy bars and pennies would serve equally well to describe each of the above situations. For example, in case 1 instead of candy bars, we have texts and instead of pennies, we have students.

■ **Question 3** Does the ratio $\frac{1}{3} : 1$ make sense for case 2?

Consider the statement 1. It describes a situation involving two sets: a set of texts and a set of students. The situation in question exhibits a property described by 2:6. We can say that the ratio of number of texts to number of students is 2 to 6. In short, there are 2 texts for every 6 students. Another name for this ratio is 3:9. This indicates that there are 3 texts for every 9 students. The ratio of the number of texts to the number of students is described by 1:3. However, the ratio of the number of students to the number of texts is 3:1, that is, 3 members of the set of students correspond to each member of the set of texts. Clearly, in making comparisons between numbers of texts and numbers of students it will not do to say that the ratio is 1:3 unless we understand that the first number indicated refers to the set of texts. The *order* in which the numbers are named is important.

There is another bit of terminology that is often used. Suppose we have a computer which can carry out 18,000 additions in a minute. Then the ratio of additions to minutes is 18,000:1 or $\frac{18000}{1}$. (The ratio of minutes to additions is $\frac{1}{18000}$.) We often say the same thing in different words: the computer performs at the *rate* of 18,000 additions *per* minute. Similarly, to say that an airplane is flying at the rate of 550 miles per hour is to say that the ratio of miles traveled to hours of travel is 550:1. If, on a cold day in January, it takes 7 minutes for a blob of molasses to flow 2 inches, then the ratio of inches to minutes is $\frac{2}{7}$ and the molasses is flowing at the rate of $\frac{2}{7}$ inches per minute.

■ **Question 4**   If Eddie can count out pennies at the rate of 2 per second, how long will it take him to pay for 10 candy bars?

Another term that is often used in connection with ratios is *proportion*. A proportion is a mathematical sentence which states that two ratios are equal. We stated above that the store where Eddie buys his candy bars does not give a discount for large orders. This means that Eddie's ratio of candy bars to pennies is the same whether he spends 3 cents, when the ratio would be $\frac{1}{3}$, or whether he spends 3 dollars, when the ratio would be $\frac{100}{300}$. The sentence which says these two ratios are equal is

$$\frac{1}{3} = \frac{100}{300}$$

and this is an example of a proportion.

While ratios are rational numbers, we rarely find need of carrying out arithmetic operation on them. A few examples are given at the end of the chapter, but a very common question about ratios is whether or not one ratio is less than, equal to, or greater than another. An example was given earlier when Eddie asked how his candy bar to penny ratio compared to Doug's ratio. As we pointed out, we learned how to answer such questions when we learned, in Chapter 12, how to compare two fractions.

Another common problem involving ratios is a proportion in which either the numerator or the denominator of one of the fractions is missing and we are to find out what it should be. For example:

Doug's candy-bar-to-penny ratio is 6:20. How many candy bars can he purchase with a dollar?

The proportion in this case is

$$\frac{6}{20} = \frac{N}{100}$$

where N is the number he could buy for a dollar (equals 100 pennies). Another problem might be:

Doug's candy-bar-to-penny ratio is 6:20. How much will 48 candy bars cost him?

Here the proportion is

$$\frac{6}{20} = \frac{48}{N}$$

In the problem set at the end of the chapter we will review methods of working these problems.

■ **Question 5** If 3 compact cars can park in the space needed for 2 regular cars, and if a parking lot can hold 40 regular cars, how many compact cars can be parked there?

PERCENTS Quite often situations arise where more than two ratios need to be compared at the same time. For example, some states provide financial aid to local schools on the basis of actual attendance at the schools. A superintendent of a school district might want to compare the attendance ratio (ratio of students actually at school to total enrollment) of all the schools in his district. This would let him spot the one with the worst ratio. Some special attention to that school might increase its ratio, resulting in more money from the state.

For another example, sports fans usually compare the ratio of games won to total games played for all the teams in a league. The records of the teams in the Western Conference of the National Basketball Association are shown in Table 18.1 for the morning of February 17, 1973.

Table 18.1

Team	Won	Lost	Ratio of games won to total games played
Chicago	37	23	37:60
Detroit	26	34	26:60
KCOmaha	31	34	31:65
Los Angeles	46	13	46:59
Milwaukee	43	18	43:61
Phoenix	29	30	29:59
Portland	15	44	15:59
Seattle	19	44	19:63
Golden State	37	23	37:60

Who is leading? Who is doing the worst? How do the others rank? To answer these questions, we would have to compare each pair of ratios. This would require more arithmetic than most sport fans wish to undertake, so newspapers usually carry out and print the results of such comparisons.

■ **Question 6** Compare Detroit and Seattle.

In order to simplify this task of comparing a set of ratios, a rather neat trick is used. Each ratio is a fraction. The trick is to find, for each of these fractions, an equivalent fraction (which will represent the same ratio, of course) whose denominator is 100. Once this has been done, then it is easy to compare the ratios. When all the denominators are the same, it is only necessary to compare the numerators.

Suppose for example we wanted to compare the ratios $\frac{2}{5}$, $\frac{7}{20}$, and $\frac{9}{25}$. First we find equivalent fractions.

$$\frac{2}{5} = \frac{40}{100} \qquad \frac{7}{20} = \frac{35}{100} \qquad \frac{9}{25} = \frac{36}{100}$$

now we can see easily that $\frac{7}{20} < \frac{9}{25} < \frac{2}{5}$.

This procedure is used so widely that a special term and a special symbol are customarily used. The symbol is %, and it means $\frac{1}{100}$. It is read *percent.* Thus 35 percent or 35% are new ways of writing $35 \times \frac{1}{100}$ or $\frac{35}{100}$. In the same way, $\frac{2}{100} = 2\%$ and $\frac{240}{100} = 240\%$.

We often see statements of the form "a is c percent of b." This is just the same as "the ratio of a to b is the same as the ratio of c to 100."

If a is c percent of b, then

$$a = \frac{c}{100} \times b$$

But this multiplication problem can be turned into a division problem (see Problem 19)

$$\frac{a}{b} = \frac{c}{100}$$

which tells us that the ratio of a to b is indeed the same as the ratio of c to 100.

■ **Question 7**   What is 100 percent of 1?

COMPUTATION WITH PERCENTS

What was said about arithmetic operation on ratios applies of course to percents, since they are merely a special form of a ratio. There are times when percents are multiplied, but rarely are they added or subtracted. Most arithmetic computations connected with percents deal with the situation where two of the quantities, a, b, or c, in the sentence

$$\frac{a}{b} = c\%$$

or, what is the same

a is c percent of b

are given, and the third needs to be computed. Typical questions are:

1   What percent is 10 of 25?
2   10 is 40 percent of what?
3   What is 40 percent of 25?

We will look at Example 1 in some detail, because there are some special conventions about writing out the answers which need to be pointed out.

In Example 1 we are given $a = 10$ and $b = 25$. All we need to do is calculate a/b. Simple division gives $a/b = .4$, but to find the percent, we must express .4 as a fraction with 100 as the denominator. That is easy: $.4 = .40 = \frac{40}{100}$. So the answer is:

10 is 40 percent of 25

This procedure causes no difficulty as long as the division process stops after one or two decimal places. But what are we going to do about:

What percent is 1 of 8?

This question says that $a = 1$ and $b = 8$, and asks what c is in

$$\frac{1}{8} = c\%$$

Carrying out the division, we find that $\frac{1}{8} = .125$. Now $.125 = \frac{125}{1000}$, so we cannot express this ratio as a fraction with 100 as the denominator and a whole number as the numerator. However, $\frac{125}{1000} = \frac{12.5}{100}$, so we can, and do, write

1 is 12.5 percent of 8

■ **Question 8**   1,000 is what percent of 100?

A possible source of confusion comes from another way of working this problem. Recognizing that the quotient has to be multiplied by 100 to obtain the percent, some like to multiply by 100 *before* doing the division. Then, $100 \div 8$ is calculated to be $12\frac{1}{2}$, and another way of writing the answer is

1 is $12\frac{1}{2}$ percent of 8

So far, there is no confusion. But what about the question:

1 is what percent of 80?

The first procedure above, using decimals, can be copied and we easily find that

1 is 1.25 percent of 80

The second procedure, if followed carelessly, could result in this statement:

1 is $1.2\frac{1}{2}$ percent of 80

Now, strictly speaking, $\frac{1}{2}$ in the second decimal place means

$$\frac{\frac{1}{2}}{100} = \frac{5}{1,000}$$

so

$1.2\frac{1}{2}\% = 1.205\%$

which of course is different from 1.25 percent, the correct answer.
It is best therefore not to use fractions to the right of the decimal point.
Another kind of problem arises when the division process never ends. For example:

3 is what percent of 11?

We have seen that $\frac{3}{11} = .272727 \cdots$
The best we can do is to recognize that we cannot write an exact decimal answer. But, as we saw at the end of the preceding chapter, if we say that

3 is 27.27 percent of 11

then we are within $\frac{1}{100}$ of the exact result, while

3 is 27.2727 percent of 11

is within $\frac{1}{10000}$ of the exact result, etc.

■ **Question 9**   Find what percent 3 is of 1 to within $\frac{1}{1000}$.

PROBLEMS

1   Look at the picture in Figure P18.1 of a fifth-grade class.   What is one way of writing a symbol which represents the ratio of boys to girls?

What is one way of writing the symbol which represents the ratio of boys to girls?

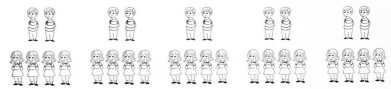

Figure P18.1

2   Copy and complete this table, where all boxes show the same ratio.

4:12	16:___	8:___	20:___	36:___	___:24	12:___	___:6

3   One store sells potatoes at 29 cents for 6 pounds, another store at 36 cents for 8 pounds.   Which is the better buy?

4   If a Fahrenheit thermometer changes 9°, a Celsius thermometer changes 5°, how much drop will a Fahrenheit thermometer record when a Celsius thermometer registers a drop of 16°?

5   On a house plan, 2 inches represent 5 feet.   A room length measures $7\frac{3}{16}$ inch on the plan.   What is the length?

6   The area of Bryant Park is 20 acres and the area of Churchill Park is 36 acres.   What is the ratio of the area of Bryant Park to the area of Churchill Park?   If the ratio is to be made three times as great as it is, and if Churchill Park is not to be changed, to what area must Bryant Park be expanded?   If, once Bryant Park has been expanded, it is desired to enlarge Churchill Park so that the original ratio is regained, what will the new area of Churchill Park be?

7   A mixture of grass seed requires by weight 5 parts of bluegrass seed and 3 parts of clover seed.   If a total of 20 pounds of seed is needed, how much bluegrass seed and how much clover seed should be in the mixture?

8   If 12 men can complete a job in 18 days, how many days will it take a crew of 30 men to do the job?

9 I weigh 160 pounds and wear a size 10 shoe. My son weighs 200 pounds. What size shoe does he wear?

10 We saw on page 257 that the question "How many candy bars can Doug purchase with one dollar?" leads to the open sentence

$$\frac{6}{20} = \frac{N}{100}$$

Similarly, the question "10 is what percent of 25?" leads to the open sentence

$$\frac{10}{25} = \frac{C}{100}$$

In each case we have two equivalent fractions, but we do not know, and we are supposed to find, the numerator of one of them. There is a way of solving such an open sentence which depends on remembering, from Chapter 7, the connection between division and multiplication. We will illustrate with the open sentence

$$\frac{6}{20} = \frac{N}{100}$$

Now $\frac{N}{100}$ is the same as $N \div 100$. So this sentence tells us that the rational number $\frac{6}{20}$ is the result that we get when we divide N by 100. But $N \div 100$ is the number which, when multiplied by 100 gives N. So

$$\frac{6}{20} \times 100 = N$$

It is easy to multiply out the numbers on the left, and we see that

$$N = 30$$

Use the same method to find what percent 10 is of 25.

11 The question "How much will 48 candy bars cost Doug?" led to the open sentence

$$\frac{6}{20} = \frac{48}{N}$$

Again we have two equivalent fractions, but this time it is the denominator of one of them that is missing. To handle this case, we use the idea of reciprocals from Chapter 15. The two fractions, $\frac{6}{20}$ and $48/N$, are equivalent, which means that they are names for the same rational number. Then their reciprocals, $\frac{20}{6}$ and $N/48$, must be two different names for the same rational number, the multiplicative inverse of the rational number named by $\frac{6}{20}$ or $\frac{N}{48}$. So $\frac{20}{6}$ and $\frac{N}{48}$ are equivalent fractions, or

$$\frac{20}{6} = \frac{N}{48}$$

Now we can use the procedure of Problem 10.

$$\frac{20}{6} \times 48 = N$$

Carrying out the multiplication on the left, we find that

$$N = 160$$

Use the same procedure to find what 5 is 20 percent of.

12 Fill in the blanks in Table P18.12.

Table P18.12

	Fraction	Decimal	Percent
(a)	$\frac{1}{4}$	.25	25%
(b)		.75	
(c)			80%
(d)		.125	
(e)			$66\frac{2}{3}$%
(f)	$\frac{12}{5}$		
(g)			
(h)		.00048	7.40%

13 Find
   (a)   3 percent of 9,612   (b)   150 percent of 28
   (c) $5\frac{1}{2}$ percent of $1,600   (d) $\frac{1}{2}$ percent of 840
   (e)   2,000 percent of 20   (f)   100 percent of 1
   (g)   1 percent of 100

14 (a)   What percent of 460 is 23?
   (b)   45 is what percent of 135?
   (c)   105 is 25 percent of what?
   (d)   What is the interest rate if $2,000 brings $135?

15 Represent each of these on a number line:
   (a) $16\frac{2}{3}$%   (b)   25%   (c)   125%
   (d) $66\frac{2}{3}$%   (e) $12\frac{1}{2}$%   (f)   105%

16 Write a mathematical sentence for each of these problems.
   (a)   A salesman receives a commission of 10 percent on his sales.   He pays an assistant 30 percent of his commission.   What net rate will this give him?
   (b)   A sign in a store advertised, "Save $4 on ladies' dresses.   30% off original price."   What was the original price of the dresses?
   (c)   The population of a city increased 30 percent in the period from 1950 to 1960.   In 1960, the population was 112,000.   Compute the 1950 population.
   (d)   A clothes dryer, regularly priced at $219, is on sale for $179.   Find the percent of markdown.

17 During 1958 the owner of a business found that sales were below normal. The owner announced to his employees that all wages for 1959 would be cut 20 percent.   By the end of 1959 the owner noted that sales had returned to the 1957 levels.   The owner then announced to the employees that the 1960 wages would be increased 20 percent over those of 1959. Were the 1960 wages the same as the 1958 wages?

18 In a certain store each customer pays a sales tax of 6 percent and is given a 10 percent discount for cash.   That is, if a customer purchased an article priced at $100 and paid cash, it would cost him $90 plus the sales tax or $95.40, since 6 percent of $90 is $5.40.   Suppose the sales tax were computed on $100 and then the 10 percent discount allowed, would the resulting net cost be the same?

19 (a)   Mr. Brown claims he paid $120 in gasoline taxes during a year.   If the tax on gasoline is 30 percent, how much did he spend on gasoline?
   The income tax collector looked at Mr. Brown's income tax return.   He inquired to find that Mr. Brown drove a Volkswagen which would go about 30 miles on a gallon of gasoline.   He also found that Mr. Brown could walk to work.

(b) If gasoline cost $.50 a gallon (including tax), how many gallons did Mr. Brown buy?

(c) How far could he drive with this amount of gasoline?

(d) What would be the average per day if he drove to work five days a week for 50 weeks?

(e) Why did the tax collector question Mr. Brown's return?

20 (a) Why do you suppose that 100 was chosen as the common denominator for comparing ratios rather than some other number, such as 10, or 1,000, or 50, or 127?

(b) When a baseball player bats 300, is he getting a hit 300 percent of his times at bat?

19.

MORE ON PROBABILITY

In Chapter 13, where some of the beginning ideas about probability were introduced, we saw that the probability of an event could be expressed as a fraction. However, we did not do much with those fractions except to compare them, since at that time we had not yet looked into the arithmetic of fractions.

Now we do know how to add, subtract, multiply, and divide fractions. We shall see that this helps us to do more with probability.

INTERSECTION OF TWO EVENTS
We start by looking again at an experiment discussed in Chapter 13. We have a box containing a red, a green, and a yellow marble. We shake the box and, without looking, take out a marble. We record its color as R, G, or Y, and replace the marble in the box. We repeat this action and record the color of the second marble along with the color of the first, to give a pair such as GY, RG, etc.

Now we can ask the probability of getting a red on both draws. We already know how to find this. In Chapter 13 we saw the tree diagram for this experiment, Figure 13.5, and it is shown again in Figure 19.1.

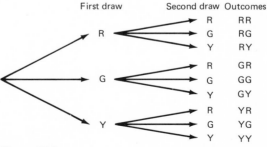

Figure 19.1

We see that there are nine equally likely outcomes, and the event RR contains only one of them, so

$$P(\text{RR}) = \tfrac{1}{9}$$

■ **Question 1** What is the probability that we will get the same color each time?

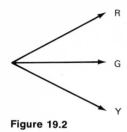

Figure 19.2

But we can look at this in another way. What is the probability of the event "red on the first draw"? The tree diagram for this simple experiment is shown in Figure 19.2 and we see that $P(\text{red on first draw}) = \tfrac{1}{3}$.

Next we ask about the probability of the event "red on second draw." The tree diagram for this is again Figure 19.2, so $P(\text{red on second draw}) = \tfrac{1}{3}$.

Now we notice two things. First, the event "red on first draw" consists of the outcomes $\{\text{RR}, \text{RY}, \text{RG}\}$ and the event "red on second draw" consists of the elements $\{\text{RR}, \text{GR}, \text{YR}\}$. The event "red on both draws" is $\{\text{RR}\}$, which is precisely the intersection of the two events.

$$\{\text{RR}, \text{RG}, \text{RY}\} \cap \{\text{RR}, \text{GR}, \text{YR}\} = \{\text{RR}\}$$

Second, we notice that

$$\tfrac{1}{3} \times \tfrac{1}{3} = \tfrac{1}{9}$$

In this case, the probability of the intersection of two events turned out to be the product of the probabilities of the events.

■ **Question 2**
What is the probability of "green on first draw"?
What is the probability of "yellow on second draw"?
What is the probability of "green on first draw and yellow on second draw"?

Let us try another example. First we toss a coin and then we roll a die. The array for this experiment was exhibited (Figure 13.7) in Chapter 13, and here it is again.

		Number on Die					
		1	*2*	*3*	*4*	*5*	*6*
Result of Coin Toss	H	(H, 1)	(H, 2)	(H, 3)	(H, 4)	(H, 5)	(H, 6)
	T	(T, 1)	(T, 2)	(T, 3)	(T, 4)	(T, 5)	(T, 6)

Suppose we ask about the probability of the event "the coin shows T and the die shows an even number." There are twelve equally likely outcomes in all, and there are three of them in this event, so its probability is $\tfrac{3}{12}$ or $\tfrac{1}{4}$.

But the event "the coin shows T" has probability $\tfrac{1}{2}$ and the event "the die shows an even number" is $\tfrac{3}{6}$ or $\tfrac{1}{2}$. The event "the coin shows T and the die shows an even number" is the set of outcomes $\{(\text{T}, 2), (\text{T}, 4), (\text{T}, 6)\}$. This is just the intersection of the event "the coin shows T" and the event "the die shows an even number." So again we have a case where the probability of the intersection of two events is the product of the probabilities of the events.

■ **Question 3** What is the probability that the coin shows H and the die shows 7?

INDEPENDENT EVENTS

This is in fact true in general provided that the two events are *independent*, that is, if the probability of either event does not depend on the other event. In particular, whenever we have, as in the example above, two independent experiments, and an event A which depends only on the outcome of the first experiment and an event B which depends only on the second experiment, then

$$P(A \cap B) = P(A) \times P(B)$$

If we have three or more experiments, all of which are independent, then the same result holds; the probability of the intersection of events is the product of their probabilities.

Of course, if the experiments are not independent, the result need not hold. For example, in the marble experiment, if we do *not* replace the marble after the first draw, then the second draw is not independent of the first. For example, if we draw the red marble first, and do not put it back, then we could not possibly draw red the second time and $P(\text{RR}) = 0$, rather than $\frac{1}{9}$.

■ **Question 4** Suppose we make the *definition* that two events are independent if the probability of their intersection is the product of their probabilities. Toss one die and let A be the event $\{2, 4\}$ and let B be the event $\{\text{even}\}$. Are A and B independent?

We can use the fact that the probability of the intersection of two independent events is the product of their probabilities to simplify some problems. At the end of Chapter 13 we considered the probability of drawing a black marble from a box containing 2 blacks and 1 white. It worked out to be $\frac{2}{3}$. Now, if we think of the outcomes of this experiment to be B and W (rather than B_1, B_2, and W), we can draw the tree diagram in Figure 19.3, where the fractions beside the arrows indicate that the probability of the outcome B is $\frac{2}{3}$ and that of W is $\frac{1}{3}$.

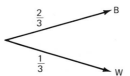

Figure 19.3

Suppose we replace the marble, after recording its color, and then repeat the experiment. Then we have the tree diagram in Figure 19.4. The outcomes are not all equally likely. The probabilities in the last column were calculated by multiplying the probabilities on the arrows leading to that outcome.

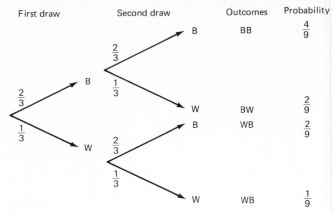

Figure 19.4

If we had used the old method of labeling the 2 black marbles B_1 and B_2, so that all outcomes would be equally likely, then the tree diagram for the two-stage experiment would have contained 12 arrows, twice as many as in Figure 19.4.

■ **Question 5** We have 3 white marbles, 2 red ones, and 1 green marble. We pull one out, record its color, and replace it. We then draw out a second marble. If we label the marbles, so that all outcomes are equally likely, how many arrows will there be in the tree diagram?

UNION OF TWO EVENTS Now let us look at another way of combining two events. In the marble experiment, what is the probability of the event "either red on the first draw *or* green on the second draw"? A look at the outputs shown in Figure 19.1 shows that this event is the set {RR, RG, RY, GG, YG}, so its probability is $\frac{5}{9}$.

But there is another way of doing the problem. Suppose we look at the simpler events, "red on first draw" and "green on second draw." The first of these is the set {RR, RG, RY}, which we call A, and the second is the set {RG, GG, YG}, which we call B. Figure 19.5 is a diagram showing these two sets.

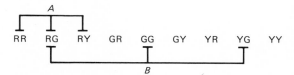

Figure 19.5

In Chapter 5, the *union* of two sets was defined to consist of those elements belonging to either the first set or the second or, perhaps, to both, and we note that $A \cup B$ is precisely the set {RR, RG, RY, GG, YG}. We also note that $P(A) = \frac{3}{9}$ and $P(B) = \frac{3}{9}$, but that $P(A \cup B)$, which is $\frac{5}{9}$, is *not* the same as $P(A) + P(B)$, which is $\frac{6}{9}$. Why is this?

This is because {RG} appears in both A and B, so it is counted once in finding $P(A)$ and it is counted again in finding $P(B)$, so it has been counted twice in finding $P(A) + P(B)$. But {RG} is counted only once in finding $P(A \cup B)$. So, in order to obtain $P(A \cup B)$ from $P(A) + P(B)$ we have to subtract $P(RG)$ once to make up for the fact that {RG} was counted twice on one side and only once on the other. So we have

$$P(A \cup B) = P(A) + P(B) - P(RG)$$

Finally, we notice that {RG} is the only outcome that is in both A and B, so

$$\{RG\} = A \cap B$$

Now we can write

$$P(A \cup B) = P(A) + P(B) - P(A \cap B)$$

■ **Question 6** Toss a coin twice. What is the probability of either a head on the first toss or a tail on the second?

Let us look at another situation. Suppose we roll two dice and ask for the probability of either an odd sum or a sum less than 6.

The set of all possible outcomes can be shown as an array, as in Figure 19.6. The event $A =$ "odd sum" consists of the outcomes which are circled. The event $B =$ "sum less than 6" consists of the outcomes within the triangular region.

Second Throw

	1	2	3	4	5	6
1	1,1	(1,2)	1,3	(1,4)	1,5	(1,6)
2	(2,1)	2,2	(2,3)	2,4	(2,5)	2,6
3	3,1	(3,2)	3,3	(3,4)	3,5	(3,6)
4	(4,1)	4,2	(4,3)	4,4	(4,5)	4,6
5	5,1	(5,2)	5,3	(5,4)	5,5	(5,6)
6	(6,1)	6,2	(6,3)	6,4	(6,5)	6,6

First Throw

Figure 19.6

Now we do some counting. The event "odd sum or sum less than 6," which is of course $A \cup B$, contains exactly 22 elements, so

$$P(A \cap B) = \frac{22}{36}$$

We find that A contains 10 elements and B contains 18 elements, so

$P(A) = \frac{10}{36}$ and $P(B) = \frac{18}{36}$

Finally, we note that $A \cap B$, which consists of those outcomes which are both in the triangular region and circled (and hence were counted twice) contains exactly six elements, so

$P(A \cap B) = \frac{6}{36}$

Putting these together, we see that

$\frac{22}{36} = P(A \cup B) = P(A) + P(B) - P(A \cap B) = \frac{10}{36} + \frac{18}{36} - \frac{6}{36} = \frac{22}{36}$

We have already seen that it is sometimes a bit tricky to find the probability of the intersection of two events when they are not independent, but there are still many real problems where we can find $P(A \cap B)$ by simple counting, just as we did above.

■ **Question 7**   We toss two dice. What is the probability of either an odd product or a product divisible by 3?

MUTUALLY EXCLUSIVE EVENTS

There is one case, however, for which it is easy to calculate $P(A \cap B)$. This is when the events A and B are *mutually exclusive,* that is, when it is impossible for both events to happen at once. Thus, for example, if we toss a coin twice, the events "exactly 1 head" and "no tails" are mutually exclusive. If we toss 2 dice, the events "sum less than 4" and "both numbers even" are mutually exclusive.

If A and B are mutually exclusive, then $A \cap B$ is the empty set, so $P(A \cap B) = 0$ and, in this case $P(A \cup B) = P(A) + P(B)$.

A special case of mutually exclusive events is when B consists of precisely those outcomes which are not in the event A. In this case, we call B the *complement* of A, and we often use the notation $\tilde{A}$ (read "not A"). For example, if we toss two dice, and if A is the event "sum less than 4," then $\tilde{A}$ is the event "sum greater than or equal to 4."

Since, by definition, each element of $\tilde{A}$ is not in A, $A \cap \tilde{A}$ is empty. Therefore

$P(A \cup \tilde{A}) = P(A) + P(\tilde{A})$

But we also see that $A \cup \tilde{A}$ is the entire set of possible outcomes of the experiment, since if any particular outcome is not in A, then by definition, it belongs to $\tilde{A}$. Since $A \cup \tilde{A}$ is the entire set of outcomes, $P(A \cup \tilde{A}) = N/N = 1$, where N is the total number of possible outcomes. Putting these two equations together,

$P(A) + P(\tilde{A}) = 1$

Finally, another way of writing this is

$P(\tilde{A}) = 1 - P(A)$

This says that the probability of A not happening is obtained by subtracting from 1 the probability of A happening.

■ **Question 8** There are 2 red, 3 white, 4 green, and 5 black marbles in a box. If we pull one out, what is the probability that it is *not* red?

SOME EXAMPLES To illustrate some of these ways of finding probabilities by calculating with fractions, let us consider the experiment of tossing a coin 10 times in a row. What, for example, is the probability of getting H 10 times in a row?

Before calculating this probability let us observe that it would be a tedious job to draw the tree diagram. Figure 13-4 shows the tree diagram for four tosses in a row, and it is clear that each additional toss adds a substantial number of new arrows. In fact, the tree diagram for this problem of 10 tosses contains exactly 2,046 arrows.

But, to go back to our problem, $P(\text{H}) = \frac{1}{2}$, so the probability of 10 heads in a row is

$$\tfrac{1}{2} \times \tfrac{1}{2} \times \tfrac{1}{2} \times \tfrac{1}{2} \times \tfrac{1}{2} \times \tfrac{1}{2} \times \tfrac{1}{2} \times \tfrac{1}{2} \times \tfrac{1}{2} \times \tfrac{1}{2}$$

Although there are 10 factors, this calculation is not hard to carry out and the result turns out to be $\frac{1}{1024}$. Therefore

$$P(10 \text{ heads in a row}) = \tfrac{1}{1024}$$

This means, incidentally, that since all outcomes are equally likely, there must be exactly 1,024 different outcomes.

Of course, H and T are equally likely, so

$$P(10 \text{ tails in a row}) = \tfrac{1}{1024}$$

■ **Question 9** Toss a coin 5 times. Then toss it 5 more times. What is the probability of 5 heads the first time? What is the probability of 5 heads the second time? What is the probability of 5 heads each time?

With a little more effort we can find the probability of exactly 9 heads out of 10. Any outcome can be thought of as a string of 10 letters, each of which is either H or T. If there are to be exactly 9 H's, then there can be only 1 T. There are only 10 places where this single T can appear (first, second, third, . . . , tenth), so there are exactly 10 outcomes with exactly 9 heads.

$$P(\text{exactly 9 heads}) = \tfrac{10}{1024}$$

If we combine the events "10 heads" and "exactly 9 heads," their union is the event "at least 9 heads." These two events are mutually exclusive, so

$$P(\text{at least 9 heads}) = \tfrac{1}{1024} + \tfrac{10}{1024} = \tfrac{11}{1024}$$

Again, since H and T are equally likely,

$$P(\text{at least 9 tails}) = \tfrac{11}{1024}$$

We can go a little further with our calculation. The complement of the event "at least 9 heads" is the event "no more than 8 heads." Therefore $P(\text{no more than 8 heads}) = 1 - \frac{11}{1024} = \frac{1013}{1024}$.
Similarily

$$P(\text{no more than 8 tails}) = \frac{1013}{1024}$$

We could calculate the probabilities of other events connected with this experiment, but these are enough to illustrate that some simple arithmetic can sometimes replace some complicated diagrams.

■ **Question 10** How much more likely is 9 heads out of 10 than 9 heads in a row?

ODDS Finally, we need to comment on another term often used in connection with probability. This term is *"odds."*

We often hear such expressions as: "the odds are 1 to 1 that a coin will fall heads," or "it is 5 to 1 against a die showing 6."

These two examples give a hint as to the relationship between the "odds in favor of an event" and the "probability of an event."

The odds in favor of an event E are simply a to b, where a and b are any two numbers in the ratio of $P(E)$ to $P(\tilde{E})$. ($\tilde{E}$, you recall, is the event "E does not occur.")

For example, if the probability of E is $\frac{2}{3}$, so that the probability of $\tilde{E}$ is $\frac{1}{3}$, then the odds in favor of E are $\frac{2}{3}$ to $\frac{1}{3}$ or, more simply, 2 to 1. (They are also 10 to 5, etc.)

Evidently, if the odds in favor of an event E are a to b, then the odds against E—that is, in favor of $\tilde{E}$—are b to a. In the preceding illustration, the odds against E are 1 to 2.

For example, if two coins are tossed and A is the event "2 heads," what are the odds in favor of A? They are 1 to 3 in favor of 2 heads on the toss of 2 coins, and 3 to 1 against.

We know that if two fractions have the same denominator, then the ratio of these fractions is the same as the ratio of their numerators. Therefore if, for any event E, we are given $P(E)$ and $P(\tilde{E})$ as fractions with the same denominator, then we can immediately state the odds in favor of E. We simply read off the numerators.

For example, the ratio of $\frac{2}{3}$ to $\frac{1}{3}$ is the same as the ratio of 2 to 1 (the odds in favor of E in the illustration at the beginning). The ratio of $\frac{1}{4}$ to $\frac{3}{4}$ is the same as the ratio of 1 to 3 (the odds in favor of A in the second example).

If we know that $P(E) = p$, then the odds in favor of E are in the ratio p to q, where $q = 1 - p$. Suppose we know the odds in favor of E; can we find $P(E)$?

Yes, because we are looking for two fractions which have to add up to 1 $[P(E) + P(\tilde{E}) = 1]$ and we know what the ratio of these two fractions is. For example, suppose we know that the odds in favor of E are 3 to 2. We need to find two fractions, with the same denominator, which add up to 1 and whose numerators are in the ratio of 3 to 2. Since we add numerators when we add fractions, the simplest thing is to let the denominator be $3 + 2 = 5$. Then $P(E) = \frac{3}{5}$ and $P(\tilde{E}) = \frac{2}{5}$.

■ **Question 11** What are the odds against tossing 3 tails in a row?

THE LONG RUN We have now met with many of the important mathematical ideas about probability and we have learned how to calculate the probabilities of some fairly complicated events. However, before leaving this topic, it is necessary to return to a remark made early in Chapter 13:

> "Probability is the part of mathematics that looks at the kind of regularity nature displays in the long run."

But what do the probabilities we have learned how to calculate have to do with nature? For example, we calculated above that the probability of no more than 8 heads, when 10 are tossed, is $\frac{1013}{1024}$. Now suppose you are holding 10 pennies in your hand. You are going to toss them on the floor, but before doing it you take another look at the fraction $\frac{1013}{1024}$. What does it tell you about the result you will get when you toss your coins?

In one sense it tells us very little, because we know we cannot predict, in advance, what the result will be. But in another sense, it tells us a lot. The kind of information it gives us was hinted at in the last five problems at the end of Chapter 13.

In each of those problems, we repeated an experiment a large number of times. In each case we compared two fractions. One fraction was the probability of a particular event, which we calculated in advance. The other fraction had for its numerator the number of times the event actually took place and for its denominator the total number of trials.

We saw that in general the two fractions were close together. Thus we would expect that if we were to toss our 10 coins a large number of times, then no more than 8 heads would be the result most, but not quite all, of the time.

The reason why the probability ideas we have been studying are useful and important is that they really do describe nature. Whenever we have a real experiment (as long as it has only a finite number of outcomes, all of which are equally likely) which we perform many times, then if A is an event (some set of outcomes) then the two fractions

$$P(A) \quad \text{and} \quad \frac{\text{number of occurrences of } A}{\text{total number of trials}}$$

will be close together. Now the fraction on the right is the one we are most interested in, since it tells us what really happened. But the one on the left can be calculated in advance, as we have seen.

Unfortunately, we cannot in general say just how close the two fractions will be, nor can we say how many times we would have to repeat the experiment in order to be sure that they would be within a specified distance of each other, although we will see in Chapter 25 that something can be said. Nevertheless, to know that the probability of an event is a good approximation to the frequency of that event is to know something about the way nature really works.

■ **Question 12** If you were to toss 4 coins 30 times, would you be surprised if 4 heads came up twice?

APPROXIMATE PROBABILITIES

In actual use, however, most probabilities are approximations. This is true even in those situations in which the probabilities are "given." When we say a coin is "fair," we assume $p = \frac{1}{2}$. No amount of testing of a particular coin could *prove* that p is $\frac{1}{2}$.

Suppose that a friend tells you that he has a spinner colored red and yellow. You cannot see the spinner. Your friend spins the pointer 30 times and tells you the result of each spin, which you record as follows:

RRYYR   YRRYR   RRRYR

YYRYR   RRYRR   RRYYR

What probability can you estimate for the outcome of red on each spin? You would probably guess that two-thirds of the spinner is colored red and one-third yellow. But you could not be sure of this, even if your friend were to spin the spinner 300 times and get 202 reds and 98 yellows.

The problem that we have been discussing—where we know the results of performing the experiment a certain number of times but do not know the exact design of the spinner—is illustrative of many real life situations. There are many examples where decisions are made on the basis of estimated probabilities. These estimates, in turn, are based on past experience. Here are two examples.

1   In a baseball game Robinson comes in as a relief pitcher. The opposing manager than orders Jones to bat for Smith. His decision is based on previous games, where experience has shown that Jones has had better success than Smith against Robinson. Regardless of the result in the present game, the manager may well claim that he is "playing percentage baseball."

2   A doctor decides not to operate on Mrs. N. His decision is based on the fact that, in medical experience, a large percentage of the patients with her symptoms have been cured without undergoing expensive (and perhaps dangerous) surgery.

In a particular situation the confidence that is placed in a decision based on the results of previous trials depends both on the nature of the results and on the number of trials.

■ **Question 13** If you toss a coin 20 times and it shows tails each time, what do you guess it will show the 21st time?

In the coin, marble, dice, and spinner examples we have had situations for which we had reasonable methods of estimating probabilities before any experiments were performed. Often, in practice, we are not able to make accurate predictions in advance. The following situation provides an example.

If a rivet (or a tack or a flat-headed screw) is tossed onto a flat surface, it may fall "up" like this: ⊥ or it may fall "down" like this: ⤬

Tossing such a rivet provides an example of a situation in which we have no way of determining the probability of each outcome by inspection. Certainly you would expect a broad-headed tack to fall "up" more often than you would a long screw with a small head. For a given rivet, whatever guess you make is not likely to be close.

TWO EXPERIMENTS

Finally, let us end up with two physical experiments, just as we did at the end of Chapter 13. We can do both at once. We need a cup, 5 pennies, and 5 nickels.

Put all the coins in the cup, shake well, and pour them out on the table. Record the number of pennies showing H in the first row and second column of a table like Table 19.1 and the number of nickels showing H in the first row and third column. Add the numbers in the second and third columns and record the results in the fourth column.

Table 19.1

Number of experiment	Heads		
	Pennies	Nickels	Total
1			
2			
3			
4			

We can think of this as doing the experiment of tossing 5 coins, and counting the number of heads, twice, once with the nickels and once with the pennies. At the same time, we have done the experiment of tossing 10 coins once.

Repeat this 99 times, so that 100 rows of the table are filled in. The 5-coin experiment has been done 200 times and the 10-coin experiment 100 times.

Now fill out two more tables. For the first one, fill in how many times, out of the 200, no heads appeared, how many times just 1 head appeared, and so on. The table has six rows, like Table 19.2. The sum of the entries in the right-hand column should be 200.

Then do the same for the 10-coin experiment. This table will have 11 rows. The sum of the entries in the right-hand column should be 100.

Be sure to save these tables. You will use them again in Chapter 25.

Table 19.2

Number of heads	Number of times
0	
1	
2	
3	
4	
5	

PROBLEMS

1 A game is played with a die and a spinner. The spinner shown in Figure P19.1 contains regions which have equal areas that are colored red, yellow, and green. In the game, you spin once and throw the die once. What is the probability that you will spin red and then get 6 on the die?

2 There are 3 boys and 2 girls in a group. One of them is chosen to buy soft drinks for a party. Another is chosen to buy the food.

(a) What is the probability that 2 boys are selected?

(b) What is the probability that a boy and a girl are picked?

(c) What is the probability that at least 1 boy is selected?

Figure P19.1

3 Suppose that you have a bag containing 5 red marbles and 4 white marbles.

(a) What is the probability of drawing 2 white marbles in succession from the bag, if the first marble drawn is replaced before the second drawing? (Hint: We can consider 9×9 outcomes.)

(b) What is the probability of drawing 2 white marbles if the first marble is not replaced before the second drawing? (Hint: Are there still 9×9 outcomes?)

4 On a single toss of an ordinary die, find the probability that the number is either greater than 3 *or* an even number.

5 On a single toss of 2 dice, find the probability that the sum of the numbers thrown is either an odd number *or* a prime number.

6 Let us think of throwing 2 dice, 1 red and 1 green. If A is the event "the sum of the numbers is 6," and if B is the event "both numbers in the pair are even numbers," find

(a) $P(A \cup B), P(A), P(B),$ and $P(A \cap B)$.

(b) Check that $P(A) + P(B) - P(A \cap B)$ agrees with your result in part a for $P(A \cup B)$.

7 The principal of Jones Junior High School said: "60 percent of the students in my school are boys; 15 percent of the students in the school play in the band. The number of boys who play in the band is 10 percent of the total number of students." If a student is chosen at random, what is the probability that the student is either a boy or a band member?

(a) If A is the event "boy is chosen," then $P(A) =$ _____ .

(b) If B is the event "band member is chosen," then $P(B) =$ _____ .

(c) If $A \cap B$ is the event that a boy band member is chosen, then $P(A \cap B) =$ _____ .

(d) The probability that either a boy or a band member is chosen is $P($ _____ $)$.

(e) $P(A \cup B) =$ _____ .

8 In each of two bureau drawers you have some socks, not sorted into pairs. One drawer contains 9 socks, of which 5 are black and 4 are blue. The other drawer contains 7 black and 8 blue socks. If you pick 1 sock from each drawer without looking, what is the probability that:

(a) both are black?

(b) both are blue?

(c) 1 is black and 1 is blue?

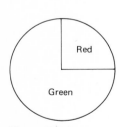

Figure P19.9

9 Suppose that you have a spinner like the one shown in Figure P19.9, with one-fourth of the area red and three-fourths green. Make a tree diagram showing the possible outcomes of two spins. We assume that the color on the first spin and the color on the second spin are independent events. Compute $P(RG), P(GR),$ and $P(GG)$.

10 In a rural territory a county road crosses a state road at a dangerous intersection. Since the state road carries more traffic, the traffic light for the county road shows red three-fourths of the time and green one-fourth of the time. An agent on the county road drives across the state road each morning and returns each afternoon, at a random time. He wants to know the probability that on a given day he will get a green light both times, or one time, or not at all.

(a) As a model for this problem construct a spinner similar to the one in Problem 9. What important change must you make?

(b) Draw a tree diagram or a table and assign probabilities to the *not* equally likely outcomes, as was done before.

$P(GG) =$ _____ $P(RG) =$ _____

$P(GR) =$ _____ $P(RR) =$ _____

(c) How many times as likely is he to get a green light in the morning and red in the afternoon as he is to get green lights both times?

(d)   Which is more likely, that he will get a red light both morning and afternoon or that he will get *at least one* green light?

(e)   What percent of the time will he get one red light and one green light?

11   A jar contains 5 marbles which are alike except for color; 2 are red, 2 are green, and 1 is blue. An experiment consists of drawing one marble at random from the jar and then drawing another marble at random, without having replaced the first.

Consider these events:

A The first marble is green. B The second marble is green.
C The second marble is blue.

(a)   Find $P(A)$, $P(B)$, $P(A \cap B)$. Does $P(A \cap B) = P(A) \times P(B)$? Are A and B independent events?

(b)   Find $P(A)$, $P(C)$, $P(A \cap C)$. Does $P(A \cap C) = P(A) \times P(C)$? Are A and C independent events?

12   A die is tossed. Find the probability of each of these events:

(a)   "Less than 4"

(b)   "Even number"

(c)   "Even number less than 4"

Are the first two events independent?

13   Select a thumb tack or a rivet. Write down your guess as to how many times it will land "up" if you toss it 50 times. Then actually toss it 50 times and compare your result with your guess. What is your present estimate of the probability of "up"?

14   A die is tossed. B is the event "a number greater than 2." What are the odds in favor of B? What are the odds against B?

15   Swat King's batting average is .325. What are the odds in favor of a hit his next time at bat? The odds against a hit?

16   The record of a weather station shows that in the past 120 days its weather prediction of "rain" or "no rain" has been correct 89 times. What is the probability that its prediction for tommorrow will be correct?

17   A random sample of 500 patients with a certain disease were treated with a new drug. Of these, 380 were helped.

(a)   What is the estimated probability that a given person with this disease will be helped by the new drug?

(b)   If 4,000 patients were treated with the drug, about how many would you expect to be helped?

18 (a)   If a batter's probability of getting a hit is .325, what is his probability of *not* getting a hit?

(b)   If the probability that a student passes a test is .85, what is the probability that he fails?

(c)   If the probability that a certain manufactured article is defective is .017, then what is the probability that it is *not* defective?

19 (a)   Draw a tree diagram for the experiment of tossing a coin 5 times in a row. (Hint: You can just extend the tree in Figure 13.4.)

(b)   Compute the probability of the event of either 3 or 4 heads.

(c)   In the table for the 5-coin experiment you carried out in Chapter 13, count the number of times you got either 3 or 4 heads.

(d)   Write the fraction whose numerator is the number you found in part c and whose denominator is 200. How close is this to the probability you computed in part b?

20 (a)   In the table for your 10-coin experiment, count the number of times you got either no heads or else just 1 head. Write the fraction whose numerator is this number and whose denominator is 100.

(b)   How close is this fraction to $\frac{11}{1024}$, which we saw to be the probability of no more than 1 head?

20.

SOME MORE FAMILIAR GEOMETRIC FIGURES

CIRCLES In Chapter 11 we reviewed a number of simple, familiar geometric figures. They were mostly composed of *straight* line segments. But there is another very familiar figure which is different. This is the *circle.* You remember that to draw a representation of a circle you put the metal tip of a compass at the point you want for the center and, keeping it fixed and the spread of the compass unchanged, draw the curve (see Figure 20.1). The segments from the center to all the points of the curve are congruent. Thus points A and B belong to the same circle with center O if and only if $\overline{OA} \cong \overline{OB}$. More formally: A *circle* is a simple closed curve having a point O in its interior such that if A and B are any two points in the curve, $\overline{OA} \cong \overline{OB}$.

The center is not a point of the circle. In Figure 20.2 point O is the center of the circle. A line segment with one endpoint at the center of the circle and the other endpoint on the circle is called a *radius.* $\overline{OP}$, $\overline{OA}$, and $\overline{OB}$ are all radii. As we observed above, all radii (radii is the plural of radius) of a given circle are congruent. $OP \cong \overline{OA} \cong \overline{OB}$.

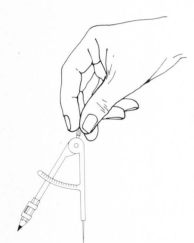

Figure 20.1

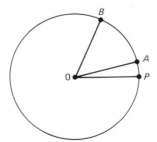

Figure 20.2

A circle is a simple closed curve. Consequently, it has an interior and an exterior. Suppose we have a circle with the center at the point

P and with radius $\overline{PR}$ (see Figure 20.3). A point, such as A, is *inside* the circle, if $\overline{PA} < \overline{PR}$, while a point, such as B, is *outside* the circle if $\overline{PB} > \overline{PR}$. Thus it is easy, in the case of the circle, to describe precisely what its interior and its exterior are. The interior is the set of all points A such that $\overline{PA} < \overline{PR}$. The exterior is the set of all points B such that $\overline{PB} > \overline{PR}$.

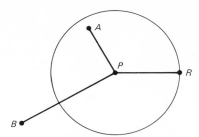

Figure 20.3

Now, shade lightly the interior of the circle as shown in Figure 20.4. The union of a simple closed curve and its interior is called a "region." The union of the circle and its interior is a *circular region*. Note that the circle is the curve only, while the circular region includes the curve and its interior.

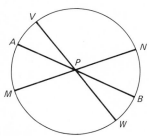

Figure 20.4

■ **Question 1** Could two different circles intersect in more than two points?

The word diameter is closely associated with the word radius. A *diameter* of a circle is a line segment which contains the center of the circle and whose endpoints lie on the circle. For the circle represented by Figure 20.5, three diameters are shown: $\overline{AB}$, $\overline{MN}$, and $\overline{VW}$. (A diameter of a circle is the longest line segment that can be drawn in the interior of a circle such that its endpoints are on the circle.)

A *tangent* to a circle is a line that intersects the circle in exactly one point. In the drawing $\overleftrightarrow{DE}$ is tangent to circle P. The single point of their intersection is T. It is called the *point of tangency*. A tangent to a circle cannot contain a point of the interior of the circle (see Figure 20.6).

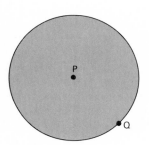

Figure 20.5

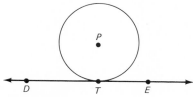

Figure 20.6

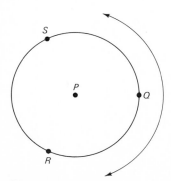

Figure 20.7

In Chapter 3 we saw that a point in a line separated the line into two half-lines. Consider a similar question with respect to a circle. Does point Q separate the circle in Figure 20.7 into two parts? If we start at

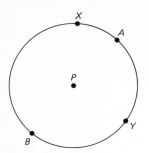

Figure 20.8

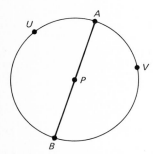

Figure 20.9

Q and move in a clockwise direction we will, in due time, return to Q. The same is true if we move in a counterclockwise direction. A single point does *not* separate a circle into two parts.

In Figure 20.8, the two points X and Y do separate the circle into two parts called *arcs*. One of the arcs contains the point A. The other arc contains B. No path from X to Y along the circle can avoid both of the points A and B. Thus, we see it takes two different points to separate a circle into two distinct parts. The arcs are written $\widehat{XAY}$ and $\widehat{XBY}$, or sometimes just $\widehat{XY}$ if it is clear from the context as to which arc is meant.

■ **Question 2** Could two different circles intersect at just one point?

In Figure 20.9, the endpoints of a diameter are A and B. These points determine two special arcs called *semicircles*. In the drawing, $\widehat{AVB}$ is one semicircle.

The endpoints of a semicircle and the center of the circle are on a straight line. For other arcs this is not true as shown in Figure 20.10. The center P is not on the straight line passing through the endpoints of $\widehat{DXE}$.

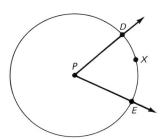

Figure 20.10

Rays $\overrightarrow{PD}$ and $\overrightarrow{PE}$ determine a *central angle*. A central angle is an angle having its vertex at the center of a circle. Such angles are measured in the same way as other angles.

Circles are not only familiar, they are very important. The world around us is full of shapes which suggest circles. And, as we are going to see, it is very handy to know something about circles when we start to think about the surface of the planet we live on, earth.

■ **Question 3** Does a semicircle separate the plane?

SPHERES What is suggested by "a set of points in space whose distances from a particular point are all the same"? This set of points would be more than a circle. It would be a surface, like the surface of a ball. Such a surface is called a *sphere*. The point from which the distances are measured is called the *center* of the sphere.

Many objects are spherical, i.e., have the shape of a sphere; rubber balls, used as toys; ball bearings, important to industry; etc.

Note that it is the surface that is called a sphere. On a ball, only that portion that could be painted represents the surface.

The sphere is also used as a mathematical abstraction of the surface of the earth. The fact that the surface of the earth is somewhat uneven and is thought to be a bit flattened at the poles is, for most purposes, not important. It is still useful to study the sphere and to regard it as an abstraction of the surface of our earth.

Just as in the case of a circle, a line segment having one end at the center of a sphere and the other end on the sphere is called a *radius* of the sphere. Similarly, a segment having both ends on the sphere which contains the center is called a *diameter*.

Any point on a sphere is one end of a diameter. The other end is said to be *diametrically opposite* to the first. We also say that either point is the *antipode* of the other.

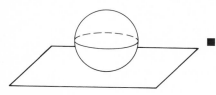

Figure 20.11

■ **Question 4** Could two spheres intersect in exactly two points?

The line passing through the poles of the earth is called the *axis* of the earth. This is approximately the line about which the earth revolves. Accepting the surface of the earth as a sphere, the diameter contained in the axis intersects the sphere at the North and South Poles. The points represented by these poles are diametrically opposite each other. The North Pole represents a point which is the antipode of a point represented by the South Pole.

Consider the intersection of a plane and a sphere. If the intersection is not empty then it might be just one point. In such a case the sphere would be tangent to the plane. This situation would be represented by a hard ball resting on a table. The surface of the ball seems to have just one point in common with the table top (Figure 20.11).

If the intersection of a plane and a sphere is not empty and contains more than one point, then it is a circle. One sees an illustration of this by slicing an orange.

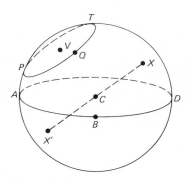

Figure 20.12

There is a distinction made as to whether the plane which intersects the sphere contains the center of the sphere. If it does, we call the intersection a *great* circle of the sphere. If the plane does not contain the center then we call the intersection a *small* circle. Note that the center of the sphere is also the center of each of the great circles of the sphere but it is *not* the center of any of the small circles of the sphere. In Figure 20.12, PQT represents a small circle with center at V. ABD represents a great circle with center at C, the center of the sphere.

■ **Question 5** Is every point on a sphere the antipode of some other point in the sphere?

There are two very important facts about great circles. The first is that any two great circles intersect in two points which are diametrically opposite. It is easy to see why this is so. Each great circle is formed by the intersection of the sphere with a plane going through the center, C, of the sphere. The two planes intersect in a line. The two points where this line meets the sphere are diametrically opposite and are on both great circles (see Figure 20.13).

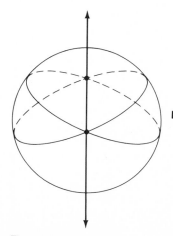

Figure 20.13

The second fact is that any two points on the sphere which are not diametrically opposite belong to exactly one great circle. The center and the two points together determine a plane which intersects the sphere in a great circle which contains the two points.

One of the interesting and important facts about spheres is that if A and B are two points of a sphere then the shortest path on the sphere between A and B is the great circle path from A to B. This fact is of great significance in navigation, both in ship-sailing routes and in airline routes.

■ **Question 6** How many great circles pass through two antipodal points?

LOCATION ON A SPHERE

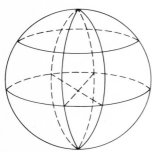

Figure 20.14

Again, think of the earth as a sphere (Figure 20.14). We can imagine many great circles of this sphere. A particular set of great circles of the earth is the set whose members pass through the North and South Poles. On the earth, half of a great circle, with the poles as endpoints, is called a *meridian.*

Figure 20.15 shows a set of parallel planes slicing the earth in horizontal sections. The top plane is tangent to the North Pole, and the bottom plane is tangent to the South Pole. The intersection of each of the remaining planes and the earth is a circle. The circles determined by these planes are all small circles except for the case where the plane contains the center of the sphere. This circle is a great circle and it is called the *equator.* All such circles are called *parallels of latitude.* They are called "parallels" because they are determined by planes parallel to the plane passing through the equator.

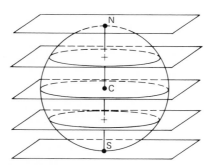

Figure 20.15

Now we are going to review the way we locate points on the surface of the earth. Suppose A is the name of some point on the sphere, as shown in Figure 20.16, perhaps the point where you are standing. There is exactly one parallel of latitude through A. There is exactly one meridian through A because the point A and the North Pole (or South Pole) determine one great circle. Since that great circle passes through the poles, the arc of the great circle containing A is a meridian. Thus, through each point of a sphere, except the poles, there is exactly one parallel of latitude and one meridian.

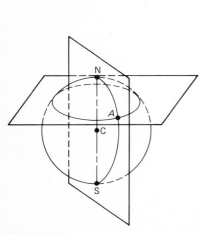

Figure 20.16

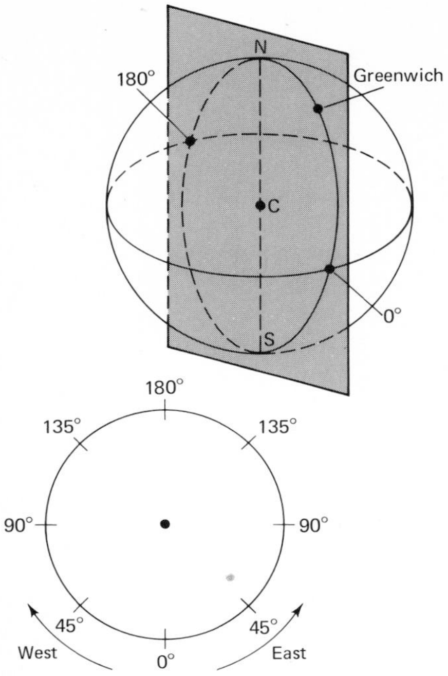

■ **Question 7** If a parallel of latitude is less than a mile in diameter, where is it?

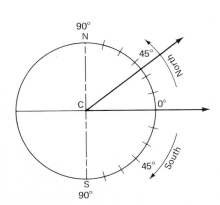

Figure 20.17

The zero meridian for the earth has been designated as the meridian which passes through a certain location in Greenwich (pronounced Gren–ich), England, near London. We sometimes refer to this meridian as the *Greenwich meridian,* even though the meridian itself passes through one particular point of the town. The meridian at Greenwich is sometimes called the prime meridian. (This has nothing to do with a prime number.)

The intersection of the Greenwich meridian and the equator is marked 0°. From this point, we follow the equator east, or west, until we reach the meridian which passes through the antipode of the Greenwich point, that is, lies on the great circle through Greenwich and the North Pole. This meridian intersects the equator at a point which is one-half way around the equator from the point labeled 0°. This point is labeled 180°. We can think of a plane intersecting the earth in this great circle. The plane separates the earth into two *hemispheres,* or half-spheres, as shown in Figure 20.17. The hemisphere on the left, as you look at the drawing, is named the Western Hemisphere. The hemisphere on the right is the Eastern Hemisphere.

The great circle, which we call the equator, is divided into 360 equal parts in a view as seen from the North Pole. The whole numbers between 0 and 180 are assigned to the points on the half equator to the left of 0°. The same is done for the points on the other half of the equator. Each of these points names the meridian passing through that point. Any point on earth may be located by the meridian passing through the point. For example, Los Angeles is approximately on the meridian 120° west of the Greenwich meridian. Tokyo is approximately on the meridian 140° east of Greenwich. We say the longitude of Los Angeles is about 120° west. The longitude of Tokyo is about 140° east.

■ **Question 8** What is the longitude of Greenwich, England?

The parallels of latitude are located in the following way. The equator is designated the zero parallel. All points above the equator are in the Northern Hemisphere, points below in the Southern Hemisphere. We choose any meridian, for instance that meridian through Greenwich.

Suppose A is a point on this meridian. Then there is a central angle determined by the ray from the center of the earth through this point and the ray from the center through the point where the meridian meets the equator. Whatever the measure of this central angle is, we use it to specify the parallel of latitude through A. Thus the part of the meridian from the intersection with the equator to the North Pole is divided into 90 equal parts. The whole numbers between 0 and 90 are assigned to these points. Each point determines a parallel of latitude. A similar pattern is followed for points on the meridian south of the equator. For any point on earth, we may locate the parallel of latitude containing the point (see Figure 20.18). For example, New Or-

Figure 20.18

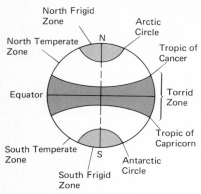

Figure 20.19

leans is approximately on the parallel 30° north of the equator. Wellington, New Zealand is approximately on the parallel 40° south of the equator. We say that New Orleans has a latitude of about 30° north. The latitude of Wellington is approximately 40° south.

Some of the parallels are given special names. The *Arctic* and *Antarctic Circles* are the parallels located about $23\frac{1}{2}°$ from the North and South Poles. The *Tropic of Cancer* is about $23\frac{1}{2}°$ north of the equator, and the *Tropic of Capricorn* is the same distance south of the equator. Portions of spheres between two parallels of latitude are sometimes called *zones.* Some of these zones are also given special names as shown in Figure 20.19.

To locate a point on earth, we name the meridian and the parallel of latitude passing through the point. Thus we name the *longitude* and the *latitude* of a point. For example: 90° west, 30° north locates a point in the city of New Orleans. We say that New Orleans is located approximately at this point on earth. Durban, South Africa is located at approximately 30° east, 30° south. Note that the longitude is always listed first.

■ **Question 9** Where is the point 90° west and 90° south?

PYRAMIDS Now we are going to take a look at some of the other familiar solid figures, those which do not lie in a plane. Our pictures of them, of course, will be in a plane—the plane of the sheet of paper you are reading—and some people find it hard to visualize a solid figure from a picture of it. We will try to draw careful figures which may help you. It would also be well for you to try to draw some of these pictures yourself. This may enable you to visualize the solid figures we are talking about. An even greater help would be to procure or make actual models of the figures we are talking about.

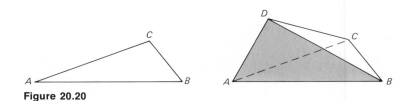

Figure 20.20

Consider the triangle ABC. It lies in a certain plane. The union of the triangle and its interior is a special case of a plane region which we call a triangular region. Select a point D which is *not* in the same plane as $\triangle ABC$. Let us suppose that D is above this page (see Figure 20.20). The line segments which can be drawn from D to points in $\overline{AB}$ all lie in the triangular region DAB. Likewise, those from D to points in $\overline{AC}$ and from D to points in $\overline{BC}$ respectively, lie in the triangular regions DAC and DBC. These four triangular regions ABC, DAB, DBC, and DAC form a *pyramid* whose *base* is the triangular region ABC, whose *vertices* are A, B, C, and D, whose *lateral faces* are DAB, DAC, and DBC and whose *edges* $\overline{AB}$, $\overline{BC}$, $\overline{CA}$, $\overline{DA}$, $\overline{DB}$, and $\overline{DC}$ outline the triangular regions. This particular pyramid is an example of the special class of pyramids called triangular pyramids, because its base is a triangular region. Any other polygon such as the quadrilateral *ABCD,* or the pentagon *PQRST* in Figure 20.21, may determine the base of a pyramid.

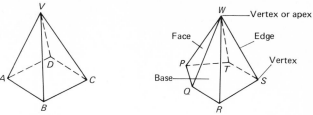

Figure 20.21

■ **Question 10**   If a pyramid were tipped upside down, would it still be a pyramid?

It should be noted that while the base of a pyramid may be any polygonal region, each lateral face is always triangular. Each region is called a *face* of the pyramid. The intersection of any two faces is a segment called an *edge* and the intersection of any three or more edges is a point called a *vertex.* The pyramid is the union of all the faces. If you think of a solid model of the situation the pyramid is the surface of the solid and not the solid itself. The distinction is much the same as the one we made before between "triangle" and "triangular region."

To summarize, a *pyramid* is a surface which is a set of points consisting of a polygonal region called the *base,* a point called the *apex* not in the same plane as the base, and all the triangular regions determined by the apex and each side of the base.

A pyramid is an example of a *simple closed surface.* There are many other simple closed surfaces which compare to our pyramid somewhat as a simple closed curve does to a triangle. We will consider some of the others such as prisms and cylinders in this unit. We have already considered the sphere.

A characteristic property of a simple closed curve in a plane was that it separated the points of the plane other than those of the curve itself into two sets, those interior to the curve and those exterior to it. In the same manner a simple closed surface divides space, other than the set of points on its surface, into two sets of points, the set of points interior to the simple closed surface and the set of points exterior to the simple closed surface. One must pass through the simple closed surface to get from an interior point to an exterior point.

■ **Question 11**   Could a pyramid intersect a sphere in exactly five points?

In a plane, we called the union of a simple closed curve and the points in its interior a *plane region.* In a similar manner, we will call the union of a simple closed surface and the points in its interior a *solid region.*

The pyramid is the surface of the solid region which it encloses. Although the word "pyramid" technically refers to the surface, it is frequently used outside of mathematics and sometimes in mathematics to refer to the solid. For instance, "the pyramids of Egypt" mean the actual stone structures and not just their surfaces. Usually the context makes clear the meaning intended.

PRISMS Consider now the surface of an ordinary closed box (see Figure 20.22a). This is a special case of a surface called a prism. The bases of this prism are the rectangles $ABCD$ and $A'B'C'D'$ which lie in parallel planes and which are congruent. The edges $\overline{AA'}$, $\overline{BB'}$, $\overline{CC'}$, etc., whose endpoints are the corresponding vertices of the bases, are all parallel to each other. They determine the *lateral faces* $ABB'A'$, $BCC'B'$, etc. Figure 20.22b shows another surface whose bases PQR and $P'Q'R'$ are parallel and lie in parallel planes. Again the edges $\overline{PP'}$, $\overline{QQ'}$, and $\overline{RR'}$ are all parallel to each other. This surface is also an example of a prism, as is that in Figure 20.22c. In Figures 20.22a and c the lateral faces are rectangles, but in Figure 20.22b they are only parallelograms. These examples lead to the general definition of a prism.

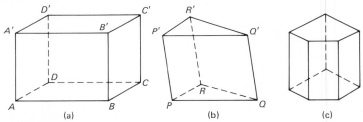

Figure 20.22

A prism is a surface consisting of the following set of *points:* Two polygonal regions bounded by congruent polygons which lie in parallel planes; and a number of other plane regions bounded by the parallelograms which are determined by the corresponding sides of the bases.

Each of the plane regions is called a *face* of the prism. The two faces formed by the parallel planes mentioned in the definition are called the *bases* and the other faces are called *lateral faces*. The intersection of two adjacent faces of a prism is a line segment, called an *edge*. The intersections of lateral faces are called lateral edges. Each endpoint of an edge is called a *vertex*. Some further examples of prisms are illustrated in Figure 20.23.

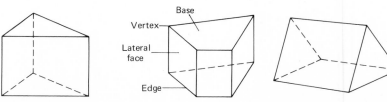

Figure 20.23

■ **Question 12** Can a prism have an odd number of vertices?

All the prisms in Figure 20.23 are examples of a special type of prism called a *right prism* in which the lateral edges are perpendicular to the base. Hence all the lateral faces are rectangles. A prism which is not a right prism is shown in Figure 20.22b.

If the polygon outlining the base is a triangle, the prism is called a *triangular prism.* A prism is a quadrangular prism if the polygon is a quadrilateral, and pentagonal if the polygon is a pentagon. The special quadrangular right prism in which the quadrilateral is a rectangle is called a rectangular prism. If the base is a square and *each* lateral face is also a square, we get the familiar *cube.* These last two are shown in Figure 20.24.

Figure 20.24

Another way to think of a prism is this. Consider any polygon such as $ABCDE$ in Figure 20.25a which lies in the horizontal plane MN. Take a pencil to represent a line segment $\overline{PQ}$, put one end of it at A, and point it upward. Move the pencil along $\overline{AB}$ keeping it always parallel to the original position as in Figure 20.25b. Then move it along $\overline{BC}$ still parallel to the original position and so on around the polygon. The pencil itself determines a surface and the upper tip of it outlines a polygon $A'B'C'D'E'$ congruent to $ABCDE$. These two polygonal regions and the surface determined by the moving pencil make up the prism.

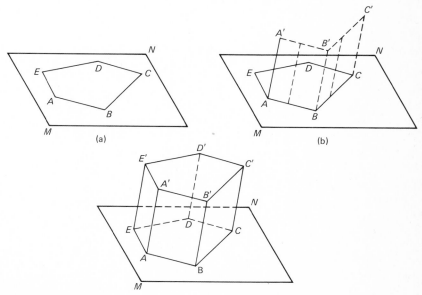

Figure 20.25

CYLINDERS AND CONES

A cylinder is defined in a manner very similar to the way we defined a prism, except that the bases are regions bounded by congruent simple closed curves instead of polygons. Thus, in a general sense, the prism is just a special case of a cylinder. A line segment which connects two corresponding points in the curves bounding the bases is called an *element* of the cylinder.

In Figures 20.26b and 20.26c, the simple closed curve is a circle and the cylinder is called a circular cylinder. If an element is perpendicular to the plane containing the curve, we get a *right* cylinder. Common examples are, of course, a tin can or a hat box. A can of beans is a good model of a right circular cylinder while a can of sardines is a good model of a right cylinder which is not usually circular.

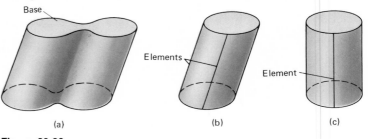

Figure 20.26

Figure 20.27

■ **Question 13**   In reference to Figure 20.27, could this region be the base of a cylinder?

CONES

A cone is related to a cylinder as a pyramid is to a prism. A *cone* is a surface which is a set of points consisting of a plane region bounded by a simple closed curve, a point called the vertex not in the plane of the curve and all the line segments of which one endpoint is the vertex and the other any point in the given curve. This differs from the definition of a pyramid only in the fact that we have changed "polygon" to "simple closed curve." Thus, in a general sense, a pyramid is a special type of cone.

If the simple closed curve is a circle, we get a *circular cone*. A cone has a base, a lateral surface, and a vertex. The most familiar cone is the one (illustrated in Figure 20.28b) whose base is bounded by a circle and in which a line drawn from the vertex to the center of the base is perpendicular to the plane in which the base lies.

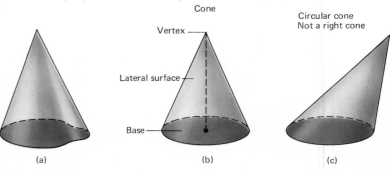

Figure 20.28

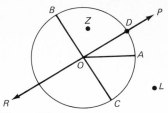

Figure P20.1

PROBLEMS

1 Given the circle in Figure P20.1:
  (a) Name two radii of the circle.
  (b) Name a diameter of the circle.
  (c) Is point Z in the interior or exterior region?
  (d) Is point L in the circular region?
  (e) Are $\overline{OA}$ and $\overline{OC}$ congruent?
  (f) Name three different arcs with endpoint A.

2 How many tangents do you find in each part of Figure P20.2?

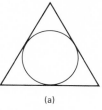

    (a)   (b)   (c)

Figure P20.2

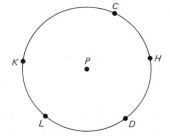

Figure P20.7

3 O is the center of a circle and $\overline{OA}$ is a radius. If $\overline{OA} > \overline{OB}$ does B lie in the circle or in the interior of the circle?

4 Two circles are given both with center O. $\overline{OA}$ is a radius of one and $\overline{OB}$ a radius of the other. If $\overline{OA} < \overline{OB}$, what is true about the circles?

5 Fold a piece of paper twice to get a model of a right angle. How many right angles can you fit together side by side around a point in the plane?

6 For any one circle, what is the intersection of all its diameters?

7 Use Figure P20.7 in answering the following:
  (a) Point L separates $\overset{\frown}{CLH}$ into two arcs. Name these two arcs.
  (b) Does point L separate the circle into two arcs?

8 Use your answers in Problem 7, above, to answer the following:
  (a) Does a point on an arc separate the arc into two arcs?
  (b) On an arc, must a point which is not an endpoint of the arc be between two points of the arc?
  (c) Does an arc have a "starting" point and a "stopping" point? If so, what are they called?
  (d) For an arc, are the notions of betweenness and separation more like those of a line segment or of a circle?

9 (a) Make a drawing of a sphere.
  (b) Label four points of the equator in diametrically opposite pairs.
  (c) Dot in the segments joining the diametrically opposite pairs in part b.
  (d) Draw two small circles, one of which intersects the equator and one of which does not. Label their centers.

10 (a) Is there an antipode of any given point on a sphere?
  (b) Is there more than one antipode of any given point on a sphere?

11 (a) How many great circles pass through a given point, such as the North Pole, of a sphere?
  (b) How many small circles pass through a given point of a sphere?
  (c) Can a small circle pass through a pair of antipodal points on a sphere?

12 (a) On a sphere, does every small circle intersect every other small circle?
  (b) On a sphere, does every great circle intersect every other great circle?

13 Identify the circle of latitude and circle of longitude for P on the globe in
Figure P20.13. Indicate which angle measures the latitude of P. If NQS
is the prime meridian indicate the angle which measures the longitude of
P.

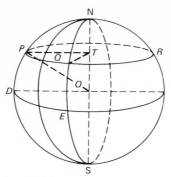

Figure P20.13

14 (a) In how many points does each meridian cut the equator?
  (b) In how many points does each meridian cut each parallel of latitude?
  (c) Does a parallel of latitude intersect any other parallel of latitude?
15 Greenwich, England is located approximately on the parallel of latitude
  labeled 52° north. Without obtaining further information, write the location
  of Greenwich.
16 Chisimaio, Somalia, in eastern Africa, is located on the equator (or very
  near the equator). It is about 42° east of Greenwich. Without using a reference, write the location of Chisimaio.
17 (1) Which of the drawings in Figure P20.17 are of pyramids?
  (2) Indicate the base, a face different from the base, and a vertex for each
    pyramid.

(a)   (b)   (c)

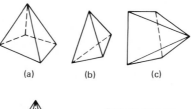

(d)   (e)

Figure P20.17

18 (a) Sketch a prism with a square base.
  (b) Sketch a triangular prism.
  (c) Sketch a cylinder which is not a right cylinder.
  (d) Sketch a pyramid with a base which is a quadrilateral.
  (e) Sketch a cone.
  (f) Sketch a sphere.

19 Name the solids pictured in Figure P20.19.

20 Point out one example, somewhere in the above figures, of each of the following: base, edge, vertex, center, radius, element, lateral face, diameter.

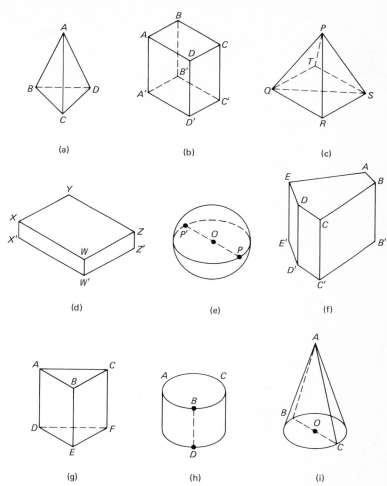

(a)

(b)

(c)

(d)

(e)

(f)

(g)

(h)

(i)

Figure P20.19

21.

CONGRUENCE

In Chapter 4 we talked about congruence of line segments and in Chapter 10 about congruence of angles. In this chapter we shall extend the notion of congruence from congruence of line segments and angles to congruence of triangles. We discuss the question: Under what conditions are two triangles congruent?

CONGRUENCE OF GEOMETRIC FIGURES

Two figures are *congruent* if a copy of one can be fitted exactly on the other. For example, take a sheet of tracing paper and make a tracing of the left-hand figure in Figure 21.1. Then see if this tracing can be exactly fitted on the right-hand figure. Remember that "fitted exactly" means that each point of the tracing covers a point of the second figure and there are no points left over.

In practice, of course, we can never make a perfect copy of a figure. Whenever we conclude that two figures are congruent by use of a tracing we mean that this is true as nearly as we can tell with our imperfect equipment. Sometimes we emphasize this by saying that two figures "seem" to be congruent.

Certain pairs of figures are always congruent: any two points are congruent, and so are any two lines and any two planes. Some figures are never congruent: a tracing of a segment can never fit exactly a ray or a line or a circle. Other figures are sometimes congruent and sometimes not: some pairs of segments are congruent and some are not, some pairs of angles are congruent and some are not.

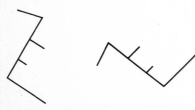

Figure 21.1

Figure 21.2

■ **Question 1** In Figure 21.2, is $\overrightarrow{PQ}$ congruent to $\overrightarrow{QR}$?

Much of classical (Greek) geometry is concerned with finding out whether or not two figures can be fitted without actually having to make a copy of one of them. Many applications of geometry to architecture, engineering, surveying, navigation, and other subjects depend upon the idea of "fitting figures."

One application is illustrated by the following story. In ancient Greece there were two towns separated by a mountain. It was desired to build a *straight* road joining the towns by tunneling through the mountain. The towns agreed to start tunneling on both sides of the mountain. The first problem was to locate points P and Q so that the points A, P, Q, and B were collinear (on the same straight line). The towns A and B, the mountain, and possible locations of P and Q are shown in Figure 21.3.

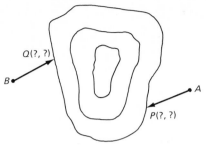

Figure 21.3

The ancient Greeks used their knowledge of geometry to solve this problem in the following way. First they picked a point C from which both towns A and B were visible. Then line segments were drawn connecting C with A and B.

On the line $\overleftrightarrow{AC}$ they located point D so that the distances from C to D and from C to A were the same. Similarly they located point E on the line $\overleftrightarrow{BC}$ so that the distances from C to E and from C to B were equal. The segment $\overline{DE}$ was drawn as shown in Figure 21.4.

The ray $\overrightarrow{AP}$ was drawn so that $\angle A$ was a true copy of $\angle D$. Likewise, the ray $\overrightarrow{BQ}$ was drawn so that $\angle B$ was a true copy of $\angle E$.

Two construction teams started tunneling at P and Q on the same level in the directions indicated by the rays $\overrightarrow{AP}$ and $\overrightarrow{BQ}$. Sure enough, they met in the middle of the mountain.

Later in this chapter we shall discuss the geometric ideas which led the ancient Greeks to this solution of this problem.

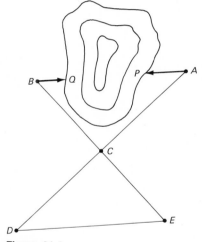

Figure 21.4

■ **Question 2** Which parts of Figure 21.4 are copies of which other parts?

We shall use the idea of congruence so often that it will be convenient to have a shorthand way of expressing it. In the drawing below you can verify, by use of tracing paper[1], that Figure 21.5a seems congruent to Figure 21.5b. We introduced the symbol "≅" as a shorthand for the phrase "is congruent to" back in Chapter 4. Thus, instead of writing out in words the statement, A is congruent to B, we can write $A \cong B$.

(a)   (b)

Figure 21.5

---

[1] Notice that you have to turn the tracing paper over, reversing top and bottom. In Figure 21.1, you did not need to do so.

In Figure 21.6 you can verify that segment AB seems to be congruent to segment $\overline{PQ}$. This information can be expressed by the statement $\overline{AB} \cong \overline{PQ}$. Now notice that when two segments are congruent, they are congruent in two ways. Let $\overline{A'B'}$ be a tracing of $\overline{AB}$. If we have a fitting that places A' on P and B' on Q this is a congruence matching A with P and B with Q. By turning the tracing around, however, we can also fit the tracing on $\overline{PQ}$ in such a way that A' fits on Q and B' fits on P so this second fitting is a congruence matching A with Q and B with P. When we write the statement $\overline{AB} \cong \overline{PQ}$ we will agree that we are thinking of the particular congruence which matches A with P and B with Q. That is, the statement $\overline{AB} \cong \overline{PQ}$ means

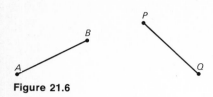
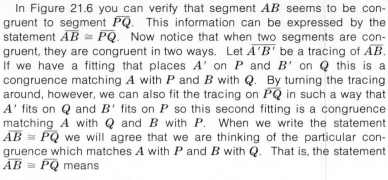

$\overline{AB}$ is congruent to $\overline{PQ}$ and $A \leftrightarrow P$, $B \leftrightarrow Q$

It would be equally correct to write $\overline{AB} \cong \overline{QP}$, but this time we would be thinking of the congruence which matches A with Q and B with P. This agreement about which matching is intended will be of importance later when we consider triangles.

Figure 21.6

Figure 21.7

■ **Question 3** The two angles in Figure 21.7 are congruent. In how many ways can they be matched?

CONSTRUCTING CONGRUENT FIGURES

So far we have been deciding whether two given figures are congruent by fitting a tracing of one on the other. It often happens that instead of having two figures we have one figure and want to *draw* a congruent copy of it in another position. How can this be done? You can, of course, make a tracing of the given figure and move it to the desired location. By going over the tracing firmly with a pencil you should be able to see the slight dents in the paper below and can then go over these with a pencil to get the desired congruent copy. Try this process of making a congruent copy of the pentagon shown in Figure 21.8.

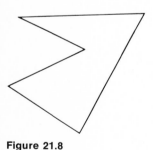

Figure 21.8

Perhaps it has occurred to you that for a polygon, or for any figure consisting only of segments, there would be a more satisfactory process. After moving the tracing to its new position one could take a pin and punch down through into the paper below at each corner. It would then only be necessary to use a straightedge to draw the segments joining the pinholes. Try this method of producing a congruent copy of the polygon above. Of course no procedure actually gives a perfect congruent copy, but it is clear what we have in mind.

The figures that we most often want to reproduce are segments and angles. It is worthwhile to consider special ways of making congruent copies of them.

■ **Question 4** Do we need to make a tracing of a circle in order to draw a congruent copy of it in another position?

Figure 21.9

Let $\overline{AB}$ be a given segment and suppose we want a congruent copy of $\overline{AB}$ with one end at the given point P (see Figure 21.9). Take a straightedge of cardboard or paper. (If you have none available, you can make one simply by making a fold in a sheet of paper.) Lay the straightedge along segment $\overline{AB}$ as shown in Figure 21.10 and make marks at A and B. If the straightedge is moved so that one of the marks is at P, it is only necessary to trace along the edge between the two marks to obtain the desired congruent image PQ of $\overline{AB}$. Here the piece of the straightedge between the marks plays the role of the tracing in our original procedure.

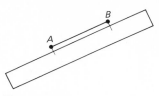 

Figure 21.10

In Chapter 4 we used a compass to compare two segments. We can also use a compass to make a copy of $\overline{AB}$. We place the point of the compass at A and adjust the opening until the pencil point is at B (see Figure 21.11). Now draw any line through P. Without changing the opening of the compass, place the point at P and draw an arc which intersects the line at another point Q. Since the opening of the compass was not changed, $\overline{AB}$ and $\overline{PQ}$ are congruent (see Figure 21.12).

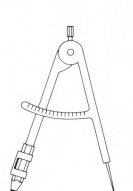

Figure 21.11

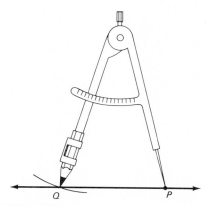

Figure 21.12

We can also devise an instrument for copying angles, just as we used the straightedge for copying segments. On a piece of light cardboard draw a circle and a line through the center, dividing the circular region into two congruent pieces. Cut out the figure so you have half a circular disk as shown (in Figure 21.13). Now let $\angle ABC$ be an angle for which we want a congruent image. Let the half-disk be placed on the angle in such a way that the center of the disk is at the vertex B of the angle. Make marks on the disk at the points where rays $\overrightarrow{BA}$ and $\overrightarrow{BC}$ meet the circle (see Figure 21.14). If then the disk is

Figure 21.13

moved to some new position it only is necessary to make dots by the center of the disk and by the two marks on the disk and draw the indicated rays. In this case the center and the two marks on the disk serve the same purpose as the tracing in our original procedure.

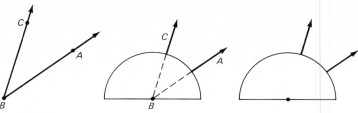

Figure 21.14

■ **Question 5** In Figure 21.9, how many segments can be drawn which are congruent to $\overline{AB}$ and have P as one end?

We saw in Chapter 11 that the idea of congruent angles and congruent segments may be used to define the familiar terms, rectangle and square. A rectangle is a quadrilateral with all four angles congruent. A square is a rectangle having all four sides congruent.

CONSTRUCTING CONGRUENT TRIANGLES In the problem of the ancient Greeks building a tunnel, we saw an application of making a copy of a triangle. We shall now study the problem of copying triangles and see just how much information about a triangle must be known in order to make a copy of it.

Use a tracing of $\triangle ABC$ to make a copy of $\triangle ABC$ in a new position. $\triangle A'B'C'$ as shown in Figure 21.15 is a copy of $\triangle ABC$ and $\triangle ABC \cong \triangle A'B'C'$. We shall consider, however, the problem of constructing a congruent copy of a triangle *without* making a tracing.

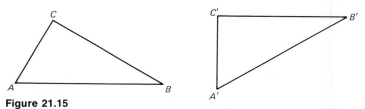

Figure 21.15

Let us start with the same triangle $\triangle ABC$ and a segment $\overline{A'B'}$ where $\overline{A'B'} \cong \overline{AB}$ (see Figure 21.16). To complete the construction of the triangle $\triangle A'B'C'$ we need simply to locate point C' because when the location of C' is found, the triangle $\triangle A'B'C'$ can be drawn.

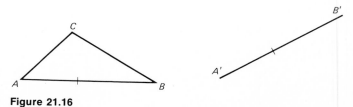

Figure 21.16

Three methods for locating C' will be shown. These methods involve making congruent copies of some, but not all, of the sides and angles of $\triangle ABC$.

■ **Question 6** In Figure 21.15, how many parts, segments, and angles of $\triangle A'B'C'$ are congruent to parts of $\triangle ABC$?

FIRST METHOD

The first method for locating C' consists of drawing a ray from A' so that $\angle A' \cong \angle A$. Point C' must be on this ray (see Figure 21.17). We locate C' on this ray so that $\overline{A'C'} \cong \overline{AC}$. Then $\triangle A'B'C'$ is completed by drawing $\overline{B'C'}$.

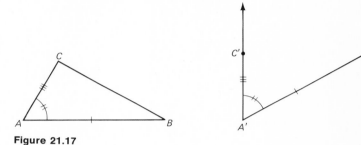

Figure 21.17

To copy a triangle, then, it is only necessary to be able to make congruent copies of two sides and the included angle. $\angle CAB$ is said to be included between the sides $\overline{AC}$ and $\overline{AB}$ because both these sides are contained in the angle.

Since this information is enough to determine the congruent copy $\triangle A'B'C'$, the following property is now stated.

SAS (Side-Angle-Side) Congruence Property
If two sides and the included angle of one triangle are congruent to two angles and the included side of another triangle, then the triangles are congruent.

SECOND METHOD

Start again with $\triangle ABC$ and $\overline{A'B'}$. Draw from A' a ray so that $\angle A' \cong \angle A$. Then draw from B' a ray so that $\angle B \cong \angle B'$ (see Figure 21.18). Point C' must lie on both rays and therefore is the point of intersection of the two rays.

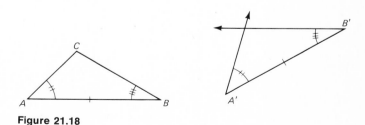

Figure 21.18

In this case, to copy a triangle it is only necessary to make congruent copies of two angles and the included side. $\overline{AB}$ is said to be included between $\angle A$ and $\angle B$ because both these angles contain the side. Since this information is enough to determine the congruent copy $\triangle A'B'C'$, the following property is now stated.

ASA (Angle-Side-Angle) Congruence Property
If two angles and the included side of one triangle are congruent to two angles and the included side of another triangle, then the triangles are congruent.

■ **Question 7** In Figure 21.18, will the ray from B' always meet the ray from A'?

THIRD METHOD This time we shall use congruent copies only of the sides of $\angle ABC$ and none of the angles.

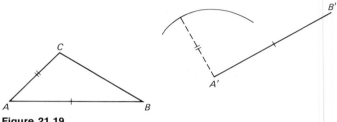

Figure 21.19

In Figure 21.19, point C' must lie on the circle with center at A' and radius congruent to $\overline{AC}$. Similarly, point C' must lie on the circle with center at B' and radius congruent to $\overline{BC}$. Since C' lies on both these circles, then it must be the point of intersection of the arcs shown in Figure 21.20.

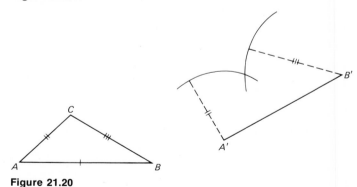

Figure 21.20

Using this method to copy a triangle, it is only necessary to be able to make copies of the three sides. For this reason the following property is now stated.

SSS (Side-Side-Side) Congruence Property
If the three sides of one triangle are congruent to the three sides of another triangle, then the triangles are congruent.

You will have noticed that we have confined our construction to one particular half-plane determined by line $\overleftrightarrow{A'B'}$.

By using the opposite half-plane we could, of course, construct a triangle congruent to $\triangle ABC$ which is a mirror image of $\triangle A'B'C'$.

For the second method it is essential that we use only one half-plane. Otherwise the constructed rays would not intersect (see Figure 21.21).

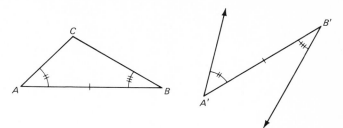

Figure 21.21

■ **Question 8** In each of these three methods, how many parts of the original triangle need to be copied in order to draw a congruent triangle?

CORRESPONDING PARTS OF CONGRUENT TRIANGLES

We have been using A', B', C' as names for the vertices of a triangle that is congruent to $\triangle ABC$ in a specially convenient way. If a copy of $\triangle ABC$ were fitted on $\triangle A'B'C'$, then the vertices A and A', B and B', C and C' would match or correspond. They are called *corresponding vertices*.

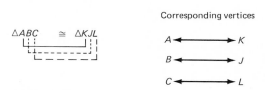

Figure 21.22

The corresponding vertices of two congruent triangles are named by the order they are written in a congruence statement as shown in Figure 21.22. From the corresponding vertices, we can determine the corresponding sides and angles as follows:

$$\triangle ABC \cong \triangle KJL$$

Corresponding sides	*Corresponding angles*
$\overline{AB} \cong \overline{KJ}$	$\angle A \cong \angle K$
$\overline{AC} \cong \overline{KL}$	$\angle B \cong \angle J$
$\overline{BC} \cong \overline{JL}$	$\angle C \cong \angle L$

Thus a congruence statement gives us six facts at once. It is helpful to show the six corresponding congruent parts by the markings as shown in Figure 21.23.

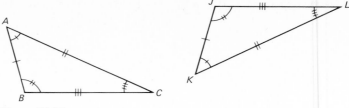

Figure 21.23

We have here agreed that when we write $\triangle ABC \cong \triangle KJL$ it means that A corresponds to K, B corresponds to J, and C corresponds to L. Remember that we made the same agreement for segments and angles. For example, to write $\overline{LM} \cong \overline{PQ}$ means that the segments are congruent with L corresponding to P and M to Q. In the case of segments this was not very important, for if a tracing of $\overline{LM}$ fits on $\overline{PQ}$ with L on P and M on Q, it is only necessary to turn the tracing around to have a fitting in which L fits on Q and M on P. That is, whenever $\overline{LM} \cong \overline{PQ}$ is true, then $\overline{LM} \cong \overline{QP}$ is also true, so it really does not matter which we write. The same applies to angles.

However, in talking about triangles it is vital to get the matching right. The situation in the figure above is correctly represented by the statement $\triangle ABC \cong \triangle KJL$. But suppose instead we write $\triangle ABC \cong \triangle JKL$. What would this say? Among other things it would assert that $\overline{AC} \cong \overline{JL}$ and $\overline{BC} \cong \overline{KL}$, both of which are false statements. Thus it will be important to write exactly what we mean.

■ **Question 9** Are there any triangles for which $\triangle ABC \cong \triangle ACB$?

SOME APPLICATIONS

Here are two applications of what we have learned about congruence of triangles. In the first application, the problem is to find how far it is from a point A on the shore of a lake to a buoy B in the lake. We can do this by selecting a point C on the shore as shown in Figure 21.24. Next we can locate a point B' such that $\triangle ABC \cong \triangle AB'C$. Once B' is found, then we know $\overline{AB} \cong \overline{AB'}$. Since $\overline{AB'}$ is conveniently on dry land, we can measure it at our leisure.

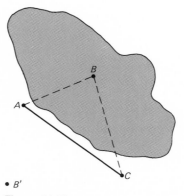

Figure 21.24

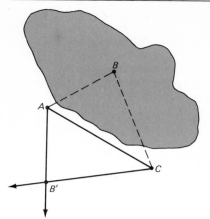

Figure 21.25

To find B', from A draw a ray which forms with $\overrightarrow{AC}$ an angle congruent to $\angle BAC$.

From C draw a ray which forms with $\overrightarrow{CA}$ an angle congruent to $\angle BCA$.

The intersection of these two rays is at B' (Figure 21.25).

$\triangle ABC \cong \triangle AB'C$ by the ASA property.

The second application is to the problem of drawing a copy of an angle. So far whenever you have copied an angle you have made a tracing or used your semicircular disk. Let us see now whether it is possible to get along without either of these. The game is to see if we can copy an angle without using any instruments but a compass and straightedge.

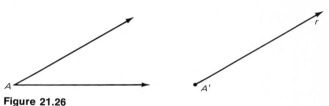

Figure 21.26

The problem is to make a congruent copy of $\angle A$ with vertex at A' and ray r as one side, using only your compass and straightedge (see Figure 21.26). Open your compass to an arbitrary amount and draw an arc with center at A intersecting the sides at B and C. Use the same compass setting and draw an arc with center at A' intersecting ray r at B' as shown in Figure 21.27. Now adjust the compass so that the tips coincide with B and C. With this setting, draw an arc with center at B'. The intersection of the two arcs is at C' (see Figure 21.28). Complete the construction by drawing the ray $\overrightarrow{A'C'}$. The SSS congruence property shows that $\triangle ABC \cong \triangle A'B'C'$ and therefore $\angle BAC \cong \angle B'A'C'$.

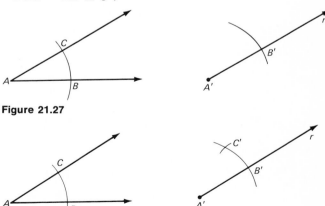

Figure 21.27

Figure 21.28

MOTIONS Now that we have learned how to check two triangles for congruence without using tracings of them, let us go back to congruence of more complicated figures. Our original idea about congruence was that we could compare two geometric figures, to see if they were congruent, by making a tracing of one and then checking to see if the tracing could be fitted exactly on the other. We said nothing, however, about *how* the tracing was moved from one figure to the other. Of course, in the end it does not matter. We could, for example, merely carry a tracing from one side of the room to the other or, instead, we could first send it around the world by airmail. As long as the tracing can be exactly fitted over the second figure when we get it there, the two figures are congruent.

Nevertheless, there are three very simple ways of moving a tracing from one part of a plane to another that are worth discussing.

SLIDES

To describe this first type of motion, we start with a line segment in the plane with one end of it specially marked. We will use an arrowhead to indicate which end has been marked, and we will call the result an *arrow*. Figure 21.29 shows an arrow and also (broken) the line, ℓ, that contains the segment.

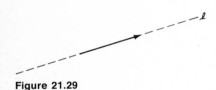

Figure 21.29

Suppose we have placed a large piece of tracing paper on the plane. Make a tracing of the arrow. Then slide the tracing paper along, keeping the tracing of the arrow on top of ℓ until the tail of the arrow trace is on the head of the original arrow while the tip of the arrow trace is still on the broken line. This completely describes a way of moving the tracing paper. This motion is called a *slide*.

■ **Question 10** If a point is not on the line containing the arrow, can its trace end up on the line after the slide?

Let us see what this operation does to a figure in the plane (see Figure 21.30). Consider the left-hand figure consisting of a circle and a pair of segments. When we lay the tracing paper on its original position, we copy this figure and point T on the tracing paper. When the tracing paper is slid along $\overrightarrow{TT'}$ to its new position, the tracing covers the right-hand figure. Obviously the two figures are congruent. We call the second figure the *image* of the first.

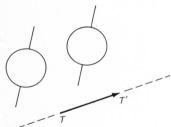

Figure 21.30

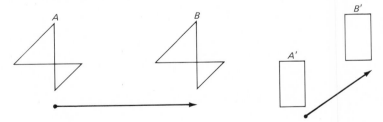

Figure 21.31

In Figure 21.31 we have two more pairs, each of which consists of congruent figures as can be seen from the slides specified by the arrows. We can see that if we make a trace of a point, and then perform a slide which takes the tracing of P to a point P', the image of P, then the line segment joining P to P' is parallel to the arrow which

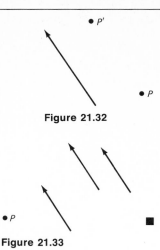

Figure 21.32

Figure 21.33

specified the slide, and $\overline{PP'}$ and the arrow have the same length (see Figure 21.32).

We know all about a slide as soon as we are given an arrow, so we can say a slide is described by an arrow. The length of the arrow gives the distance the tracing of any point moves and the direction of the arrow shows the direction of the motion.

An arrow should not be confused with a ray, even though we have similar representations of them. They both have a starting point and a direction but an arrow has a terminal point while a ray does not.

Arrows which have the same length and the same direction show the same slide. Thus, each of the arrows in Figure 21.33 determines the same slide. Check this by locating where the tracing of P lands for the slides specified by each of these arrows.

■ **Question 11** A slide carried the tracing of A to the point A' (see Figure 21.34). Where did the tracing of B land? Where was the point whose tracing landed at C?

Figure 21.34

TURNS

For the second kind of motion, we start with an angle together with a curved arrow pointing from one ray of the angle to the other. Figure 21.35 shows three such configurations. Now we suppose that the plane is covered with tracing paper. We trace a copy of the angle. Then we stick a pin through the tracing paper into the vertex of the angle. Next we turn the tracing paper around the pin in the direction, clockwise or counterclockwise, indicated by the arrow. We turn until the tracing of the ray from which the arrow starts ends up on the other ray.

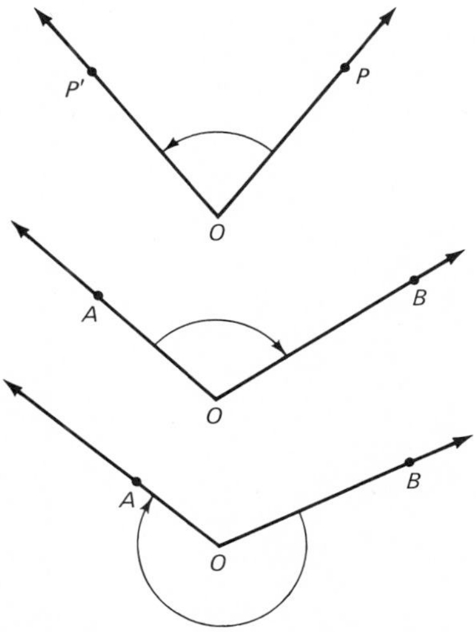

This kind of motion is called a *turn* or a *rotation*. If we start with a figure in the plane and make a tracing of it, after a turn the tracing will cover another figure in the plane, which is congruent to the original one. We call the second figure the *image* of the first. Figure 21.36 shows a figure (solid lines), a turn, and the congruent image (broken lines) where the tracing of the original figure lands. Note that no slide could carry a tracing of the figure into the congruent figure. (A slide always takes a line segment into a parallel line segment.)

Figure 21.35

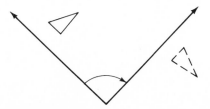

Figure 21.36

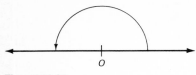

Figure 21.37

When we defined an angle, we required the two rays to be on different lines. For a turn, we do not insist on this. Thus the configuration shown in Figure 21.37 defines a turn.

We saw in Figure 21.33 that different arrows could specify the same slide if they were parallel, congruent, and pointed in the same direction. A similar relationship holds for turns. Two configurations specify the same turn if the angles have the same vertex, if they are congruent, and if the curved arrows point in the same direction, as in Figure 21.38.

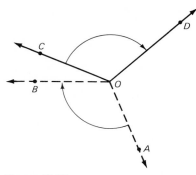

Figure 21.38

■ **Question 12** The two angles in Figure 21.39 are congruent. Do these configurations specify the same turn?

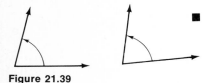

Figure 21.39

In the case of a slide, we saw that the copy on the tracing paper of a point traces out a line segment parallel to the arrow. In the case of a turn, the copy on the tracing paper of any point traces out an arc of a circle whose center is at the vertex of the angle.

FLIPS The third kind of motion is called a *flip*. We start with a line in the plane and a tracing of the line on our tracing paper. We then pick the paper up, turn it over, and put it back down on the plane with the tracing of the line back where it started. However, it is not enough that the tracing of the line ends up where it started. The tracing of each point in the line should end up where it started. Marking one particular point on the line makes it easy to check on this.

Another way of thinking about a flip is to imagine that this page of this book represents part of our tracing paper. When you turn this page you are flipping it around the spine of the book.

The line around which we do the flipping is called the *axis* of the flip. Figure 21.40 shows a figure (solid lines), the axis (broken line), and the position of the tracing, or image, (broken lines) after the flip.

As was the case for the other two kinds of motions, any figure in the plane is congruent to the figure covered by the tracing after it has been flipped.

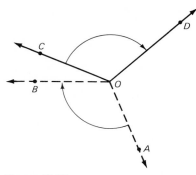

Figure 21.40

■ **Question 13** Can two different lines ever specify the same flip?

The location of the axis of a flip does make a difference. In Figure 21.41 we see a figure, an axis, and the location of the tracing of the figure after a flip around the line.

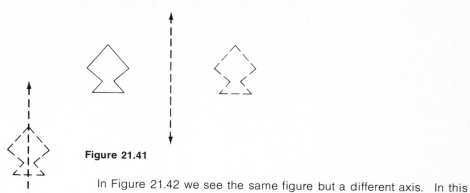

Figure 21.41

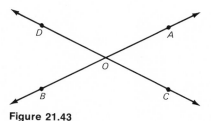

Figure 21.42

In Figure 21.42 we see the same figure but a different axis. In this case the tracing after the flip matches the original figure exactly. Whenever this happens, we say that the figure is *symmetric* with respect to the line, and we say that the axis of the flip is a *line of symmetry* for the figure.

CONGRUENT ANGLES

Finally, there are two facts about the congruency of angles which fall out of our discussions of slides and turns. Suppose we have two lines, $\overleftrightarrow{AB}$ and $\overleftrightarrow{CD}$ which intersect, as in Figure 21.43, at a point O. There are four angles which have O for a vertex, and they can be arranged in two pairs of "opposite" angles. One pair is

$\angle AOC$ and $\angle BOD$

Another pair is

$\angle AOD$ and $\angle BOC$

The technical term for pairs of "opposite" angles such as these is *vertical* angles.

Figure 21.43

Notice in this figure that the two angles in a pair seem to be congruent. You should check this by making a tracing of one of the angles in a pair and matching the tracing with the other angle of the pair. Do this also with a few other pairs of intersecting lines.

In fact, it is true that the two angles in any pair of vertical angles are always congruent. You will see in Problem 9 below how this fact is

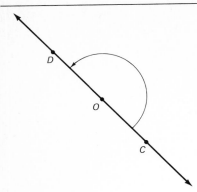

Figure 21.44

needed to justify the procedure the ancient Greeks used to locate their tunnel. To see why this is so, think of the turn shown by Figure 21.44. For this turn, the image of $\angle COA$ is $\angle DOB$, which confirms that these "vertical" angles are congruent. Similarly, the image of $\angle AOD$ is $\angle BOC$, so these two angles also are congruent.

The other fact about the congruency of angles has to do with the situation in which a line intersects each of two parallel lines, as in Figure 21.45. (Parallel lines were discussed in Chapter 3.) In this diagram it looks as though any pair of corresponding angles at O and O' (for example, the upper angles on the right, as marked) are congruent. This is in fact the case, but you should check this by comparing tracings of the angles not only in this diagram, but in some others as well.

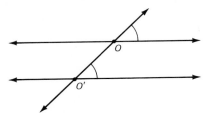

Figure 21.45

To justify this observation, think of the arrow from O to O'. We see that the image, for this slide, of the marked angle at O is the marked angle at O', which confirms that these two angles are congruent.

As we shall see, this fact about parallel lines and congruent angles has a number of uses.

■ **Question 14** What is another angle with vertex at O which is congruent to the marked angle at O' in Figure 21.45?

MOTIONS AND CONGRUENCE

We have noticed that any motion of any of these three kinds will carry a tracing of a geometric figure onto another figure that is congruent to the original one. Suppose we carry out any number of these motions, one after the other. Then the tracing, in its final position, will again cover a figure congruent to the original.

What is perhaps more interesting is that the above statement can be turned around. In fact, if two figures in the plane are congruent, then a tracing of one can be moved over to where it exactly fits the other in just *two* steps, each of which is either a slide, a turn, or a flip.

Unfortunately, the job of demonstrating how this can be done is quite lengthy. The fact that it is so, however, explains why these three kinds of motions are worth studying.

PROBLEMS

1 Each pair of triangles in Figure P21.1 has markings indicating congruent parts. If the triangles are congruent, state the congruence and identify the congruence property used.

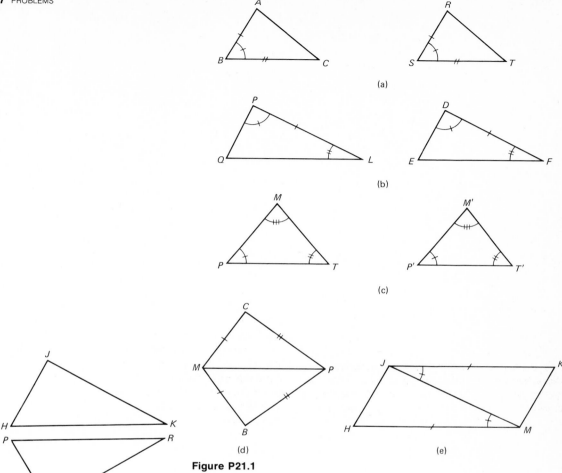

Figure P21.1

Figure P21.2

2 Given that $\triangle JHK \cong \triangle QPR$ in Figure P21.2, use marks to indicate the six pairs of corresponding congruent parts.

3 In each part of Figure P21.3, name an additional pair of congruent parts, such that if you know that these parts are congruent, you could say that the triangles are congruent. Write the congruence and identify the congruence property used. (There may be more than one correct solution.)

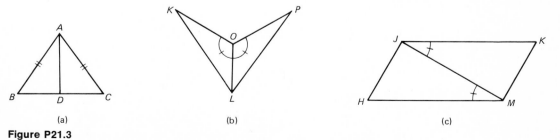

Figure P21.3

4   In each case, answer the questions on the basis of the given information in reference to Figure P21.4.

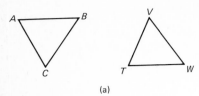

(a)

Figure P21.4

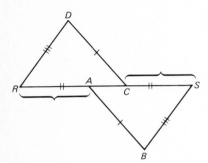

Figure P21.7

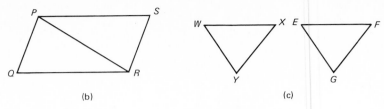

(b)

(c)

(a) $\overline{AC} \cong \overline{VW}$ $\angle A \cong \angle W$ $\overline{AB} \cong \overline{TW}$
Are the triangles congruent?   Why?
(b) $\overline{PQ} \cong \overline{SR}$ $\overline{QR} \cong \overline{PS}$
Are the triangles congruent?   Why?
(c) $\angle W \cong \angle F$ $\angle X \cong \angle E$ $\overline{WX} \cong \overline{EF}$
Are the triangles congruent?   Why?

5   We know that two segments are either congruent or a copy of one is wholly contained in the other. Is this true for triangular regions? For square regions? For circular regions?

6   Sometimes sets of points may be congruent in more than one way. In how many different ways can you fit a tracing of a point to a point? A line to a line? A ray to a ray? A segment to a congruent segment?

7   The markings (Figure P21.7) indicate corresponding congruent segments. Are the triangles congruent?   Why?

8   (a)   In reference to Figure P21.8a, can you find $\triangle ABC$ whose three sides are congruent to the given segments? If not, what seemed to go wrong?

_____

_____

_____

Figure P21.8a

(b)   Figure P21.8 was constructed by drawing the angle at A, choosing the point B so that the segment AB has a certain length, and then drawing a circle, with a given radius and with its center at B. The circle intersects the other side of the angles in the two points C and D. Notice that

$$\angle A \cong \angle B \qquad \overline{AB} \cong \overline{AB} \qquad \overline{BC} \cong \overline{BD}$$

Do you think that there is an Angle-Side-Side Congruence Property?

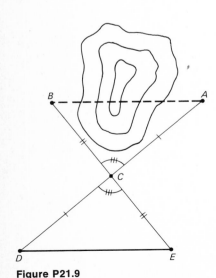

Figure P21.9

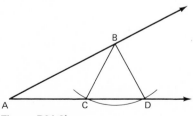

Figure P21.8b

9   Return to the tunneling problem of the ancient Greeks and show why their procedure worked (see Figure P21.9). (Note: In the figure, it is indicated that $\angle ACB$ and $\angle DCE$ are congruent. Why?)

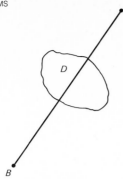

Figure P21.10

10 It is desired to measure the length of AB but a deep ditch at D intervenes. Show how to use congruent triangles to solve this problem (see Figure P21.10).

11 In Figure P21.11, draw an arrow for the slide in which.
(a) figure B is the final position of the tracing of figure A.
(b) figure A is the final position of the tracing of figure B.

Figure P21.11

12 For the slide shown by the arrow (Figure P21.12)
(a) Find the image of polygon P.
(b) Find the polygon which has polygon P as its image.

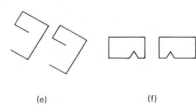

Figure P21.12

13 For which pairs of figures (Figure P21.13) is there a slide in which the second figure is the image of the first? If there is, show the slide with an arrow.

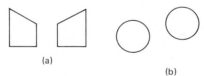

(a)

(b)

(c)

(d)

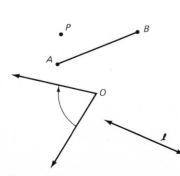

(e)

(f)

Figure P21.13

14 For the turn shown at O (see Figure P21.14), find
(a) P', the image of P (b) $A'B'$, the image of AB
(c) ℓ', the image of ℓ
Draw the arcs of circles which connect P with P', A with A', and B with B'.

Figure P21.14

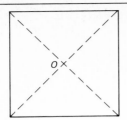

Figure P21.15

15 Given the square shown in Figure P21.15, with point O at the intersection of the diagonals:

   Describe three different turns about O for which the square is its own image.

16 Find the images of Figures P21.16a, b, c, and d for the flip whose axis is ℓ.

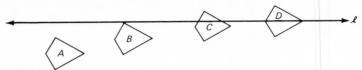

Figure P21.16

17 Find all the lines of symmetry for each part of Figure P21.17.

    (a)   (b)   (c)   (d)

Figure P21.17

18 How many lines of symmetry does a circle have?

19 (a) In Figure P21.19a, show exactly how to map figure 1 into figure 2 by two motions, slides, turns, or flips, or a combination. You should find at least three answers.

    (b) Do the same for Figure P21.19b.

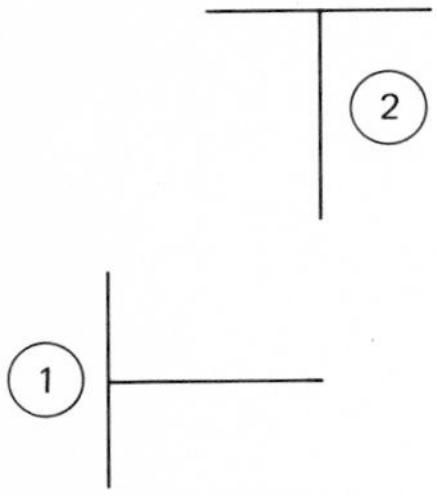

Figure P21.19a

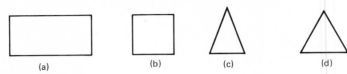

Figure P21.19b

20 Think of each figure in Figure P21.20 as a design on a wall and assume that it extends infinitely in both directions. Describe a motion or pair of motions for which the image of the figure is exactly the same figure. Find two answers.

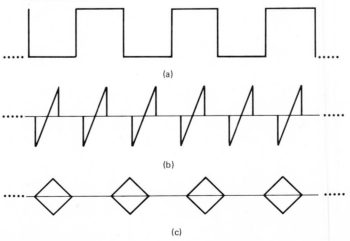

(a)

(b)

(c)

Figure P21.20

310

22. SIMILARITY

SIMILAR FIGURES Any two geometric figures which are congruent certainly have the same size, and the same shape. But there are many cases in the world around us of pairs of figures which seem to have the same shape but different sizes. A very familiar example would be a photograph and an enlargement of it, as sketched in Figure 22.1.

Figure 22.1

In this case, we have a 2× enlargement, which means that the width of the enlarged picture is twice that of the original and the height of the enlargement is also twice that of the original. In other words, the ratio of the width of the original to the width of the enlargement and the ratio of the height of the original to the height of the enlargement are both the same, namely $\frac{1}{2}$.

Now the important thing is that we get the same ratio whenever we compare the distance between any two points in the original and the distance between the matching points in the enlargement. Check this by measuring the distance between the center of the reel and the end of the fishing pole in the original and in the enlargement. The first will come out to be half of the second. Check again with the length of the fishline between the water and the end of the pole. Check again with the distance from the fisherman's nose to his right big toe. In all cases, the ratio is $\frac{1}{2}$. (Of course, going in the other direction, the ratios of distances in the enlargement to the corresponding distances in the original are all the same, namely 2.)

■ **Question 1** How does the area of the enlargement compare with the area of the original?

But now let us compare angles in the original and in the enlargement. Do we find a ratio of $\frac{1}{2}$? No. In fact, any angle in the original is actually congruent to the corresponding angle in the enlargement. Using some tracing paper, check this with the angle between the pole and the fishline. Check again with a few other pairs of corresponding angles in the two pictures.

Whenever we have a situation such as this, it is certainly plausible to say that the two figures have the "same shape" even though they have different sizes. The standard terminology is to say that the two figures are *similar*.

It will be helpful to look at this idea in more detail. Fortunately, we can find out all we need to know about the similarity relationship by studying only the case where both the figures are triangles. Our definition in this case says that

Two triangles are similar if there is a correspondence between their parts such that:

1 corresponding angles are congruent and
2 the ratios of the lengths of corresponding sides are equal.

In Chapter 10, we used the symbol m($\overline{AB}$) to denote the length of the segment $\overline{AB}$. In this chapter we will very frequently have to refer to the lengths of various segments. To simplify matters, from now on we will denote the length of the segment $\overline{AB}$ by AB instead of m($\overline{AB}$).

Now let us consider an example and explain a bit more precisely just what that second condition above means. Consider two triangles, $\triangle ABC$ and $\triangle XYZ$, and the correspondence $A \leftrightarrow X, B \leftrightarrow Y, C \leftrightarrow Z$. Then $\overline{AB}$ and $\overline{XY}$ are a pair of corresponding sides; $\overline{BC}$ and $\overline{YZ}$ are another; $\overline{AC}$ and $\overline{XZ}$ are the third. Now the length of $\overline{AB}$ is denoted by AB and that of $\overline{XY}$ by XY (see Figure 22.2). Then

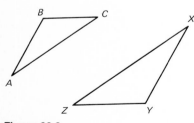

Figure 22.2

$$\frac{AB}{XY}$$ is a ratio of the lengths of two corresponding sides

Also,

$$\frac{AC}{XZ}$$ is a ratio of the lengths of two corresponding sides

$\dfrac{BC}{YZ}$ is a ratio of the lengths of two corresponding sides

$\dfrac{XY}{AB}$ is a ratio of the lengths of two corresponding sides

$\dfrac{XZ}{AC}$ is a ratio of the lengths of two corresponding sides

$\dfrac{YZ}{BC}$ is a ratio of the lengths of two corresponding sides

Do we mean to say that *all six* of these are to be the same? No. Consider the photo and its enlargement in Figure 22.1. Let A, B, and C be the tip of the pole, the tip of the fisherman's nose, and the knob on the reel, respectively, in the smaller picture, and let X, Y, and Z be the corresponding points in the enlargement. Then,

$$\frac{AB}{XY} = \frac{AC}{XZ} = \frac{BC}{YZ} = \frac{1}{2} \quad \text{and} \quad \frac{XY}{AB} = \frac{XZ}{AC} = \frac{YZ}{BC} = 2$$

We see that these ratios are naturally separated into two sets:

1 ratios whose numerators refer to sides of $\triangle ABC$ and whose denominators refer to sides of $\triangle XYZ$
2 ratios whose numerators refer to sides of $\triangle XYZ$ and whose denominators refer to sides of $\triangle ABC$

The three ratios in the first set are all equal to each other; the three ratios in the second set are also equal to each other. We can see, of course, that if the ratios in one set of three are equal, then so are the three in the other set. (A ratio in one set is a reciprocal of one in the other.)

We could restate our definition as follows:

$\triangle ABC \sim \triangle XYZ$ (read "triangle ABC is similar to triangle XYZ") if

1 $\angle A \cong \angle X$, $\angle B \cong \angle Y$, $\angle C \cong \angle Z$ and

2 $\dfrac{AB}{XY} = \dfrac{AC}{XZ} = \dfrac{BC}{YZ}$

or (which is the *same* thing)

$$\frac{XY}{AB} = \frac{XZ}{AC} = \frac{YZ}{BC}$$

Note that we use the same convention as that used in stating congruence, namely $\triangle ABC \sim \triangle XYZ$ means that the two triangles are similar under the correspondence $A \leftrightarrow X$, $B \leftrightarrow Y$, and $C \leftrightarrow Z$.

■ **Question 2** If we know that $\triangle ABC \sim \triangle XYZ$, can we say that $AB/XZ = BC/YZ$?

SCALE FACTORS When $\triangle ABC \sim \triangle XYZ$, then

$$\frac{AB}{XY} = \frac{AC}{XZ} = \frac{BC}{YZ}$$

Calling this common ratio k, we have:

$$\frac{AB}{XY} = k \quad \text{so that} \quad AB = k \times XY$$

$$\frac{AC}{XZ} = k \quad \text{so that} \quad AC = k \times XZ$$

$$\frac{BC}{YZ} = k \quad \text{so that} \quad BC = k \times YZ$$

Thus, in "going from" $\triangle XYZ$ to $\triangle ABC$ each side's length is multiplied by the factor k. Let us refer to this factor as "the scale factor in going from $\triangle XYZ$ to $\triangle ABC$."

In "going the other way"—from $\triangle ABC$ to $\triangle XYZ$—the scale factor would be $1/k$, of course, since if $AB = k \times XY$ then $XY = 1/k \times AB$ and similarly for the other two sides. Thus, in any similarity there are two scale factors, one for each "direction," and *they are the reciprocals of each other*.

The idea of similarity and scale factors is at the heart of many maps and other "scale drawings." In making such a map of a region that is, say, 5 miles square, what we want is a figure consisting of a square and certain points in it marked so that for any two points H and S in the region (say your home and your classroom) the distance between the corresponding points on the map H' and S' will be some fixed number times the distance between H and S. That is, $H'S' = k \times HS$ and (most importantly) the same number, k, serves for *all* pairs of points. The number k is the scale of the map.

If your map is 10 inches square, then the 10-inch-square region on the map corresponds to a 5-mile-square region on earth. So 10 inches is k times the length, 5 miles. Now

5 miles = 5 × 5,280 feet = 5 × 5,280 × 12 inches = 316,800 inches

so the scale k (in going from the region to the map) is given by

$10 = k \times 316,800$

That is,

$$k = \frac{10}{316,800}$$

The topographical maps made by the U.S. Geological Survey have scales of $\frac{1}{24000}$ and of $\frac{1}{62500}$.

If 1 inch on a map corresponds to 1 mile, what is the scale of the map?

1 mile = 5,280 feet = 63,360 inches

so the scale is $\frac{1}{63360}$.

On a $\frac{1}{24000}$ map, how many miles are represented by one inch? One inch on the map represents 24,000 inches, and

24,000 inches = $\frac{24000}{63360}$ miles $\approx$ .38 miles

■ **Question 3** Could you hang on your wall a map showing both San Francisco and Los Angeles if the scale factor were $\frac{1}{10000}$?

How can we tell whether two triangles are similar? In the case of congruence, we had three congruence properties which we could use to compare two triangles. For similarity, we also have three ways of comparing. These are:

SAS Similarity Property
If the ratios of the lengths of two pairs of sides are equal and the included angles are congruent, then the triangles are similar.

SSS Similarity Property
If the ratios of the lengths of corresponding sides of two triangles are equal then the triangles are similar.

AA Similarity Property
If two angles of one triangle are congruent, respectively, to two angles of another triangle, then the triangles are similar.

In the problems at the end of the chapter we will see how to justify these properties.

Now let us look at some examples of applications of similarity ideas to some surveying problems.

SOME APPLICATIONS

A first example concerns the problem facing a man who wishes to bridge a stream (see Figure 22.3). The bridge must reach from point A on one bank to point B on the other. Suppose he chooses a point C on the same bank as A and measures side $\overline{AC}$ and angles A and C of $\triangle ABC$. He now retires to his desk and carefully constructs a triangle $A'B'C'$ similar to $\triangle ABC$. Then $AB/A'B' = AC/A'C'$ and $AB = AC/A'C' \times A'B'$. Since all the segments $\overline{AC}$, $\overline{A'C'}$, and $\overline{A'B'}$ can be measured, AB can be calculated.

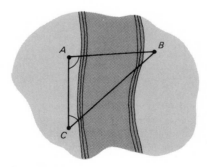

Figure 22.3

If $\overline{AC}$ is 100 feet long, $\overline{A'C'}$ is 6 inches long, and $\overline{A'B'}$ is 10 inches long, how long must the bridge be (in feet)?

$$AB = \tfrac{100}{6} \times 10 = \tfrac{500}{3} \qquad \text{or} \qquad 166\tfrac{2}{3}$$

The length of the bridge is $166\tfrac{2}{3}$ feet, or 166 feet 8 inches.

Another example is the problem of drilling through a hill along the line joining two points, A and B, on opposite sides, which we first encountered in Chapter 21. The problem is, in what direction should we start digging from A in order to come out at B? Choosing a point C from which both A and B can be seen, we measure $\angle ACB$ and the lengths of $\overline{AC}$ and $\overline{BC}$. Again, we now retire to our desk and make a drawing of a triangle $A'B'C'$ similar to $\triangle ABC$. Then $\angle A \cong \angle A'$; this gives us the direction for digging (see Figure 22.4).

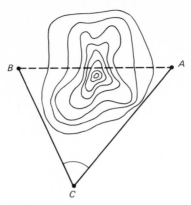

Figure 22.4

■ **Question 4** Another way of working this problem was given in Chapter 21. Are the two answers the same?

ANOTHER SIMILARITY PROPERTY

Figure 22.5

Suppose we have two similar triangles: $\triangle ABC \sim \triangle DEF$. Part of the definition for similarity requires that if $DE = k \times AB$, then $EF = k \times BC$ and $DF = k \times AC$ (see Figure 22.5). Another way to express this requirement is:

$$\frac{DE}{AB} = k \qquad \frac{EF}{BC} = k \qquad \frac{DF}{AC} = k$$

Thus, we have three equal ratios. Each ratio compares one side of *one* triangle with the corresponding side of the *other* triangle.

The reciprocals of the above give us three additional ratios which are equal:

$$\frac{AB}{DE} = \frac{1}{k} \qquad \frac{BC}{EF} = \frac{1}{k} \qquad \frac{AC}{DF} = \frac{1}{k}$$

Each of these ratios again compares one side of *one* triangle with the corresponding side of the *other.*

Each ratio in the chain

$$\frac{DE}{AB} = \frac{EF}{BC} = \frac{DF}{AC}$$

is of the form

$$\frac{\text{A side of } \triangle DEF}{\text{The corresponding side of } \triangle ABC}$$

Each ratio in

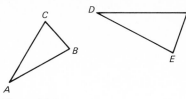

Figure 22.6

$$\frac{AB}{DE} = \frac{BC}{EF} = \frac{AC}{DF}$$

is of the form

$$\frac{\text{A side of } \triangle ABC}{\text{The corresponding side of } \triangle DEF}$$

From these two chains, we may get *other* equal ratios. We show three more such pairs that are of particular interest.

Let $\triangle ABC \sim \triangle DEF$ (see Figure 22.6). Then

$$\frac{AB}{DE} = \frac{BC}{EF}$$

That is,

$$AB \times EF = BC \times DE$$

Dividing by $BC \times EF$, we have

$$\frac{AB \times EF}{BC \times EF} = \frac{BC \times DE}{BC \times EF} \qquad \text{or} \qquad \frac{AB}{BC} = \frac{DE}{EF}$$

Notice that each ratio in the equality

$$\frac{AB}{BC} = \frac{DE}{EF}$$

compares one side of a particular triangle with another side of the *same* triangle. Likewise, we can show

$$\frac{BC}{AC} = \frac{EF}{DF} \qquad \text{and} \qquad \frac{AC}{AB} = \frac{DF}{DE}$$

In general, we have the following result:

If two triangles are similar then the ratio of the lengths of any two sides of one triangle equals the ratio of the lengths of the corresponding sides of the other.

■ **Question 5** Can we turn this around and say that if the ratio of the length of any two sides of one triangle equals the ratio of the length of the corresponding sides of another triangle, then the triangles are similar?

Here is an application of this (see Figure 22.7). We have a flag pole. We wish to determine its height. Not wishing either to climb the pole or to cut it down, we proceed as follows:

1 Measure the length of its shadow, $\overline{AB}$;
2 Measure the length of a friend's shadow, $\overline{A'B'}$; and
3 Measure the friend's height $\overline{B'C'}$.

Now *if* the two triangles are similar ($\triangle ABC \sim \triangle A'B'C'$), we can easily solve our problem, since

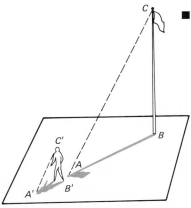

Figure 22.7

$$\frac{BC}{B'C'} = \frac{AB}{A'B'}$$

But we have measured AB and $A'B'$, so we know their ratio and so we know the ratio of BC to $B'C'$. Then we can find BC. But are the triangles similar? Yes, they are, for all practical purposes. The shadows are cast by the sun's rays. Because of the great distance of the sun from the earth we can consider these rays as practically parallel. Now a horizontal line forms angles at A and A' which are congruent (as we saw in Chapter 21) since the rays (broken lines in the figure) are parallel. Also, the angles at B and B' are congruent because both the flag pole and the man are vertical and form right angles with the horizontal.

This is enough to show that the triangles are indeed similar, which is all that is needed for the problem.

GEOMETRIC REPRESENTATIONS OF RATIOS

In this chapter we have been working a great deal with *ratios*. In our instances it was lengths of segments whose ratios were considered, but since lengths are positive numbers we have indeed been working with ratios of numbers. Going the other way, any statement that the ratio of two (positive) numbers, a/b, equals another ratio of two numbers, c/d, *can be interpreted geometrically* as in Figure 22.8, where the two segments, $\overline{PQ}$ and $\overline{RS}$, are parallel. ($\angle C$ can be of any size.)

Now in the statement $a/b = c/d$, if we know any three of the numbers, a, b, c, d, we can solve for the fourth number; this is equivalent to the geometric fact that if we know any three of the points, P, Q, R, S, we can construct the fourth.

Often we express ratios in terms of percent. For example, $\frac{3}{4} = \frac{75}{100} = 75$ percent. We say that 3 is 75 percent of 4; also that the ratio of 3 to 4 is the same as the ratio of 75 to 100. All of the following statements express the same thing:

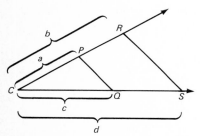

Figure 22.8

1 a is p percent of b
2 p percent of b is a
3 The ratio of a to b is the same as the ratio of p to 100
4 $\dfrac{a}{b} = \dfrac{p}{100}$

All of these can be represented as in Figure 22.9. (Note that since the ratio of the lengths of two segments is independent of the unit of measure used in their measurement, we can choose the unit on each ray as we please.)

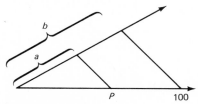

Figure 22.9

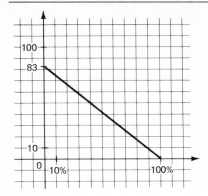

Figure 22.10

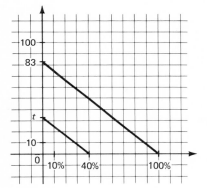

Figure 22.11

Since students often find that a geometric "picture" of a numerical relation is helpful in seeing the relation and in estimating an answer, the following may prove an aid in problems involving percent.

QUESTION:
What is 40 percent of 83?

PICTURE:
1. On squared paper draw lines as shown in Figure 22.10 and select a point on the horizontal (or percent) line and mark it 100 (10 squares is often a convenient choice). This establishes the full number scale on the horizontal line.
2. With any convenient number scale on the vertical line, mark the point 83. Connect the two points.
3. The line parallel to this line through the point 40 on the horizontal line will cut the vertical line at a point with a coordinate, say t.
4. Since $t/83 = \frac{40}{100}$, the number t is the required number—it appears to be just a bit less than 35, say 33 (see Figure 22.11).

This construction is based on the idea "If 83 corresponds to 100 percent, what number t corresponds to 40 percent?"

ARITHMETIC SOLUTION:
If t is the required number, $t/83 = \frac{40}{100} = .4$. Hence, $t = (.4)(83) = 33.2$.

■ **Question 6** Why can't we work all percent problems this way?

PROBLEMS

1. If two triangles are congruent, are they necessarily similar?
2. If two triangles are similar, are they necessarily congruent?
3. In reference to Figure P22.3, suppose it is given that $\triangle ABC \sim \triangle PQR$ and $\triangle PQR \sim \triangle XYZ$. What can we say about $\triangle ABC$ and $\triangle XYZ$?

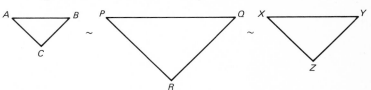

Figure P22.3

4. Lengths of the sides of three triangles are given. Which two of the triangles are similar?
   (a) 5, 6, 7   (b) 10, 11, 12   (c) $3\frac{1}{2}$, $2\frac{1}{2}$, 3
5. In Figure P22.5, $BC = \frac{1}{4}DC$ and $EC = \frac{1}{4}AC$. Are the two triangles similar? Give an explanation to justify your answer.
6. If a segment joining points on two sides of a triangle is parallel to the third side, why is the triangle formed similar to the original triangle?
7. In each part of this problem you are given some information (see Figure P22.7). Explain why the triangles are similar.

   (a) Given: $\dfrac{AE}{DE} = \dfrac{BE}{CE}$

   Why is $\triangle AEB \sim \triangle DEC?$

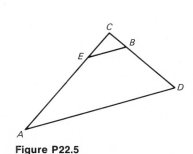

Figure P22.5

(b) Given: $AB = 15$, $AD = 12$, $AC = 20$, $AE = 16$
     Why is $\triangle ADE \sim \triangle ABC$?

(c) Given: $ST = 15$, $VW = 5$, $RS = 12$, $ZV = 4$, $RT = 9$, $ZW = 3$
     Why is $\triangle RST \sim \triangle ZVW$?

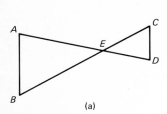

(a)

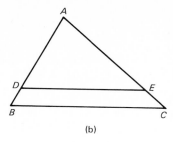

(b)

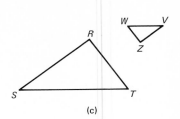

(c)

Figure P22.7

8 In each case, decide whether the given information enables you to conclude that the given pair of triangles are similar. Give a reason for your
   decision and, if they are similar, so state in the proper form:
   $\triangle$_____ $\sim$ $\triangle$_____. See Figure P22.8.

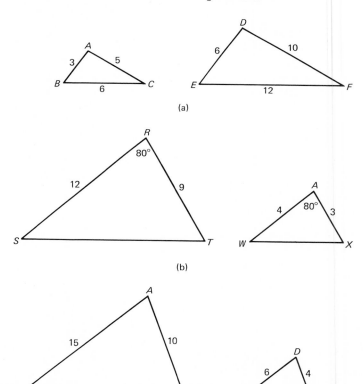

(a)

(b)

(c)

Figure P22.8

9 For each pair of similar triangles (see Figure P22.9), write one set of equal ratios. Then determine the lengths of the remaining sides from the given lengths.

(a) $\triangle PQR \sim \triangle DEF$ (b) $\triangle ABC \sim \triangle DEC$
Find AD and BE.

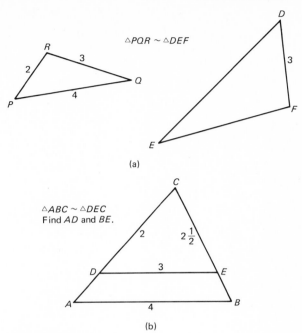

(a)

(b)

Figure P22.9

10 The three triangles in Figure P22.10 are similar, and $AB:PQ = 4:3$ while $AB:XZ = 4:5$. If in terms of a certain unit $AB = 12$, what are PQ and XZ? If $PR = 15$, what is AC and what is XY?

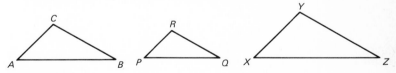

Figure P22.10

11 The two triangles in Figure P22.11 are similar.
(a) If RS corresponds to AB, find RT.
(b) State the correspondence which is a similarity.
(c) Follow the directions for Problem 11a and b if RS corresponds to AC.

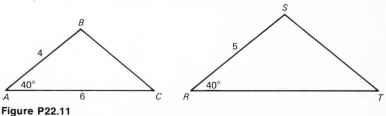

Figure P22.11

12 In an architect's drawing, the scale is given as $\frac{1}{4}$ inch to 1 foot. How big should the plan of a room be if the room is to measure 17 feet by 23 feet?

13 (a) The scale of a map is $\frac{1}{4}$ inch to 10 miles. If the distance on the map from city A to city B is $2\frac{1}{8}$ inches what is the actual distance between the cities?

(b) On the map of part a, a salesman who lives at A finds the distance from A to B is $2\frac{1}{8}$ inches, B to C is $3\frac{3}{8}$ inches, from C to D is $4\frac{1}{2}$ inches and from D to A is $1\frac{1}{16}$ inches. How far does he travel to visit all four cities and return home?

14 Use Figure P22.14 to estimate $\frac{1}{4}$ as a percent.

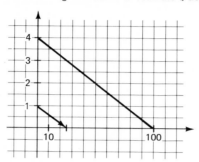

Figure P22.14

15 Estimate the following graphically.
(a) 24 percent of 75
(b) 120 percent of 95
(c) 84 is what percent of 240?
(d) 96 is what percent of 64?
(e) 87 is 75 percent of what number?
(f) 133 is 140 percent of what number?

16 On a certain day, 836 students were present at the Hillside School. If this was 95 percent of the student enrollment, what was the student enrollment? Estimate your answer graphically.

17 A large apartment house in Chicago has 180 apartments. In order to cover costs, the owner estimates that he must have 80 percent of the apartments rented. If he had 171 apartments rented during one year, did he cover costs? How many apartments above his break-even point did he have rented? Estimate your answer graphically.

18 Given $\triangle ABC$ and segment $\overline{A'B'}$ (Figure P22.18). Required: Construct C' so that

$$\triangle A'B'C' \sim \triangle ABC$$

(a) To have $\triangle A'B'C' \sim \triangle ABC$, what must be true about $\angle C'A'B'$?

(b) Hence, C' must lie on the ray $\overrightarrow{A'Y}$ constructed so that $\angle A \cong$ _____. Construct such a ray.

(c) To have $\triangle A'B'C' \sim \triangle ABC$, what must be true of AB, AC, $A'B'$, and $A'C'$?

(d) Since AB and $A'B'$ are given we may determine the number

$$k = \frac{A'B'}{AB}$$

$\overline{AC}$ is given also, so $A'C'$ is given by: $A'C' =$ _____.

(e) This means C' must lie on a circle with center at A' and with radius = _____. Construct it.

(f) Thus, C' must lie on the intersection of the ray and the circle and we have located C'.

(g) Which similarity property have we demonstrated?

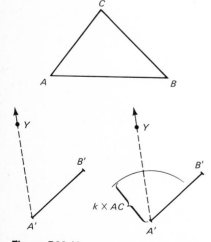

Figure P22.18

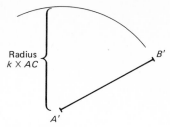

Radius
$k \times AC$

Figure P22.19c

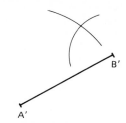

Figure P22.19e

19 Given: $\triangle ABC$ and segment $\overline{A'B'}$. Required: Construct C' so that $\triangle ABC \sim \triangle A'B'C'$.

(a) To have $\triangle ABC \sim \triangle A'B'C'$, what must be true of $A'B'/AB$, $A'C'/AC$, and $B'C'/BC$?

(b) Since AB and $A'B'$ are given, we may determine the number $k = A'B'/AB$ by measurement. AC is also given, so $A'C'$ is determined by: $A'C' = $ _____.

(c) This means that C' must lie on a circle with center at A' and with radius = _____. Construct it (see Figure P22.19c).

(d) BC is also given, so $B'C'$ is determined by: $B'C' = $ _____.

(e) This means that C' must lie on a circle with center at B' and radius = _____. Construct it (see Figure P22.19e).

(f) Since C' lies on both circles, C' is a point of intersection of the two circles. (There will be two such points, one on each side of $A'B'$.)

(g) Which similarity property have we demonstrated?

20 Given: $\triangle ABC$ and segment $\overline{A'B'}$ (Figure P22.20). Required: Construct C' so that $\triangle ABC \sim \triangle A'B'C'$.

(a) To have $\triangle ABC \sim \triangle A'B'C'$, what must be true about $\angle CAB$ and $\angle C'A'B'$? About $\angle CBA$ and $\angle C'B'A'$?

(b) Hence C' must lie on the ray $\overrightarrow{A'Y}$ constructed so that $\angle A' \cong$ _____. Construct such a ray.

(c) Also, C' must lie on the ray $\overrightarrow{B'X}$ constructed so that $\angle B' \cong$ _____. Construct such a ray on the Y side of $\overline{A'B'}$.

(d) C' lies on $\overrightarrow{A'Y}$ and C' lies on $\overrightarrow{B'X}$, so C' lies on their intersection. We now have a triangle $A'B'C'$ so that $\angle A' \cong \angle A$ and $\angle B' \cong \angle B$.

(e) Which similarity property have we demonstrated?

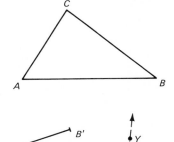

Figure P22.20

23.

MORE ABOUT MEASUREMENT

In Chapters 10 and 11, where the measurement of lengths, angles, and areas was discussed, we saw that we always had to measure to the nearest unit. We also saw that the smaller the unit, the more accurate the measurement. Thus, for example, if we choose this segment

P   Q

as the unit for measurement of length, then each of these segments, AB and CD, in Figure 23.1 has length 3.

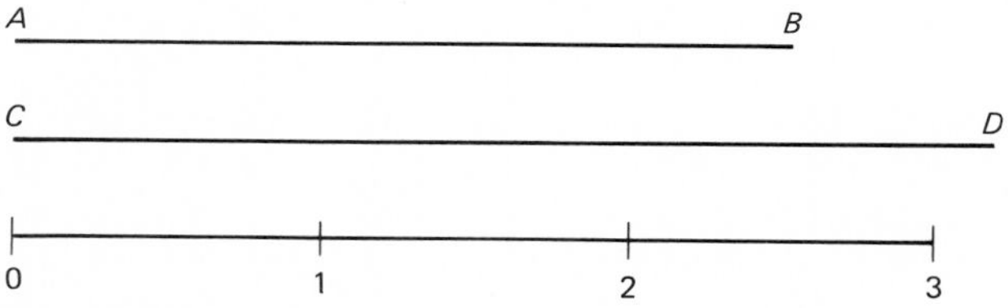

A   B

C   D

```
|-----------|-----------|-----------|
0           1           2           3
```

Figure 23.1

However, suppose we choose this smaller segment as the unit.

R   S

Then we see that $m(AB) = 13$ and $m(CD) = 17$ in terms of the unit shown in Figure 23.2. This measurement is more accurate and it certainly shows that AB and CD have different lengths.

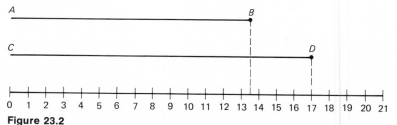

Figure 23.2

FRACTIONAL UNITS Now that we know something about fractions, we can choose these smaller units in a systematic fashion. Suppose we start with a ruler marked off in inches (see Figure 23.3). If we want to gain greater accuracy we could divide each unit segment into 2 congruent segments.

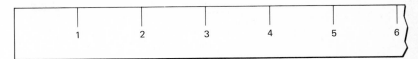

Figure 23.3

Then, just as we did in Chapter 12, we would label the new points (Figure 23.4).

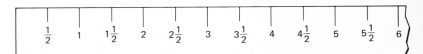

Figure 23.4

If we wanted even more accuracy, we could divide each of these new segments into 2 congruent parts and, if we wished, we could continue to repeat the subdividing as long as we wanted (Figure 23.5).

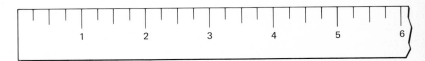

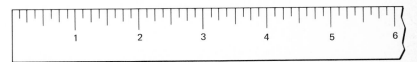

Figure 23.5

Generally, the rulers that we buy are marked off in eighths or sixteenths of an inch, but even finer subdivisions are used in certain kinds of jobs.

Why did we subdivide the original unit into 2 congruent parts, rather than 3, or 4, or some other number? Probably this was done, long ago when our present measurement system was being developed, for simplicity. There is no logical reason for choosing 2, rather than some other number, and in fact we will see later that there is a better choice.

■ **Question 1** How many times must the ruler at the bottom of Figure 23.5 be subdivided until the unit is less than $\frac{1}{100}$ inch long?

This process that we have described really allows us to keep our original unit, rather than replace it by a smaller unit, but to report measurements in terms of fractions of that unit rather than in terms of whole numbers. Since we know how to carry out arithmetical operations on rational numbers, we can now work the same kind of length problems that we did before. For example, what is the perimeter of a triangle whose sides are $\frac{1}{2}$, $\frac{3}{8}$, and $\frac{5}{8}$ inches? We know how to find $\frac{1}{2} + \frac{3}{8} + \frac{5}{8}$. In fact the sum of these three fractions is $\frac{12}{8}$ or $1\frac{1}{2}$, so the perimeter of the triangle is $1\frac{1}{2}$ inches.

FRACTIONAL UNITS AND AREAS

In addition to addition, we also used multiplication in measurement problems, in order to compute areas, for example. Can we use the same procedures if the measurements are given in fractions? For example, can we find the area of a rectangle (by multiplying the length by the width) if the length and width are fractions? Let us look again at an example we first saw in Chapter 11.

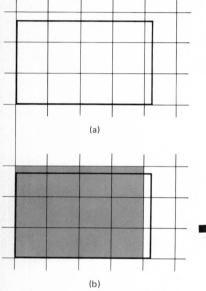

(a)

(b)

Figure 23.6

In Figure 23.6, the length and width of the heavily outlined rectangle are 4 and 3 inches respectively to the nearest unit. By counting unit regions in the superimposed grid we saw that the area was between 8 and 15 square inches. When we used the measures of the sides to the nearest inch in the formula

$$A = a \times b$$

we got $A = 3 \times 4 = 12$ and the area was 12 square inches. This was the exact area of the region shaded in Figure 23.6b, which seemed to have about the same size as the given rectangle, and it lies between the underestimate 8 and the overestimate 15 that we got using the grid. It was probably not, of course, the exact area of the original rectangle but it seemed to be pretty close.

■ **Question 2** One rectangle measures $2\frac{3}{8}$ by $3\frac{7}{8}$ inches. Another measures $2\frac{5}{8}$ by $3\frac{1}{8}$ inches. If you measure to the nearest inch, which is larger?

Now suppose the length of the rectangle was exactly $4\frac{1}{4}$ inches and the width was $2\frac{3}{4}$ inches. If we use these two fractions in the formula

$$A = b \times h$$

we find that $A = 11\frac{11}{16}$ square inches. Certainly, 12 is the nearest whole number to $11\frac{11}{16}$, so it looks as though our formula still works even with fractions.

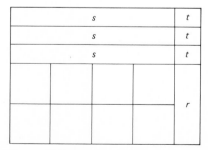

Figure 23.7

Let us take another look at the rectangle, with the parts labeled as shown in Figure 23.7. The part of the rectangle within the heavy lines is exactly 4 by 2, and its area is 8. The strip r is $\frac{1}{4}$ by 2, so its area seems to be $\frac{2}{4}$ or $\frac{1}{2}$. Each of the strips labeled s is 4 by $\frac{1}{4}$, so each strip seems to have area $\frac{4}{4}$ or 1. Each little square labeled t is $\frac{1}{4}$ by $\frac{1}{4}$. Clearly, it would take 16 of them to fill up a unit square, so the area of each of the little squares is $\frac{1}{16}$ (see Figure 23.8).

Thus the area of the original rectangle is $8 + \frac{1}{2} + 3 + \frac{3}{16} = 11\frac{11}{16}$.

This is, of course, just what we got by multiplying $4\frac{1}{4}$ and $2\frac{3}{4}$, and confirms that finding areas when the measurements are fractions is no different from finding areas when the measurements are whole numbers.

There is one thing to be careful of. When we measure to the nearest inch and put these whole numbers in the formula, we do *not* always get the answer correct to the nearest square inch. Try a rectangle $4\frac{3}{8}$ by $2\frac{7}{8}$. To the nearest inch, the measurements are 4-by-3 inches, and the formula gives 12 square inches. Now multiply $4\frac{3}{8}$ by $2\frac{7}{8}$. Is 12 the nearest whole number?

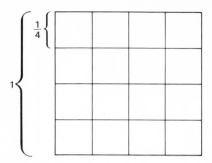

Figure 23.8

■ **Question 3** If the measures of the rectangles in Question 2 are exact, which is larger?

REPORTING MEASUREMENTS

We observed above that we can make more accurate measurements if we choose a smaller unit or if we use fractional parts of our original unit.

Now we are faced with a communication problem. Suppose we measure something. How can we write down the result in such a way that the reader will know how accurate our measurement was? Suppose that we report that the distance between Palo Alto and Tracy is 60 miles. What unit did we use? Did we measure to the nearest mile? To the nearest 10 miles? To the nearest yard? To the nearest furlong?

We need some systematic way of handling this problem. Let us start by taking another look at measuring lengths.

If the measurement of a pencil using a ruler scaled in inches is reported as 3 inches, then we interpret that measurement to mean that if one end of the pencil is matched with the zero point on the ruler, the other end of the pencil is closer to 3 than to 2 or 4; that is, the length of the pencil is somewhere between $2\frac{1}{2}$ inches and $3\frac{1}{2}$ inches. Using such a ruler, we would report the lengths of both pencils pictured in Figure 23.9 to be 3 inches.

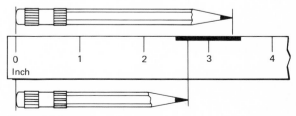

Figure 23.9

If we have a ruler scaled in $\frac{1}{2}$-inch segments, then we can measure the length of an object to the nearest $\frac{1}{2}$ inch. In Figure 23.10, a rod is measured using a ruler marked off in 1-inch segments and one marked off in $\frac{1}{2}$-inch segments. The length of the rod as measured by the first ruler would be reported as 4 inches; the length as measured by the second ruler would be reported as $3\frac{1}{2}$ inches.

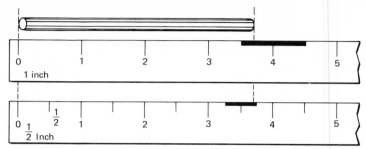

Figure 23.10

■ **Question 4** To the nearest $\frac{1}{4}$ inch, how long is the rod in Figure 23.10?

We say that the first measurement was made to the *nearest inch* while the second was made to the *nearest $\frac{1}{2}$ inch.* When a line segment is measured to the nearest inch, the actual length differs from the reported one by no more than $\frac{1}{2}$ inch. Thus the measure to the nearest inch of the rod in Figure 23.10 is sometimes reported as $4 \pm \frac{1}{2}$ inches, meaning that the actual length is somewhere between $3\frac{1}{2}$ and $4\frac{1}{2}$ inches.

The measure of the same rod, to the nearest half inch is $3\frac{1}{2}$ inches. The actual length differs from $3\frac{1}{2}$ by no more than $\frac{1}{2}$ of $\frac{1}{2}$ inch, or $\frac{1}{4}$ inch. We could report it as $3\frac{1}{2} \pm \frac{1}{4}$ inches.

Another way of saying this is that when a line segment is measured to the nearest inch, the *greatest possible error* is $\frac{1}{2}$ inch. When it is measured to the nearest $\frac{1}{2}$ inch, the greatest possible error is $\frac{1}{4}$ inch.

The standard convention is that the denominator of the fractional part of a measurement indicates the unit of measurement. Thus, if a pencil is measured to the nearest $\frac{1}{4}$ inch, and the measurement is $3\frac{2}{4}$ inches, we would not change this to $3\frac{1}{2}$ inches. To say that the length is $3\frac{2}{4}$ inches means that the greatest possible error is $\frac{1}{8}$ inch and the actual length is between $3\frac{3}{8}$ and $3\frac{5}{8}$ inches. To say that the length is $3\frac{1}{2}$ inches is to say that the actual length is between $3\frac{1}{4}$ inches and $3\frac{3}{4}$ inches. The first measurement is more precise than the second.

■ **Question 5** If we report the distance between two road markers to be 225 yards, what are we saying about the actual distance?

REPORTING DECIMAL MEASUREMENTS

Usually scientific measurements are expressed in decimal form. Suppose that a measurement of a length is reported as 4.2 inches. What precision is indicated by "4.2 inches"? We are measuring to the nearest .1 inch and we know that the actual length is closer to 4.2

inches than to 4.1 inches or 4.3 inches. Thus the greatest possible error is .05 inches and the actual length lies between 4.15 inches and 4.25 inches (since 4.15 lies halfway between 4.1 and 4.2, and 4.25 lies halfway between 4.2 and 4.3). Figure 23.11 shows a segment whose length would be reported as 4.2 inches to the nearest .1 inch.

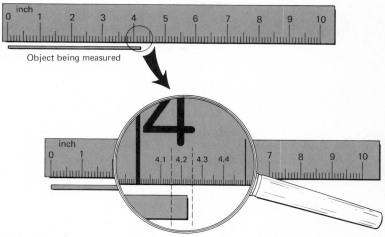

Object being measured

Figure 23.11

Unless the contrary is specified, the greatest possible error of a measurement given in decimal form is understood to be $\frac{1}{2}$ of the place value of the last digit which is used in the numeral for a purpose other than simply locating the decimal point.

Let us consider some examples.

Numeral	*Place value of last digit used for a purpose other than locating the decimal-point*	*Greatest possible error*
48.6	.1	.05
9800	100	50
.054	.001	.0005
830.00	.01	.005

Most readers probably have little question about the first and third examples. In the second, the two zeros are used simply to locate the decimal point and hence neither is the "last digit to be considered." In the fourth example, the second and third zeros are not used simply to locate the decimal point. They could be omitted. Hence they are considered used to indicate precision and the agreement gives .005 as the greatest possible error.

■ **Question 6** If the distance from the earth to the sun is stated to be 93 million miles, what is the greatest possible error?

COMPUTATION WITH MEASUREMENTS

Each number we use in measurement has a certain precision. We ask how precise or accurate the sum (or the product) of such numbers will be. The situation gets very complicated very rapidly. The best we can do here is to give some examples and suggest some reasonable "rules of thumb." Some understanding of both the nature of the problem and the limitations of our "rules" is necessary.

Suppose we want to add two numbers such as 18.6 and 23.9. The greatest possible error is .05 in both cases. Below we have made some computations revealing the greatest possible error of the sum.

Least values		*Greatest values*
18.55	18.6	18.65
23.85	23.9	23.95
42.40	42.5	42.60

Thus the sum 42.5 really has a greatest possible error of 0.1; that is, we know only that the "true" value is somewhere between 42.4 and 42.6. We could have written our computations as follows:

$$18.6 \pm .05$$
$$\underline{23.9 \pm .05}$$
$$42.5 \pm .10$$

In effect, we add the greatest possible errors of the addends to find the greatest possible error of the sum.

If we had three numbers, 18.6, 23.9, and 41.2, to add together, then the greatest possible error of the sum would be 0.15. The more numbers we add together the less precise the answer can be asserted to be. However, it is inconvenient to state explicitly the greatest possible error of the sum. In the first illustration above, we would probably write our answer as 42.5 with the standard agreement that the "greatest possible error" is .05 but with the clear understanding that we cannot be certain about this much precision. In a sense we are "caught"; we have to comprise between technical validity of our statements and giving too many details.

■ **Question 7** What is the sum and the greatest possible error in

$$.1 + .2 + .3 + .4 + .5 + .6 + .7 + .8 + .9$$

Another kind of problem arises if we want to add 86 to 18.48. Here it simply does not make sense to write the answer as 104.48 for in so doing we are implying precision to the nearest .01 whereas the 86 presumably was precise only to the nearest unit. Thus we ought to write our answer as 104 or possibly as 104.5 with the .5 interpreted more as $\frac{1}{2}$ than as $\frac{5}{10}$. The "true" value is quite likely to be somewhere between 104 and 105 and thus 104.5 seems like a reasonable answer.

The situations relative to the greatest possible error in multiplication (and division) are even less satisfactory than those in addition and subtraction.

If we multiply two measurements together, what can we say about the precision of the product? For instance, how many square feet are there in a room which is 16 feet by 18 feet? Most of us would say "288 square feet" but how precise is our answer? We assume (by our agreement) that 16 and 18 are precise to the nearest unit.

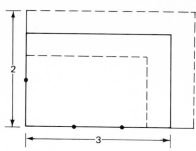

Figure 23.12

Consider the computations below.

Least values		*Greatest values*
17.5	18	18.5
15.5	16	16.5
271.25	288	305.25

In other words the "true" area can differ from 288 by as much as about 17 units. Being explicit, the best we could say is 288 ± 17.25. The size of the greatest possible error has been massively magnified in the process of multiplication. We can see this geometrically by considering Figure 23.12. The 2-by-3 region is enclosed by the heavy segments. The $1\frac{1}{2}$-by-$2\frac{1}{2}$ possibility is indicated on the inside and the $2\frac{1}{2}$-by-$3\frac{1}{2}$ on the outside.

■ **Question 8** Suppose the measurements of the room were 16.0 and 18.0. What is the greatest possible error in the area?

Going back to the 288 ± 17.25 case discussed above one might well ask, "How should the answer be written if we do not wish to indicate the greatest possible error explicitly?" There is no clear-cut answer. Some would prefer 288, but clearly this implies much greater precision than is present. Some would prefer 290. Here the "true" value would be indicated as being between 285 and 295 which, while not necessarily correct, seems not unreasonable. Again, we are "caught."

CHANGE OF UNIT The standard system of units of length used in the United States and some British countries is the British system, first discussed in Chapter 10. One basic unit of length in this system is the *inch*. The segment below is an inch segment.

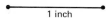

1 inch

The other units of length in the British system can be defined as follows:

1 foot is 12 inches
1 yard is 3 feet
1 mile is 1,760 yards

To measure a line segment we can choose any unit segment we like. For a given measurement, some units are more convenient than others. Miles are commonly used to measure distances between cities; feet or yards are used to measure the sizes of rooms, etc.

Frequently it is necessary to change a measurement using one unit to one using a different unit. Since a foot segment is 12 inches long, a measurement expressed in feet can be converted to one expressed in inches by multiplying the number of feet by 12. Thus a segment 4 feet in length is 4×12 inches long. This conversion can be expressed by the following conversion operation:

feet-to-inches   multiply by 12

Thus to find the number of inches in a 6-foot segment, we start with 6 and get 12×6 or 72. Hence the length of the segment is 72 inches.

■ **Question 9** Draw a flow chart which will tell a computer to print out the lengths, in inches, of any lengths that are input to the computer in terms of feet.

Similarly, to find the length in inches of a 3.6-foot segment, we start with 3.6 and get 43.2, because $12 \times (3.6) = 43.2$. Thus the segment is 43.2 inches long.

To change a measurement expressed in feet to one expressed in inches we use the operation

inches-to-feet   multiply by $\frac{1}{12}$

Thus, a segment 96 inches long is also 8 feet long because $\frac{1}{12} \times 96 = 8$. Again, the length of a 44-inch segment can be expressed as $3\frac{2}{3}$ feet since $\frac{1}{12} \times 44 = \frac{44}{12} = 3\frac{2}{3}$.

The operation used to convert a measurement expressed in yards to one expressed in feet is

yards-to-feet   multiply by 3

To convert a measurement expressed in miles to one expressed in yards we use the operation

miles-to-yards   multiply by 1,760

Table 23.1

Conversion	Operation
feet-to-inches	multiply by 12
yards-to-feet	multiply by 3
miles-to-yards	multiply by 1,760
inches-to-feet	multiply by $\frac{1}{12}$
feet-to-yards	multiply by $\frac{1}{3}$
yards-to-miles	multiply by $\frac{1}{1760}$

Table 23.1 gives the operations we have considered so far for converting a measurement of length expressed in one British unit to one expressed in another British unit.

To remember whether to multiply by 12 or $\frac{1}{12}$ when changing a measurement from feet to inches we remember that, since an inch is shorter than a foot, the number of inches in the measurement of a segment is greater than the number of feet. Thus we multiply by 12 to obtain a larger number. On the other hand, to change from inches to feet, we multiply by $\frac{1}{12}$ to obtain a smaller number.

■ **Question 10** What number would you multiply by to change a measurement from feet to yards?

From Table 23.1 we can create new "conversion" operations. For example, to to obtain a miles-to-feet operation, we "combine" the two operations

miles-to-yards   multiply by 1,760
yards-to-feet   multiply by 3

in the following way:

Start with a number, and use the first operation of multiplying by 1,760.

Then use the second operation of multiplying the result by 3. But

$$3 \times 1{,}760 = 5{,}280$$

so our miles-to-feet operation is

   miles-to-feet   multiply by 5,280

Similarly, we can construct an inches-to-yards operation by combining the operations

   inches-to-feet   multiply by $\frac{1}{12}$
   feet-to-yards   multiply by $\frac{1}{3}$

Hence the inches-to-yards operation is

   inches-to-yards   multiply by $\frac{1}{36}$

That is, a length of 54 inches is the same as a length of $\frac{1}{36} \times (54)$ yards, or $\frac{3}{2}$ yards.

The fact that 1 foot and 12 inches are measurements of the same segment is often written

  1 foot = 12 inches

Here the "equals" symbol does not denote that we have two numerals representing the same number; it denotes that the two measurements describe the same property of a segment. With this use of the symbol "=," the last sentence in the preceding paragraph can be written

54 inches = $\frac{3}{2}$ yards

It should be emphasized that 54 inches and $\frac{3}{2}$ yards do *not* represent numbers. What is true is that if the number of inches in the length of a segment is 54, then the number of yards in its length is $\frac{3}{2}$.

■   **Question 11**   What is the "inches-to-miles" operation?

THE METRIC SYSTEM   We mentioned above that there is a better way of subdividing a unit of measurement than by dividing it into halves, quarters, eighths, etc. This better way is used in another system of standard units of length. This second system, more widely used in the world than the British system, is the metric system. Its basic unit is the *meter,* which is a little more than 3 inches longer than the yard. The meter was originally intended to be 1 ten-millionth of the distance from the equator to the North Pole measured along the meridian through Paris. To standardize this unit, two marks were made on a platinum-iridium bar which was kept at the International Bureau of Weights and Measures in Paris. Until 1960, a meter was defined as the distance between these marks measured under a specific atmospheric condition. Currently the meter is defined as 1,650,763.73 wavelengths of the orange-red radiation of the element krypton 86 in vacuum. Other metric units of length are defined as follows:

one millimeter is one-thousandth of a meter (.001 meter)
one centimeter is one-hundredth of a meter (.01 meter)
one kilometer is one thousand meters (1,000 meters)

One advantage of the metric system is that it is decimal based; that is, a measurement using one metric unit of length can be changed to a measurement using any other metric unit of length by multiplying or dividing by a power of 10; that is, by moving the decimal point to the right or left.

We can use a conversion operation to convert any of these metric units to any other of them. Some of these conversion operations are given in Table 23.2.

Table 23.2

Conversion	Operation
centimeters-to-millimeters	multiply by 10
meters-to-centimeters	multiply by 100
kilometers-to-meters	multiply by 1,000
millimeters-to-centimeters	multiply by $\frac{1}{10}$
centimeters-to-meters	multiply by $\frac{1}{100}$
meters-to-kilometers	multiply by $\frac{1}{1000}$

Notice how much simpler the arithmetic operations are in this table as compared with those in the British system conversion table.

■ **Question 12** How many wavelengths of the orange-red radiation of the element krypton 86 are in one centimeter?

If we need an operation that is not in the table, we can combine those already given. For example, to construct the meters-to-millimeters operation we combine the following two operations:

meters-to-centimeters   multiply by 100
centimeters-to-millimeters   multiply by 10

Since $10 \times 100 = 1,000$, our operation is

meters-to-millimeters   multiply by 1,000

The British and the metric systems of standard units are related to each other. We have already said that a meter is about 3 inches longer than a yard. The official relationship between the two systems is given by the following:

1 inch is 2.54 centimeters

From this we get the basic conversion operations between the two systems:

inches-to-centimeters   multiply by 2.54
centimeters-to-inches   multiply by $\frac{1}{2.54}$

By combining operations in the metric conversion table with operations in the British conversion table, we can derive operations for converting a measurement in any unit of length to one in any of the others we have discussed.

Suppose, for example, that we want to express a length of $\frac{5}{2}$ yards in terms of centimeters. To obtain the operation that converts from yards to centimeters, we combine the operations

yards-to-feet	multiply by 3
feet-to-inches	multiply by 12
inches-to-centimeters	multiply by 2.54

When we carry out the arithmetic, we find that

$2.54 \times 12 \times 3 = 91.44$

Thus, our operation is

yards-to-centimeters	multiply by 91.44

Therefore, $\frac{5}{2}$ yards corresponds to

$91.44 \times \frac{5}{2}$ centimeters   or   228.6 centimeters

Some people would prefer to show the following arrangement of work for the above problem:

1 yard = 3 feet
     $= 3 \times 12$ inches = 36 inches
     $= 36 \times (2.54)$ centimeters = 91.44 centimeters
$\frac{5}{2}$ yards $= \frac{5}{2} \times (91.44)$ centimeters = 228.6 centimeters

Fortunately, when we give up the British system for the metric system, such unpleasant conversions will no longer be necessary.

■ **Question 13** In the Olympic Games, one of the events is the 100-meter dash. How many yards is this?

VOLUME OF A SOLID REGION

So far we have discussed measurement of lengths, angles, and areas. Let us now consider the measurement of solid figures.

The discussion of volumes of solid regions is more difficult than that of areas of plane regions primarily because it is hard for us to visualize solid regions when our pictures and diagrams are always in a plane. If you make cardboard models of some of the solid regions, it will help you to see what is going on.

When we discussed areas of plane regions in Chapter 11, we ran into difficulties. We found that a unit of area was a square 1 unit on a side. Putting many such squares side by side formed a grid and we could put our region on the grid and count how many squares were totally inside, say n, and how many were necessary to totally cover it, say N. Then the area was a number somewhere between n and N which we could approximate better and better by using smaller and smaller unit squares.

The problem of actually computing the area we solved only in a few cases. We found that the area of a rectangle is measured by a number which is the product of the measures of the base and the height. As we saw at the beginning of this chapter, this is true even if the unit squares do not fit exactly. The area of a parallelogram can be computed by making an equivalent rectangle. The area of a triangle is half that of a certain parallelogram and the area of any polygon can be computed by dividing it into triangles. Formulas for the areas of certain other figures can be derived, but for most plane regions the simplest method for determining their area is still to approximate them on a rectangular grid, preferably with fairly small divisions.

When we talk about solids and want to measure them, we have to think about the amount of space they occupy. One way to approach this is again through the idea of a rectangular grid, but this time in space rather than in a plane. See Figure 23.14. The rectangular grid is made up of unit solids which are cubes, one unit in each dimension. The unit of length is a line segment one unit long; the unit of area, a square one unit in each of two dimensions; and the unit of volume is a cube one unit in each of three dimensions. See Figure 23.13. The grid in Figure 23.14 is an outline of the solid cubes we imagine.

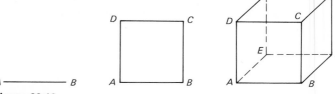

Figure 23.13

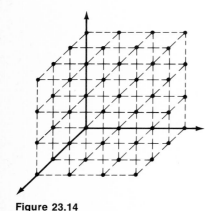

Figure 23.14

If any solid is given, a grid such as in Figure 23.14 may be imagined to surround it. A certain number n of the solid units may be completely enclosed by the solid and a larger number, N, of them will completely enclose it, thus giving an underestimate and an overestimate for V which represents the measure of the volume in terms of the chosen unit.

$$n < V < N$$

By using smaller and smaller units we can make better and better approximations to V much as was the case with measurement of length and area.

■ **Question 14** In measuring the volume of a solid could it ever be the case that $n = N$?

Let us now turn from this consideration of measuring the volume of any solid no matter how curiously shaped to consideration of the more regular and common solids. For these we will work out formulas for computing their volumes in terms of the linear measures assigned to segments associated with them, rather than approximating their measure by using space grids and counting procedures.

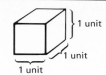

Figure 23.15

Let us ask how we can find the volume of a box. We will use a process which is much like the tiling process we used for areas. First we choose our unit of volume, a unit cube, such as the one pictured in Figure 23.15. We call its volume one *cubic unit*. If we have a rectangle whose dimensions are 3 units and 4 units, we can cover it by cubes which are congruent to the one represented in Figure 23.16.

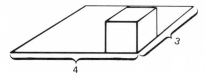

Figure 23.16

Each row will have three cubes (see Figure 23.17), and there will be 4 rows; thus, we have covered the rectangle by 12 unit cubes. We can think of the figure formed by these cubes (Figure 23.18) as being a box, 1 unit high, with a volume of 12 cubic units. The dimensions of the box are 4 units by 3 units by 1 unit.

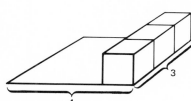

Figure 23.17

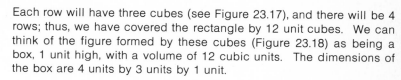

Figure 23.18

Notice that the area of the rectangle which forms the base of the box (Figure 23.19) is $3 \times 4 = 12$ square inches. The height of the box is 1 unit, and

$$3 \times 4 \times 1 = 12$$

while 12 unit cubes are needed to make this box. The box in Figure 23.20 also has a base of 4 units by 3 units. Its height is 2 units.

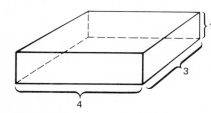

Figure 23.19

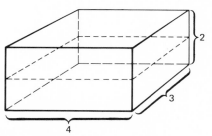

Figure 23.20

We notice that two layers of cubes are needed to make this box. There are 12 cubes in each layer, 24 in all. Notice that

$$3 \times 4 \times 2 = 24$$

The box in Figure 23.21 has a base of 4 units by 3 units and height of h units, where h is a whole number.

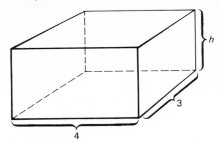

Figure 23.21

There are h layers of cubes, each layer consisting of 12 cubes. Altogether, there are

$$12 \times h$$

cubes in this box, so its volume is $12h$ cubic inches.

■ **Question 15**   If we chose a new unit, only half as long as the one shown in Figure 23.15, what would be the volume of the box in Figure 23.18 in terms of this new unit?

In each of these cases, all the layers of the box were alike, and each one contained as many unit cubes as there were square inches in the rectangular base. To find the volume of the box, we needed to find the number of unit cubes in each layer, which was the same as the area of the base, and then to multiply this number by the number of layers. The number of layers is the same as the height of the box, so we see that if A is the number of square units in the area of the base of a box, and h is the number of units in its height, then the number of cubic units, V, in its volume is $A \times h$. That is,

$$V = A \times h$$

We often say that the volume of a box is the product of the area of its base and the height. Although we have developed this formula for boxes whose dimensions are whole numbers, it also applies when the dimensions are any rational numbers, as in Figure 23.22.

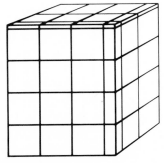

Figure 23.22

Most standard units of volume are related to standard units of length. British units of volume include the cubic inch, the cubic foot, and the cubic yard. Metric units of volume include the cubic centimeter, the cubic millimeter, and the cubic meter.

VOLUME OF A PRISM

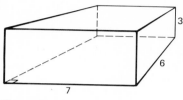

Figure 23.23

Figure 23.23 shows a box whose base is 7 units by 6 units and whose height is 3 units. The volume of this box is $7 \times 6 \times 3 = 126$. In Figure 23.24 we divide the box into two congruent figures by the plane determined by the diagonals $\overline{AC}$ and $\overline{FH}$. Each part is a triangular prism. The two parts are congruent, since the two triangles in the base are congruent, so the two prisms have the same volume, namely half the volume of the box, or 63.

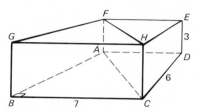

Figure 23.24

■ **Question 16** Each edge of a cube is exactly 1 inch long. What is its volume in cubic centimeters (to 3 decimal places)?

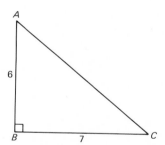

Figure 23.25

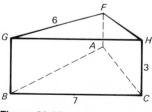

Figure 23.26

Now let us look at the right triangle ABC. This triangle is the base of one of the triangular prisms (Figure 23.25.) Its area is $\frac{1}{2}(6 \times 7) = 21$ square units. The triangular prism in Figure 23.26 is a copy of one of those above. If we multiply the area of the base by the height, we get

$21 \times 3 = 63$

which is just what we calculated the volume to be. Thus we can express the volume of the triangular prism as the product of the area of its triangular base and the height. This means that for the prism, as for a box, we have

$V = A \times h$

The following will give us another way of seeing that this is true. The triangular prism in Figure 23.27 has a height of h units. The area of its base is A square units. The prism consists of h triangular layers, each one unit high, whose bases are congruent to $\triangle ABC$.

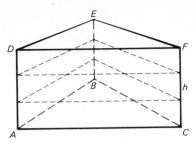

Figure 23.27

Notice that the volume of each layer is numerically the same as the area of the base. Just as in the case of the box, the total volume of the prism is h (the number of layers) times the volume of each layer. This last is just A, so again we see that

$$V = A \times h$$

■ **Question 17** Suppose another triangular prism had the same shape as the one in Figure 23.26, but that each edge was only a third as long. How would its volume compare with that of the first prism?

This particular prism happens to have a right triangle for its base. Other right prisms may have scalene triangles for their bases or quadrilaterals or in fact any polygons. Physical models of several prisms can be made out of thin plywood or cardboard with open tops so that the solid region they enclose may be filled with sand. Suppose such prisms have the same height, h, and the polygons which are their bases have the same area, B, even though they may not have the same shape (see Figure 23.28). Filling one model with sand and then pouring the sand from it into the other models demonstrates very convincingly that they all have the same volume. This may be written:

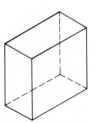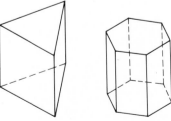

Figure 23.28

For *any* right prism: the volume is equal to the area of its base times its height.

$$V = B \times h$$

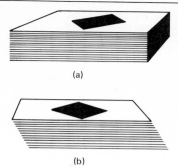

(a)

(b)

Figure 23.29

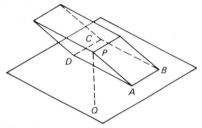

Figure 23.30

If the prism is not a right prism, a good physical model of the situation is a deck of cards which has been pushed into an oblique position as in Figure 23.29b. Putting the deck back into a straight stack does not seem to change the amount of space occupied by them. It looks as though the formula $V = B \times h$ is a valid formula to find the volume of any prism and in fact it is. What we must be careful to note, however, is that the h in this formula represents the actual height of the prism and not a lateral edge. In Figure 23.30 the base of the prism is the parallelogram $ABCD$ and the height is the measure of $\overline{PQ}$, a segment from a point P in the upper base perpendicular to the plane in which $ABCD$ lies. $\overline{PQ} < \overline{PA}$ where $\overline{PA}$ is a lateral edge. We can now say:

$$V = B \times h \qquad \text{for } any \text{ prism}$$

■ **Question 18** The area of the base of one prism is 1 square centimeter and of another is 1 square inch. What should the ratio of their heights be if they have the same volume?

MEASUREMENT OF CIRCLES

Figure 23.31

So far, we have talked about lengths, areas, etc., only for "straight" or "flat" geometric figures.

The length of a curve has not been discussed before (except for curves made up of one or more segments). However, the process for determining the length of a simple curve is to approximate the curve as closely as we please by a broken straight line and measure the length of the line. By using more and more shorter and shorter segments, we can approximate the lengths of the curve from A to B as closely as we please (see Figure 23.31). But, except in very special cases, there is no simple "formula" for the length.

The most important of these special cases is the circle. Think of a circle placed so that it touches a line at just one point of the circle, call that point A; then think of rolling the circle along the line until point A touches the line again. The starting point on the line and the point at which the point A touches the line again determine a segment. The length of this segment is the *circumference* of the circle. In Figure 23.32, the diameter of the circle is used as the unit segment. We see that in terms of this unit segment the circumference of the circle is a little more than 3 units; that is, the circumference is a little more than 3 times the diameter.

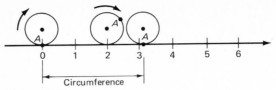

Figure 23.32

If we use a circle with a larger diameter we will get a longer segment than we did before. In Figure 23.33 the same unit segment is used as in the previous figure, but the diameter of the circle has been doubled. It appears that the circumference has been doubled also. Again the circumference is a little more than 3 times the diameter.

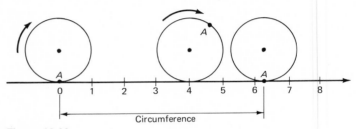

Figure 23.33

■ **Question 19** In Figures 23.32 and 23.33 the circles were rolled from left to right. What would happen if they were rolled from right to left?

Actually, the ratio of the circumference of a circle to its diameter is the same for all circles. It is a number which has been given the name π (pronounced "pie").

The number π is a number whose decimal expression never ends. It starts out

$\pi = 3.1415926535 \cdots$

For most practical purposes we can use, as an approximation to π, the first few decimal places, and write $\pi \approx 3.14$ or, if we need more accuracy, $\pi \approx 3.1416$.

For those who prefer to work with fractions, $\frac{22}{7}$ is a fair approximation to π and is often used. In fact, the ancient Greek mathematician Archimedes knew that

$\frac{22}{7} > \pi > \frac{223}{71}$

Chinese mathematicians, more than a millenium ago, used $\frac{355}{113}$ as an approximation to π.

However, for most of our purposes, we will use either 3.14 or $\frac{22}{7}$ as a satisfactory approximation of π.

To find the circumference of a circle, when we know the diameter, we use the operation:

Multiply by π

THE CIRCUMFERENCE IS THE PRODUCT OF π AND THE DIAMETER.

If we denote the number of units in the radius by r, then we can replace d in the above formula by $2r$, and obtain

$$C = 2\pi r$$

THE CIRCUMFERENCE IS THE PRODUCT OF 2π AND THE RADIUS.

What is the circumference of a circle whose raidus is 5 inches? If we use $\frac{22}{7}$ as an approximation for π, we get

$$C = 2\pi \times 5 \approx 2 \times \tfrac{22}{7} \times 5 = \tfrac{220}{7}$$

and if we use the approximation 3.14 for π, we get

$$C = 2\pi \times 5 \approx 2 \times 3.14 \times 5 = 31.4$$

So one approximation to the circumference of the circle is $\frac{220}{7}$ inches and another is 31.4 inches. Sometimes the answer is left in terms of π, in which case the circumference of the circle above would be reported as 10π inches, since

$$2\pi \times 5 = 5 \times 2\pi = 10\pi$$

In fact, 10π inches is a better answer than either $\frac{220}{7}$ inches or 31.4 inches because both of them are approximations to the number of inches in the length of the circumference while 10π is the exact number of inches.

■ **Question 20** The value of π can be found to as many decimal places as we please by calculating enough terms of the expression

$$4(1 - \tfrac{1}{3} + \tfrac{1}{5} - \tfrac{1}{7} + \tfrac{1}{9} - \tfrac{1}{11} + \cdots)$$

where the denominators are the odd numbers and we alternately subtract and add. Calculate each of the fractions (to four decimal places) and find the result of using the first two terms in the expression, the first three, etc., up to the first six.

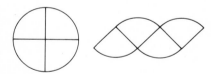

Figure 23.34

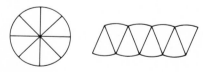

Figure 23.35

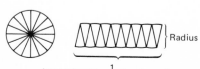

Approximately $\frac{1}{2}$ the circumference

Figure 23.36

The area of a circle may be estimated on a grid just as we estimated the area of an irregular simple closed region. A formula for this area, however, is more difficult to obtain than those for polygonal regions. All our formulas have been derived for polygonal regions which have line segments as their boundaries. The circle has no segments on its boundary. We can make some progress, however, if we cut up the circular region into a number of congruent parts and rearrange them. Thus if we cut it into 4 parts, we get Figure 23.34. If we cut it into 8 and 16 parts, we get Figures 23.35 and 23.36. We see that as we divide the circle into more and more parts, the resulting figure looks more and more like a parallelogram. The base of this figure is nearly equal in length to half the circumference as it is made up of the arcs of half the pieces. The height of the figure is more and more nearly equal to the length of the radius. Thus the area of the figure can be approximated as the product of one-half the circumference by the radius. Since the area of this figure is the same as that of the circle, we get the formula

$$A = \tfrac{1}{2} \times C \times r$$

This formula can be checked by using models of circles such as the bases of tin cans. Drawing a circle on a grid such as that of a piece of graph paper with units $\frac{1}{10}$ inch in length, the area can be estimated fairly accurately. If the circumference and radius are represented by a piece of string and these in turn measured on the graph paper scale, the results obtained by formula and by estimate can be compared.

Combining this formula for the area of a circle with the formula for the circumference of a circle, we may express the area as

$$A = \tfrac{1}{2} \times C \times r = \tfrac{1}{2} \times 2 \times \pi \times r \times r$$

or

$$A = \pi \times r \times r$$

■ **Question 21** Measure the diameters of a dime and of a quarter. Is the area of the quarter twice as great as the area of a dime?

VOLUME OF A CYLINDER

Now that we know how to find the area of a circle, we can ask about the volume of a circular cylinder. We saw in Chapter 20 that a prism is a special case of a cylinder. Can we hope that the formula

$$V = B \times h$$

where B is the area of the base, holds for cylinders as well as prisms?

In fact, any cylinder can be approximated by prisms whose heights are the same as that of the cylinder and whose bases are polygons which approximate the bases of the cylinder (see Figure 23.37). We guess that the formula $V = B \times h$ still holds and indeed it does. Of course, B stands for the area of the base of the cylinder and h for its height. Thus

$$V = A \times h \qquad \text{for } any \text{ cylinder}$$

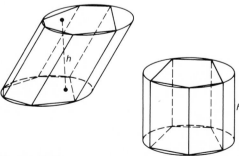

Figure 23.37

VOLUME OF SPHERES AND CONES

The last common solid to consider is the sphere. If the radius of a sphere is r, think of a right circular cone and a right circular cylinder each with the same radius r and each with height equal to the diameter of the sphere which of course is $2 \times r$. Consider hollow models of each, shown in Figure 23.38.

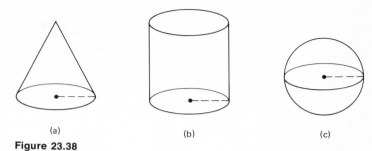

(a)   (b)   (c)

Figure 23.38

If the cone is filled with sand and this sand is poured out of it and into the cylinder, the cylinder will be one-third full. If now the sphere is also filled with sand and then emptied into the cylinder which is already one-third full with the sand from the cone, the cylinder will be completely full. Thus the volume of the sphere is just two-thirds that of the corresponding cylinder and just twice that of the corresponding cone. Since the radius of the base of the cylinder is r and its height is $2 \times r$ the volume, which is $B \times h$, is

$$V = (\pi \times r \times r) \times (2 \times r)$$

Therefore, the volume of the sphere is

$$V = \tfrac{2}{3} \times (\pi \times r \times r) \times (2 \times r)$$
$$V = \tfrac{4}{3} \times \pi \times r \times r \times r$$

Volumes of solid regions bounded by pyramids and cones are hard to find formulas for in the way we have been proceeding. But, an experiment gives the formulas quite easily. If you take a certain pyramid and make a model of it and of a prism with base and height congruent to those of the pyramid, you will find that if you fill the pyramid with sand and pour it into the prism, the prism will be filled after three such pourings. The same is true for a cone and the corresponding cylinder (see Figure 23.39).

(a)   (b)

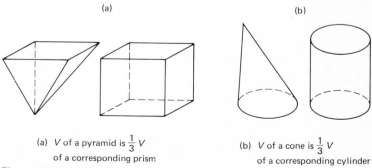

(a) V of a pyramid is $\tfrac{1}{3} V$ (b) V of a cone is $\tfrac{1}{3} V$
        of a corresponding prism   of a corresponding cylinder

Figure 23.39

This leads to the formula:

$$V = \tfrac{1}{3} \times B \times h \qquad \text{for any pyramid or cone}$$

where B is the area of the base.

■ **Question 22** A stack of seven dimes is almost exactly the same height as a stack of five quarters. How does the volume of a dime compare with the volume of a quarter?

OTHER MEASURES

We have studied measurements of length, angle, area, and volume. Of course, there are many other quantities that we can measure: weight, pressure, temperature, speed, elapsed time, voltage, electric current—to name a few.

In the case of the measurement of length, angle, area, and volume,

1 It is possible to choose a unit that can be *copied or duplicated.*
2 These units can be *put together* in a way that corresponds to counting.
3 It is possible to divide the given unit into a sufficiently large number of interchangeable subunits to reach any desired accuracy.

Measurements with these characteristics will be called *direct* measurements.

Let us consider the measurement of *weight* with these ideas in mind. There are, of course, many familiar units of weight, for example, the pound weight and the kilogram.

1 Can a weight unit be copied or duplicated? Yes, of course. We can construct as many pound weights as we please. To test whether two pieces of metal (whose weights are supposed to be units) are equal in weight, it is sufficient to put them in opposite pans of an equal-arm balance and then check to see whether the bar is horizontal (see Figure 23.40).
2 To weigh an object, we can count the number of unit weights in one pan that balance the object in the other pan.
3 In general, we must use subunits to bring about such a balance. But we can easily make subunits. For example, we can use ounce weights, 16 of which balance a pound weight.

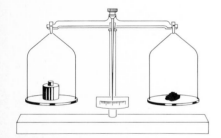

Figure 23.40

■ **Question 23** If 16 ounces weigh 1 pound and 2,000 pounds weigh 1 ton, what is the "ounces-to-tons" conversion operation?

The measurement described is a *direct* measurement because it determines the weight of the object by using a certain number of units of weight. But there are other ways of determining the weight. For example, we can use a spring balance. That is, we can find out how much the object stretches a spring. A pointer attached to the spring moves along a scale. This scale may be marked in pounds, but we are apparently measuring the weight by a stretch, that is, a distance. This is an *indirect* measurement of weight. How can we say that it

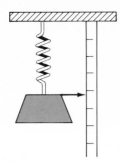

Figure 23.41

measures weight at all? Because of the discovery that a spring behaves in a certain way (see Figure 23.41). Robert Hooke (1635–1703) found that the stretch is proportional to the weight (if we do not overload the spring). For example, if we double the weight, we double the stretch, if we triple the weight, we triple the stretch, and so on. This property of springs has been thoroughly tested. We can be sure that if, for example, the spring balance reads 3.5 pounds, the stretch is 3.5 times the stretch for 1 pound. Therefore, *indirectly* we are actually measuring the desired weight in terms of a unit *weight*.

Many measurements commonly made are of this indirect sort. We follow Galileo in measuring temperature changes by changes in the length of a column of liquid. Galileo used oil; we use mercury.

Speed is a quantity that cannot be measured directly; that is, we do not measure speed by combining units (or subunits) of speed. We can measure speed by measuring distance, say, in feet, and time, say, in seconds. In fact, we divide the distance measure by the time measure. Suppose, for example, an object moves 120 feet in 5 seconds (we assume that the speed remains constant), then the speed is $\frac{120}{5}$, or 24, feet per second. This is commonly written as 24 feet/seconds. This notation does not mean that 1 unit (foot) is divided by another unit (second), but rather that a length measure in feet has been divided by a time measure in seconds.

■ **Question 24** What is the "feet/seconds-to-miles/hour" conversion operation?

It might be asked: When we read a speedometer on a car, don't we measure the speed directly? The answer is "no." This is like the case of weight and the spring. It is based upon the way in which the reading of the speedometer is related to speed. Actually the deflection of the needle is proportional to the speed of turning the front wheels.

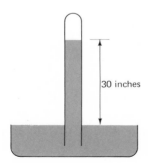

30 inches

Figure 23.42

Another example of an indirect measurement is the common practice of measuring the atmospheric pressure by the height of a column of mercury that it will support. Figure 23.42 shows this form of barometer invented by Galileo's pupil, Torricelli. The space above the mercury is called the Torricellian vacuum. Pressure is defined to be force (which can be expressed in pounds, or in grams) per unit area. It is, therefore, not a length. Nevertheless, the *length* of the mercury column is a measure of pressure since the two are proportional. Approximately 30 inches of mercury corresponds to 15 pounds per square inch (usually written 15 pounds/square inch), approximately 60 inches of mercury corresponds to 30 pounds per square inch, and so on.

We have now looked at some of the basic ideas of measurement, indeed all of those appropriate for elementary school children. There is much more to this general topic of measurement, but it can safely be postponed to the later years of school.

PRACTICAL CONSIDERATIONS

One final comment. This is a mathematics text, so we have confined our attention to the mathematics of measurement. We have paid no attention so far to the practical problems involved in making accurate measurements. But these problems are so important that we cannot neglect them entirely.

There are two facts that we should keep in mind. The first is that most of the measurements that are important to us are carried out by human beings (quite often ourselves). The second is that human beings are fallible. Consequently, the lengths, areas, volumes, times, weights, etc., that we deal with in everyday life are rarely exactly correct.

To emphasize this, you are asked to do an experiment. This is to measure the distance around the block you live on (or some other convenient block if necessary). The unit of length is to be your average step. Choose a mark in the sidewalk, for example a crack, which you will recognize when you return to it. Put your right toe against this mark and step out with your left foot. Walk at a steady pace, taking normal length steps. When you return to your starting point, record the number of steps you took, to the nearest half-step.

Do this once a day for 5 days. Record your answers. They will be used in one of the problems at the end of Chapter 26.

PROBLEMS

1 The chart in Table P23.1 contains measurements of lengths. Complete the chart.

Table P23.1

Measurement	Precise to the nearest	Actual length lies between
$4\frac{1}{2}$ miles	$\frac{1}{2}$ mile	$4\frac{1}{4}$ miles and $4\frac{3}{4}$ miles
$1\frac{1}{4}$ yards		$\cdots$ and $\cdots$
$3\frac{1}{8}$ inches		$\cdots$ and $\cdots$
52.3 millimeters		$\cdots$ and $\cdots$
142.0 kilometers		$\cdots$ and $\cdots$
93 million miles		$\cdots$ and $\cdots$
.0004 centimeters		$\cdots$ and $\cdots$

2 State the greatest possible error for each of these measurements.
   (a) 52 feet   (b) 4.1 inches
   (c) 7.03 inches   (d) 54,000 miles
3 Suppose that the area of a rectangle is reported as 64 square feet. Find the difference between the smallest and largest possible areas if the reported dimensions of the rectangle are
   (a) 64 feet and 1 foot   (b) 8 feet and 8 feet
   Assume that each of these measurements is correct to the nearest foot.
4 In each of the following use a conversion operation to complete the statement.
   (a) $4\frac{1}{6}$ feet = _____ inches
   (b) $4\frac{1}{6}$ feet = _____ yards

(c) $\frac{1}{4}$ mile = _____ yards
(d) $\frac{1}{4}$ mile = _____ feet
(e) 14,400 inches = _____ feet
(f) 14,400 inches = _____ yards

5 A common size for a sheet of paper is $8\frac{1}{2}$ inches wide and 11 inches long. Express each dimension in centimeters, to the nearest centimeter.

6 (a) A foreign sports car has a speedometer that registers speed in kilometers per hour. The car is traveling at a rate of 80 (according to its speedometer) in a 55-mile-per-hour zone. Is the car exceeding the speed limit?

(b) A boy who lives in the United States says he is 4 feet, 8 inches tall. A boy from France says his height is 144 centimeters. Which boy is taller?

7 The basic unit of weight in the metric system is the *gram:* 1 kilogram (equals 1,000 grams) weighs 2.205 pounds.
(a) How many ounces in 1 gram?
(b) How many kilograms in 1 ton?

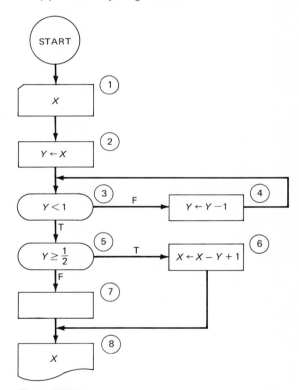

Figure P23.8

8 Figure P23.8 shows a flow chart for reporting the length of an object to the nearest inch if you know its length to the nearest $\frac{1}{4}$ inch.
(a) Trace through the flow chart to find the output for an input of $2\frac{3}{4}$.
(b) Explain what the loop consisting of boxes 3 and 4 does.
(c) Complete box 7 and check it by tracing through the flow chart for an input of $4\frac{1}{4}$.

9 Find the volume of each box whose dimensions are given below.

(a) Base is 6.2 inches by 2 inches; height is 4 inches.

(b) Base is 7.3 centimeters by 3.1 centimeters; height is 2 centimeters.

(c) Base is $\frac{1}{3}$ foot by 6 inches; height is 5 inches.

10 (a) If the altitude of a prism is doubled, its base unaltered and all angles unchanged, how does this affect the volume?

(b) If all edges of a rectangular prism are doubled and its shape left unchanged, how is the volume affected?

(c) The side of the square base of a pyramid is doubled. The height of the pyramid is halved. How is the volume affected?

(d) The sides of a rectangular prism are all doubled. How is the total area affected?

11 Compute the decimal expression for each of these fractions to six decimal places. Which is closest to π?

(a) $\frac{22}{7}$ (b) $\frac{223}{71}$ (c) $\frac{355}{113}$

12 In each part of this problem use the approximate value $\frac{22}{7}$ for π.

(a) What is the circumference (approximately) of a circular table whose diameter is 7 feet?

(b) About how much fencing is needed to enclose a circular garden with a radius of 35 meters?

(c) Find the circumference of a circular pavilion which has a diameter of 140 yards.

13 During the Apollo 11 mission, the command ship *Columbia* circled the moon at a distance of about 69 miles from its surface while the lunar module *Eagle* made its moon landing. If the radius of the moon is 1,080 miles, find the distance *Columbia* traveled during one orbit. (Use $\pi \approx 3.14$.) (Written on Participation Day, July 21, 1969.)

14 Using 3.14 as an approximation for π, find the area of a circle with *diameter*

(a) 2 inches   (b) 20 inches

(c) 7 centimeters   (d) 28 centimeters

15 The region between two circles with the same center is called a *ring* (see Figure P23.15).

(a) Find the area of the ring if the radius of the larger circle is 2 centimeters and the radius of the smaller circle is 1 centimeter.

(b) How does the area of the smaller circle compare with the area of the ring?

Figure P23.15

16 The Pyramid of Cheops in Egypt is 480 feet high, and its square base is 720 feet on a side. How many cubic feet of stone were used to build it? (Assume that the pyramid is solid.)

17 For each sphere whose radius is given below, find the volume of the corresponding spherical solid. Use 3.14 as the approximation for π.

(a) $r = 3$ centimeters   (b) $r = 6$ centimeters

(c) $r = 2$ yards   (d) $r = 4$ yards

(e) Divide the answer to part b by the answer to part a.

(f) Divide the answer to part d by the answer to part c.

18 A silo (with a flat top) is 30 feet high and the inside radius is 6 feet. How many cubic feet of grain will it hold? (What is its volume?) Use $\pi \approx 3.14$.

19 Complete the following conversion operations for units of time.

Example.

The days-to-hours operation is: Multiply by 24 (because there are 24 hours in 1 day).

(a) hours-to-minutes: _____

(b) minutes-to-seconds: _____

(c) hours-to-days: _____

(d) minutes-to-hours: _____

(e) seconds-to-hours: _____

20 A box of 100 electronic resistors has a reported weight of .6 pounds.

(a) The actual weight of one box of 100 resistors lies between _____ pounds and _____ pounds.

(b) Find the smallest and largest possible weights of a shipment of 3,000 resistors, 100 in a box.

24. NEGATIVE NUMBERS

At the very beginning of this book we investigated the whole numbers and we saw that they were useful for describing certain special aspects of the world around us. Later we studied the system of rational numbers, and we have seen in the last few chapters that there are many ways in which they can be used in dealing with real problems in the real world.

Now we are going to investigate still a third system of numbers. This system also arises from a practical need, out of a physical situation which needs a mathematical interpretation.

NEGATIVE NUMBERS   This is a situation where counting or measuring is done with respect to a fixed reference point from which the *direction* of counting or measuring is important. Examples of such situations are measuring temperature in degrees, or altitude in feet. One talks about a temperature of 33.5° above zero or 33.5° below zero, an altitude of 300 feet above or 300 feet below sea level. A business firm may have a credit balance or a debit at its bank. In each case a number by itself will measure the size or the magnitude involved but, without mention of the direction, full information on the physical situation is not given.

■ **Question 1** In finding the length of a building, does it matter if we measure from right to left rather than from left to right?

In each of these physical situations there are essentially only two directions involved. We indicate one direction with the superscript "+" and the other with the superscript "−." The choice as to which direction is labeled $+$ and which $-$ is purely arbitrary, although often the physical situation indicates the best choice. Thus, usually we indicate temperature above zero, distances to the north from a fixed starting point, altitude above sea level and a credit balance at the bank as being in the positive direction in each case. On the other hand, a deep-sea diver might want to indicate the depths of his dives as being in the positive direction and the bank might want to consider a firm's credit balance as money it owes to a depositor and so as being in the negative direction as far as the bank's assets and liabilities are concerned.

A combination such as $+\frac{1}{3}$ or -3 of one of the superscripts and one of the numerals for a rational number gives us an effective name for a number which indicates both direction and size. Such a number is also a rational number. Actually $+\frac{1}{3}$ is merely another name for the number $\frac{1}{3}$ which is one of the rationals studied in previous chapters. For these *positive rationals* we will use or omit the superscript $+$ as happens to be convenient and instructive at the moment. However, -3 is a name for a new kind of number. It is a *negative rational* number, one of the set of such numbers which we are now introducing and are going to study in this chapter. Together with 0 and the positive rationals, these form the set which from now on we will call the set of rational numbers. Sometimes we call these the *signed* rationals or just the signed numbers when we want to emphasize the directions involved; -2, $-\frac{1}{2}$, $+1$ are names which are read "negative two," "negative one-half," and "positive one."

The whole numbers together with their negatives are usually called the *integers*.

■ **Question 2** Is -0 different from $+0$?

Using these signed numbers we can say, having agreed that temperatures above zero are considered to be in the positive direction, "The boiling point of water is $+212°F$" or "The elevation of a town is $+600$ feet" if it is above sea level and we have chosen this as the positive direction.

Since from now on we will be talking mostly about rational numbers, we will use the simple "numbers" as short for "rational numbers." If any other system is intended, we will say so.

A PHYSICAL MODEL

When we studied whole numbers and positive rational numbers, we found that a good physical model such as a number line was a great help to our understanding. Once again a number line will serve our purpose admirably. We start as before by picking a point to represent 0 which we call the origin and label 0. We then pick a direction on the line for the positive direction and a unit length. The point one unit from 0 in the positive direction we label 1, or $+1$, the point 2 units in the positive direction from 0 we label 2. In general, any point which on the number line was originally labeled with the number a/b may now also be labeled

$$\frac{+a}{b}$$

See Figure 24.1.

So far the points of our number line which are on one of the rays from 0 represent only the positive rational numbers and 0. But now, using the ray in the opposite direction from 0, and calling it the negative direction we can represent the negative rational numbers by

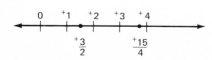

Figure 24.1

points on it. Thus, we label the point 1 unit in the negative direction from 0 with ⁻1, the point 2 units in the negative direction, ⁻2, the point halfway between, $-\frac{3}{2}$, etc.

(When using the number line, we sometimes talk loosely about the "point" or the "number" rather than the "point representing the number." This should not confuse you. It certainly saves time and effort.)

■ **Question 3** How far is it from ⁻2 to 2?

Figure 24.2 makes it apparent that the positive numbers may be thought of as extending indefinitely to the right of 0, and the negative numbers indefinitely to the left of 0.

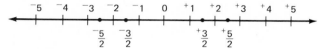

Figure 24.2

OPPOSITES Pairs of numbers such as ⁻1 and ⁺1, $+\frac{3}{2}$ and $-\frac{3}{2}$ are said to be *"opposites."* Thus ⁻2 is the opposite of ⁺2 and $-\frac{3}{2}$ is the opposite of $-\frac{3}{2}$. Pairs of opposite numbers such as ⁻4 and ⁺4 are represented by pairs of points the same distance to the right and left of 0. We write ⁻4 = opp (4) and 4 = opp (⁻4), etc. We find it useful to agree that the opposite of 0 is 0 itself. So every rational now has an opposite.

The statement "⁻2 is the opposite of 2," read "negative 2 is the opposite of 2," is cumbersome to write. If we use the lower dash, "—," to mean "the opposite of," then we can write the above statement as follows:

"⁻2 = −2"

Since "negative 2," ⁻2, "the opposite of 2," and −2, all represent the same number, it makes no difference at what height the dash is drawn. This being the case, we can use either symbol. Therefore,

"−3" is read "negative 3" or "the opposite of 3."
"−(−3)" is read "the opposite of negative 3" or "the opposite of the opposite of 3."

(It should be noted that "−(−3)" is usually *not* read "negative, negative 3.")

■ **Question 4** What is ⁻[⁻(⁻1)]?

When we attach the lower dash "—" to a number, say "7," we shall always read this "the opposite of 7." You can think of this as though we are performing on 7 the operation of "determining the opposite of 7." Do *not* confuse this with the *binary* operation of subtraction, which is performed on *two* numbers, such as 13 − 7, meaning "7 subtracted from 13."

Let us return to the number line model for rationals and let A name the point which corresponds to the number a, etc. When using the number line for positive rational numbers, we saw that if $a > b$, then on a number line in which 1 lay to the right of 0, point A lay to the right of point B. Also, if point C lay to the right of point D then $c > d$.

We say exactly the same thing for the rational numbers and their number line (see Figure 24.3). If 1 is to the right of 0 then 2 is to the right of $^-4$, $^-3$ to the right of $^-10$ and we say 2 is greater than $^-4$, $^-3$ is greater than $^-10$, or in mathematical symbols, $2 > ^-4$, $^-3 > ^-10$. We may also say $^-4$ is less than 2 and $^-10$ is less than $^-3$, writing $^-4 < 2$ and $^-10 < ^-3$.

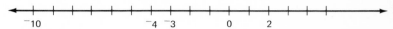

Figure 24.3

■ **Question 5** If we know that $a < b$, what can we say about ^-a and ^-b?

ANOTHER MODEL   We come now to the question of operations on signed numbers. Can we add, subtract, multiply, and divide them, and if we can, do the operations have the properties of closure, commutativity, and associativity? Is multiplication distributive over addition?

We already know about the nonnegative rationals since they have been studied extensively. The real question is: Can we extend the definitions of addition, multiplication, etc., from these numbers to all rational numbers so that all the familiar properties still hold? The answer is yes. To help us see how, we construct a second physical model of rational numbers.

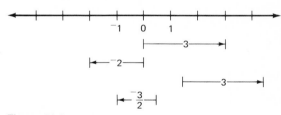

Figure 24.4

In our previous chapters we used several different physical models for whole numbers and rational numbers to help illuminate different mathematical characteristics. Up to now a number, say 3, has been represented by a *point* on the number line. Our new model for the number 3 consists of a directed line segment instead of a point. See Figure 24.4. We represent this directed segment by an arrow which points in the direction from the point 0 to the point 3 and whose length is 3 units. If the positive direction of the number line is from left to right, the model for 3 will be an arrow 3 units long pointing to the right. Similarly, the model for $^-\frac{3}{2}$ is an arrow one and a half units long pointing to the left. In general, if the number r has been represented in our first model by the point R on the number line, the new model is an arrow pointing in the direction from 0 to R and whose length is equal to the measure of $\overline{OR}$. Notice that we have not said where the arrow begins, just how long it is, and in which direction it points. Either arrow in Figure 24.4 may represent $^+3$.

■ **Question 6** In Figure 24.4, what number is represented by the arrow from 1 to $^-1$?

ADDITION OF RATIONALS   This new model will help us greatly in thinking about addition of numbers. What number do we associate with $^+5$ and $^+3$ as their sum? Since these are just new names for 5 and 3, we know that their sum is 8. Thus, $^+5 + {}^+3 = {}^+8$. If we thought of $^+5$ as representing $5.00 earned one day and $^-3$ as $3.00 lost the second day, we would be only $2.00 better off. We could represent this situation as $^+5 + {}^-3$ and the answer seems to be $^+2$. This sum is easily seen using our arrows as models to represent numbers. In our model, addition of two numbers is represented by drawing an arrow representing the first and then from the head of this arrow drawing an arrow representing the second. The sum is then represented by the arrow which goes from the tail of the first to the head of the second. Thus in Figure 24.5 the arrows represent $^+5 + {}^+3 = {}^+8$, while in Figure 24.6 the arrows represent $^+5 + {}^-3 = {}^+2$.

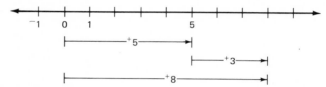

Figure 24.5

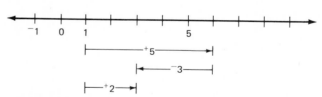

Figure 24.6

■ **Question 7** Draw a representation of $^-2 + 0$.

In these models the arrows could all be drawn in the same line but the figure would be hard to follow, so we have drawn them in separate lines. Also notice that in Figure 24.5 the arrow for $^+5$ started opposite the point 0 on the number line while in Figure 24.6 the arrow started opposite a different point, in this case the point 1. It is important to realize that in this model for addition the first arrow may start anywhere on the number line. Actually, the number line is important in the figures only to indicate the positive direction and the scale of our model. It is not where the arrow starts but its direction and length which tell us what number it represents. Thus in Figure 24.6 the sum of $^+5$ and $^-3$ is an arrow which goes from 1 to 3 and is therefore 2 units long and headed in the positive direction. It represents $^+2$. In later figures we will draw just enough of the number line to indicate the positive direction and the scale.

Figure 24.7

Thus, Figure 24.7 is a model for $^+3 + {}^-5$. The sum is modeled by an arrow 2 units long and directed to the left. The sum is, therefore, $^-2$. Figure 24.8 shows the addition of two opposite numbers, $^+2$ and $^-2$. The sum is 0 and so it will always be for any two opposite numbers.

Figure 24.8

We can equally well add two negative numbers and see the answer using arrow models. See Figure 24.9. In this case our number line is drawn as a vertical line.

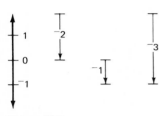

Figure 24.9

■ **Question 8** Does Figure 24.9 illustrate this subtraction?

$^-3 - ({}^-2) = {}^-1$.

The operation of addition may become clearer if you think of it as moving from home in a certain direction a given distance and then *following this* with another walk in the same or the opposite direction for another given distance and finally determining where you are with respect to the starting point. Thus, a walk $3\frac{3}{4}$ blocks east followed by one $5\frac{1}{2}$ blocks west brings you to a point $1\frac{3}{4}$ blocks west of your starting point. $3\frac{3}{4} + {}^-5\frac{1}{2} = {}^-1\frac{3}{4}$. The arrows in Figure 24.10 show you starting at a point 3 blocks east of City Hall and ending at the point $1\frac{3}{4}$ west of where you started, that is, at a point $1\frac{1}{4}$ blocks east of City Hall.

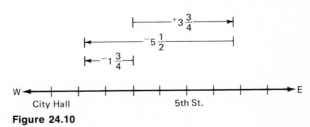

Figure 24.10

Certain general features of the model of addition presented here should be noted.

1 In an addition $b + c = s$ where b and c denote numbers and s the unknown sum, the arrow for the first addend b is *always* drawn first.

2 The arrow for the second addend c is drawn next. Its tail starts at the head of the arrow for b and its direction depends on whether the number c is positive or negative.

3 The arrow giving the unknown sum s is then drawn. Its tail is always in line with the tail of the arrow for b and its head in line with the head of the arrow for c. The length of this arrow and its direction determines s.

4 If the arrow for b is drawn with its tail at 0, the sum s is always the number on the number line opposite the head of the sum arrow.

■ **Question 9** Draw a figure, following steps 1 to 4 above, to show $1 + {}^-2 + 3 + {}^-4$.

From our model it can also be observed that addition of rational numbers is commutative and associative. Figure 24.11 illustrates commutativity. Although this one example does not prove the property, it is in fact true.

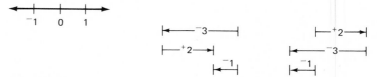

Figure 24.11

Associativity may be shown by examples, such as

$$({}^-3 + {}^-5) + {}^+4 = {}^-8 + {}^+4 = {}^-4$$

and also

$${}^-3 + ({}^-5 + {}^+4) = {}^-3 + {}^-1 = {}^-4$$

This property in fact holds for the addition of any three rationals.

AN ALGORITHM FOR ADDITION OF RATIONALS

While the model of addition given above does a good job of making clear the general nature of addition of negative numbers, it is not very useful when we have some actual computations to do. For example, would you want to work this problem by means of arrows and a number line?

$${}^-4,281 + 2,468 + {}^-10,951 + 7,869$$

Clearly, it would be helpful to have an algorithm for addition when negative numbers are involved. Now we already have an algorithm for adding positive whole numbers (Chapter 6) and for adding positive rational numbers (Chapter 15). It turns out that we can make good use of these. We will not write out all the details of an addition algorithm which will handle negative numbers, but we will describe the procedure in such a way that the details could easily be filled in.

First we note that, in terms of the arrow model of addition, there are three cases that might arise. We will consider them one by one. In each case, we are given two rationals to add, a and b.

■ **Question 10** Draw a figure showing $(^-2 + 3) + ^-4 = ^-2 + (^-4 + 3)$.

In each case, we first show arrows representing a and b. Then we show an arrow representing $a + b$. Next, and this is the basic idea of our algorithm, we consider the opposite of certain of the arrows. These opposites will point to the right and hence will stand for positive numbers, which we already know how to add or subtract. Then we will be able to find $a + b$ by, if necessary, taking one more opposite.

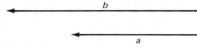

Figure 24.12

CASE 1:
Both numbers negative (see Figure 24.12).

First, we draw the arrow for $a + b$, using the model for addition outlined above. The result is shown in Figure 24.13.

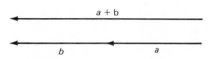

Figure 24.13

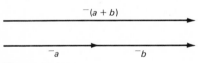

Figure 24.14

Next, we turn these arrows around in the opposite direction, both the arrow for $a + b$ and the composite arrow of a followed by b. The result is shown in Figure 24.14. Because we have changed the direction, the top arrow is a model for $^-(a + b)$. Similarly, the two bottom arrows represent ^-a and ^-b. We see from the way they are arranged that the bottom composite arrow represents $^-a + {}^-b$. Since this bottom composite arrow is the same length as the top arrow and points in the same direction, we now see that

$$^-(a + b) = {}^-a + {}^-b$$

This helps, because ^-a and ^-b are both positive numbers and we know how to add them. So we know what $^-(a + b)$ is. But what we started out to find was $a + b$ itself. We have found its opposite. Now we remember that the opposite of the opposite of a number is the number itself.

$$^-[^-(a + b)] = (a + b)$$

We saw above how to find $^-(a + b)$, so now we see that

$$(a + b) = {}^-[^-(a + b)] = {}^-(^-a + {}^-b)$$

In other words, we add two negative numbers by first adding two positive numbers, the opposites of the two negative numbers, and then forming the opposite of that sum.

As an example, let us find

$^-5 + ^-3$

As in Figure 24.14,

$^-(^-5 + ^-3) = ^-(^-5) + ^-(^-3)$

We know that $^-(^-5) = 5$ and $^-(^-3) = 3$, and of course we know that $5 + 3 = 8$. So

$^-(^-5 + ^-3) = 8$

Now, taking opposites, we see that

$^-[^-(^-5 + ^-3)] = (^-5 + ^-3) = ^-8$

CASE 2:
The longer arrow pointing right and the shorter arrow left.

Figure 24.15

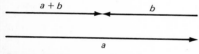

Figure 24.16

Figure 24.15 illustrates such a case and Figure 24.16 shows $a + b$ together with a and b. If we change the direction of the b arrow, we have the situation shown in Figure 24.17. In this figure, each arrow represents a positive number, and we see that they inform us that

$a = (a + b) + ^-b$

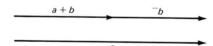

Figure 24.17

We remember from Chapter 5 that any such addition statement can be rewritten as a subtraction statement:

$a + b = a - ^-b$

Again, we have computed $a + b$ by performing standard operations on positive quantities.

As an illustration, let us compute

$5 + ^-3$

From Figure 24.17, we see that

$(5 + ^-3) + ^-(^-3) = 5$

We know that $^-(^-3) = 3$, so we see that

$(5 + ^-3) + 3 = 5$

or

$(5 + ^-3) = 5 - 3 = 2$

CASE 3:
The longer arrow points left and the shorter arrow right.

Figure 24.18

Figure 24.19

Figure 24.18 shows such a case and Figure 24.19 shows $a + b$. In Figure 24.20 we reverse the directions of both $(a + b)$ and a. Figure 24.20 shows that

$$b + {}^-(a + b) = {}^-a$$

Figure 24.20

Notice that each quantity in this statement is positive. This can be rewritten as

$${}^-(a + b) = {}^-a - b$$

Now we remember that the opposite of the opposite of a number is the number itself, so

$${}^-({}^-(a + b)) = (a + b) = {}^-({}^-a - b)$$

and once more we can find the sum by performing standard operations on positive numbers and then by forming the negative of a number.

As an illustration, we compute

$${}^-5 + 3$$

From Figure 24.20, we see that

$$3 + {}^-({}^-5 + 3) = {}^-({}^-5)$$

We know that ${}^-({}^-5) = 5$, so

$$3 + {}^-({}^-5 + 3) = 5$$

or

$${}^-({}^-5 + 3) = 5 - 3 = 2$$

Taking opposites again,

$${}^-[{}^-({}^-5 + 3)] = ({}^-5 + 3) = {}^-(2) = {}^-2$$

■ **Question 11** Why did we not consider the case where both arrows point to the right?

Thus, whenever one or both of two numbers are negative, we can find their sum by taking opposites and by adding or subtracting only positive numbers, using the algorithms in Chapter 6 or in Chapter 15.

There is no need to remember the details of the three different cases. Whenever you are faced with an addition problem involving negative numbers, draw a sketch, representing the numbers by arrows. The sketch will tell you what to do. (Or, if you wish, you can use the flow chart in Problem 19.)

SUBTRACTION OF RATIONALS

We have considered addition of rationals very carefully. What can we say about subtraction? We can consider the subtraction of r and s in two ways: first directly via the physical models, the arrows which represent r and s; and second as the inverse of addition, but again using the arrows to represent the proper addition problem. In the first case we have to consider what sort of combination of the two arrows is a model for subtraction; in the second case we say that $r - s$ is the number which answers the question "What number added to s will give r?" It is important, of course, that both methods are consistent, i.e., give the same answer. In the first case, consider $^+5 - {}^+2$. Since $^+5$ and $^+2$ are simply different names for the numbers 5 and 2 we know the answer must be $^+3$. What could we do to the arrow representing $^+2$ so that we could combine it with the one for $^+5$ and get the one for $^+3$? The answer is quickly seen to be "Reverse the direction of the $^+2$ arrow and add it to the $^+5$ arrow." But reversing the direction of a $^+2$ arrow simply gives us an arrow representing the opposite of $^+2$ or $^-2$. This gives us the hint we need. Our physical model of the subtraction of a rational number will be to *reverse its arrow and add.* Thus:

$$^+5 - {}^+2 = {}^+5 + {}^-2 = {}^+3$$

and

$$^+2 - {}^+5 = {}^+2 + {}^-5 = {}^-3$$

and

$$^+2 - {}^-5 = {}^+2 + {}^+5 = {}^+7$$

■ **Question 12** Is $^-5 - ({}^-3 - 1) = ({}^-5 - {}^-3) - 1$?

Actually making physical models of these arrows and using them in this fashion is an extremely useful and illuminating experiment. Again

$$^-7 - {}^-15 = {}^-7 + {}^+15 = {}^+8$$

and in general for any two rational numbers r and s

$$r - s = r + \text{opposite of } s$$

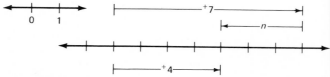

Figure 24.21

The second look at subtraction is as the inverse of addition. Thus $^+4 - {}^+7$ asks us for a number n, if any, such that $^+7 + n = {}^+4$. But the arrow model in Figure 24.21 shows that there is an arrow from the head of $^+7$ to the head of $^+4$ and clearly this arrow by size and direction represents $^-3$.

The subtraction of any rational number may be replaced by the addition of its opposite. Unlike the systems we have had before, the rational numbers are closed under subtraction, i.e., for any two rational numbers, p and q, there always exists a unique rational number, s, such that $p - q = s$. This was not true for any of the systems we have studied before. There was no whole number n such that $n = 4 - 7$, nor rational number r such that $r = \frac{4}{5} - \frac{17}{3}$.

Our rational number system is closed under subtraction, but suddenly we find we do not need subtraction any more as any subtraction can be replaced by an addition of the opposite number. This parallels closely what happened when we examined division for the positive rationals. We found that division by any number except zero could be replaced by multiplication by its reciprocal. In some ways enlarging the number system pays off in increased simplicity of operation as well as in applicability to many new kinds of physical situations.

■ **Question 13** Is the rational number system closed under the operation of "determining the opposite"?

MULTIPLICATION AND DIVISION OF RATIONALS

What about multiplication of rational numbers? Since the signed numerals such as $^+7$ and $^+5$ are simply different names for the positive rational numbers 7 and 5, their product is already known. Thus we may write

$$^+7 \times {}^+5 = 7 \times 5 = 35 = {}^+35$$

and

$$^+\tfrac{1}{3} \times {}^+\tfrac{2}{5} = \tfrac{1}{3} \times \tfrac{2}{5} = \tfrac{2}{15} = {}^+\tfrac{2}{15}$$

The product $^+3 \times {}^-2$ may be thought of as corresponding to the situation that if we lose \$2.00 a day for 3 days, we have lost a total of \$6.00. Mathematically, the number sentence would be $^+3 \times {}^-2 = {}^-6$. We hope multiplication of signed numbers is going to be commutative and therefore that $^-2 \times {}^+3$ will also be $^-6$. But physically this might be thought of as corresponding to the question "If I am making \$3.00 a day now, how did my financial situation 2 days ago compare to it today"? Obviously, I was \$6.00 poorer. And it seems true that $^-2 \times {}^+3 = {}^-6$. On the other hand, if you are spending \$3.00 a day, how did your situation compare 2 days ago with it today? Obviously, you were \$6.00 better off. This corresponds to $^-2 \times {}^-3 = {}^+6$.

Another way to consider multiplication of positive and negative numbers is to consider the pattern in the following sequnce. We know

$$^+2 \times {}^+3 = {}^+6 \qquad {}^+1 \times {}^+3 = {}^+3 \qquad 0 \times {}^+3 = 0$$

Now what should

$$^-1 \times {}^+3 \qquad \text{and} \quad {}^-2 \times {}^+3 \qquad \text{be?}$$

If the pattern of dropping 3 units each time on the right side of these equations is to be preserved, the answers should be $^-3$ and $^-6$ respectively. Thus

$$^-1 \times {}^+3 = {}^-3 \quad \text{and} \quad {}^-2 \times {}^+3 = {}^-6$$

If the commutative property is to hold, we should also have

$$^+3 \times {}^-2 = {}^-6$$

But then another pattern will develop as we see that

$$2 \times {}^-2 = {}^-4 \quad 1 \times {}^-2 = {}^-2 \quad 0 \times {}^-2 = 0$$

Now it seems right that

$$^-1 \times {}^-2 = {}^+2 \quad \text{and} \quad {}^-2 \times {}^-2 = {}^+4$$

■ **Question 14** Here is another explanation of why $^-2 \times 3 = {}^-6$. Give the reason for each step.

(a) $2 + {}^-2 = 0$
(b) $(2 + {}^-2) \times 3 = 0$
(c) $0 = (2 + {}^-2) \times 3 = 2 \times 3 + {}^-2 \times 3$
(d) $^-2 \times 3$ is the opposite of 2×3
(e) $^-2 \times 3 = {}^-(2 \times 3)$

Corresponding to the physical model and these patterns, we can define multiplication of rational numbers as follows. Suppose r and s represent *positive* rationals. We already know from Chapter 12 that $r \times s$ is a positive rational. We define

$$r \times {}^-s = {}^-(r \times s)$$
$$^-r \times s = {}^-(r \times s)$$
$$^-r \times {}^-s = {}^+(r \times s) = r \times s$$

and finally

$$^+r \times 0 = 0 \times {}^-r = 0$$

Thus

$$^-2 \times {}^-5 = {}^+(2 \times 5) = {}^+10 \quad \text{and} \quad {}^-\tfrac{1}{2} \times {}^+\tfrac{2}{3} = {}^-(\tfrac{1}{2} \times \tfrac{2}{3}) = {}^-\tfrac{1}{3}$$

and

$$^+3 \times {}^-\tfrac{1}{3} = {}^-1 \quad \text{and} \quad 0 \times {}^-\tfrac{2}{3} = 0$$

Does multiplication thus defined distribute over addition? Let's look at an example.

$$^-5 \times ({}^+6 + {}^-2) = {}^-5 \times {}^+4 = {}^-20$$

and

$$({}^-5 \times {}^+6) + ({}^-5 \times {}^-2) = {}^-30 + {}^+10 = {}^-20$$

So in this case it does distribute. We will not attempt a general proof of this or other properties, but will assure you that indeed multiplication for the signed numbers is closed, commutative, associative, and distributive over addition. The familiar properties of 0 and 1 also hold.

Division by any rational number r $(r \neq 0)$ is, as before, defined as the inverse of multiplication by r. Thus, $^+8 \div {}^-4 = {}^-2$ since $^-4 \times {}^-2 = {}^+8$, and $^-\frac{3}{5} \div {}^-\frac{3}{4} = {}^+\frac{4}{5}$ since $^-\frac{3}{4} \times {}^+\frac{4}{5} = {}^-\frac{3}{5}$. In each case $r \div s$ is that number which multiplied by s will produce r. We see that the same rules for determining the sign of the answer apply as in multiplication. Also, if the reciprocal of r is that number $\dfrac{1}{r}$ which multiplied by r produces 1, it can be seen that division by any rational number $\neq 0$ is equivalent to multiplying by its reciprocal. Thus $^+3 \div {}^-5 = {}^+3 \times {}^-\frac{1}{5} = {}^-\frac{3}{5}$.

■ **Question 15** Does $^-0$ have a reciprocal?

Now that we have explored this new number system which includes negative as well as positive numbers, we will see, in the next few chapters, a number of ways in which we can use negative numbers.

PROBLEMS

In each of the problems below, you may use a number line to help you if you wish.

1 Find these sums:
(a) $9 + {}^-5$ (b) $^-8 + 11$ (c) $3\frac{1}{2} + {}^-2\frac{1}{4}$
(d) $13 + {}^-(2 + 7)$ (e) $6\frac{1}{5} + {}^-8\frac{2}{3}$ (f) $^-14 + 9 + {}^-7$
(g) $^-7\frac{1}{2} - {}^-3\frac{7}{8}$ (h) $^-(^-18) + {}^-17$ (i) $^-[^-(^-7)] + {}^-(^-3)$
(j) $^-(^-2 + 4)$

2 Find the missing addend:
(a) $\underline{\quad} + {}^-8 = {}^-6$ (b) $^-4 + \underline{\quad} = {}^-5$
(c) $^-(3 + \underline{\quad}) = 8$ (d) $^-5 = \underline{\quad} + 3$
(e) $^-(\underline{\quad} + 2) = 4$ (f) $^-[^-(2 + \underline{\quad})] = {}^-(1 + {}^-4)$

3 Fill in the rest of this addition table (Table P24.3).

Table P24.3

+	-4	-3	-2	-1	0	1	2	3	4
-4									
-3									
-2									
-1			-3						
0									
1									
2							4		
3									
4		1							

4 Work these subtractions:

(a) $2 - 35$ (b) $^-11 - ^-13$

(c) $\frac{2}{3} - ^-\frac{4}{5}$ (d) $^-1 - (^-3 + ^-4)$

(e) $^-\frac{9}{4} - \frac{7}{4}$ (f) $72 - ^-28$

(g) $^-3 - 10 - ^-6$ (h) $^-(^-3 - 6) - ^-10$

(i) $^-5 - (^-7 - 4) - ^-8$ (j) $^-1 - ^-(^-2) - ^-[^-(^-3)]$

5 Fill in the missing number:

(a) $\underline{\quad} - ^-9 = ^-15$ (b) $^-20 - \underline{\quad} = ^-15$

(c) $1 - \underline{\quad} = ^-15$ (d) $^-6 - \underline{\quad} = 3$

(e) $\underline{\quad} - 7 = ^-3$ (f) $\underline{\quad} - ^-7 = 3$

6 Fill in the rest of this subtraction table (Table P24.6).

Table P24.6

−	⁻4	⁻3	⁻2	⁻1	0	⁺1	⁺2	⁺3	⁺4
⁻4		⁻1							
⁻3									
⁻2									⁻6
⁻1									
0									
1			3						
2									
3						2			
4									

7 Complete the following sentences by writing the correct sign of operation.

(a) $^-5 \underline{\quad\quad} ^-7 = ^+2$ (b) $^+1 \underline{\quad\quad} ^-1 = 0$

(c) $^+13 \underline{\quad\quad} ^-10 = ^+23$ (d) $^-9 \underline{\quad\quad} ^+7 = ^-2$

(e) $^-1 \underline{\quad\quad} ^+6 = ^+5$ (f) $^+18 \underline{\quad\quad} ^-3 = ^+21$

(g) $^-2 \underline{\quad\quad} ^-4 = ^+2$ (h) $^+26 \underline{\quad\quad} ^+11 = ^+15$

8 Find these products:

(a) $^-2 \times 8$ (b) $\frac{3}{5} \times ^-\frac{5}{2}$

(c) $^-6 \times ^-5 \times \frac{1}{8}$ (d) $^-5 \times ^-3$

(e) $^-5 \times ^-3 \times ^-1$ (f) $\frac{2}{3} \times ^-\frac{2}{5}$

(g) $^-4 \times 3 \times ^-\frac{1}{12}$ (h) $^-(^-5 \times 2) \times ^-(6 \times ^-\frac{1}{2})$

(i) $^-(\frac{1}{7} \times ^-\frac{3}{2})$ (j) $^-[^-(2 \times ^-3 \times ^-\frac{1}{4})]$

9 Find these quotients:

(a) $^-12 \div 3$ (b) $6 \div ^-2$

(c) $^-15 \div ^-5$ (d) $\frac{1}{4} \div ^-\frac{1}{2}$

(e) $^-\frac{1}{2} \div \frac{1}{4}$ (f) $\frac{4}{7} \div ^-14$

10 Fill in the missing numbers:

(a) $\underline{\quad\quad} \times 6 = ^-12$ (b) $^-2 \div \underline{\quad\quad} = 4$

(c) $^-3 \div 2 = \underline{\quad\quad} \times ^-1$ (d) $^-(^-2) \times \underline{\quad\quad} = ^-8$

(e) $36 \times \underline{\quad\quad} = ^-6$ (f) $^-2 \times 3 \times \underline{\quad\quad} = ^-4$

(g) $^-(\frac{1}{2} \times ^-\frac{1}{3}) = ^-4 \div \underline{\quad\quad}$ (h) $^-6 = 2 \times ^-1 \times \underline{\quad\quad}$

11 Fill in the rest of this multiplication table (Table P24.11).

Table P24.11

×	⁻4	⁻3	⁻2	⁻1	0	1	2	3	4
⁻4									
⁻3			6						
⁻2									
⁻1								⁻3	
0									
1							2		
2									
3		⁻9							
4									

12 Copy each number, write its opposite, and then circle the greater of the number and its opposite.
(a) 3   (b) 0
(c) 17   (d) −5
(e) −(−2)   (f) $^-[^-(^-\frac{1}{2})]$

13 Copy and complete the following sentences by writing the correct sign between the numbers, ">," "<," or "=."
(a) $^+3$ _____ $^+5$ (b) $^-1\frac{2}{7}$ _____ $^-\frac{4}{5}$
(c) $^-\frac{8}{3}$ _____ $^+\frac{6}{4}$ (d) $^+1$ _____ $^-19$
(e) $^-16$ _____ $^-32$ (f) $^+479$ _____ $^+421$
(g) $^+89$ _____ $^+95$ (h) $^-26$ _____ $^-26$
(i) $^-\frac{3}{7}$ _____ $^-\frac{5}{12}$ (j) 0 _____ $^-7$

14 (a) What is the smallest integer greater than $^-5$?
(b) What is the largest integer less than $^-9$?

15 Choose the smallest integer from each set.
(a) $P = \{^-29, ^+3, ^+31, ^-50, ^-1\}$ (b) $T = \{^+5, ^+1, ^-2, ^-1, ^-5\}$
(c) $W = \{^+23, ^-41, ^-30, ^+29, ^+20\}$ (d) $F = \{0, ^-3, ^-7, ^-2, ^-6\}$
(e) $G = \{^-4, ^+10, ^+15, ^-9, ^+1\}$

16 The temperature reading at 5 P.M. was 3°F below zero. The temperature then dropped 10°F. What was the new temperature reading?

17 The lowest temperature ever recorded in the United States was 70°F below zero at Roger's Pass, Montana. The highest temperature ever recorded in the United States was 134°F at Death Valley, California. How many degrees higher was the temperature recorded at Death Valley than the temperature recorded at Roger's Pass?

18 In four successive plays from scrimmage, starting at its own 20-yard line, Franklin High makes a gain of 17 yards, then a loss of 6 yards, next a gain of 11 yards, and finally a loss of 3 yards.
(a) Represent the gains and losses in terms of positive and negative numbers.
(b) Where is the ball after the fourth play?
(c) What is the net gain after the four plays?

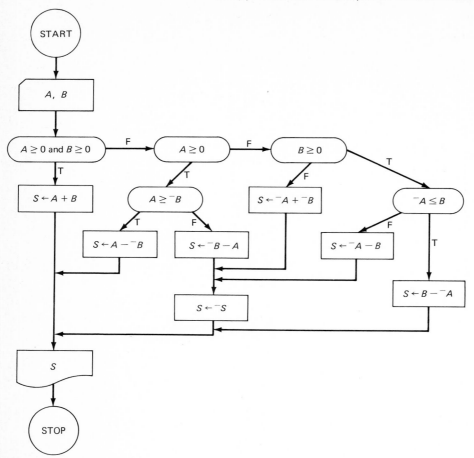

Figure P24.19

(1) 1, 2   (2) 3, 2   (3) ⁻1, 2
(4) ⁻3, 2   (5) 1, ⁻2   (6) 3, ⁻2
(7) ⁻1, ⁻2   (8) ⁻3, ⁻2

(b) What does this flow chart do?

20 (a) The flow chart in Figure P24.20 is supposed to multiply any two numbers, A and B, whether they are positive or negative. What should be in boxes 5 and 7 to make it work properly?

(b) Use these pairs of numbers to check out the flow chart:
(1) $\frac{1}{2}, \frac{3}{5}$ (2) ⁻4, $\frac{2}{3}$
(3) 1.70, ⁻6   (4) ⁻5, ⁻8

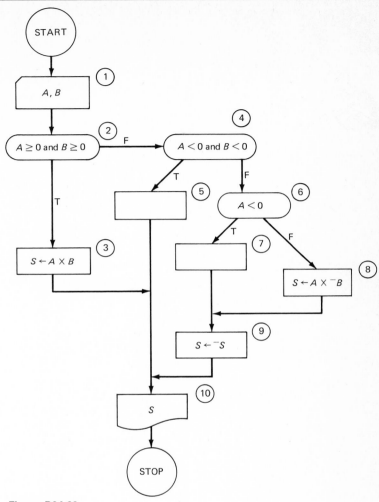

Figure P24.20

25.

SOLVING MATHEMATICAL SENTENCES

We looked, in Chapter 9, at number sentences such as

$$7 + 4 = 11$$
$$17 - 3 = 4$$
$$n + 8 = 22$$
$$3 \times 13 \,\square\, 10 \times 4$$

We saw there that many problems which asked "how many" questions could be translated into open sentences and we saw that by solving the sentences we could often find answers to the problems.

However, we did not develop then any specific techniques for solving sentences, and in fact, the only procedure that was suggested was that of trial and error.

We did notice, however, that there were some sentences that we were not able to solve at that time. An example is $n + 4 = 2$. We can see now that if we replace n by $^-2$, the resulting sentence is true, but when we studied Chapter 9, we only had the nonnegative whole numbers to work with.

We did not mention it at the time, but the open sentence $2 \times n = 3$ is another example of a sentence we could not solve at that time. Now that we have studied fractions, we see that the solution of the sentence is $\frac{3}{2}$.

EQUATIONS   In this chapter we are going to see that there is a systematic way to go about trying to solve some kinds of mathematical sentences. In order to use this method, we have to know how to work with negative numbers. We will start with a specific situation:

A student has saved $7.00 and he plans to use his summer vacation doing odd jobs in order to earn additional money for the purchase of a bicycle. Let us assume that he can save $2.00 a week from the money earned doing odd jobs.

If we ask, "How much money will he have at the end of his summer vacation?" we can give no answer. We must know how many weeks he will work during the summer; one week? two weeks? nine weeks? The chart below shows the savings for various numbers of weeks.

Number of weeks worked	Savings
1	$2 \times 1 + 7$
2	$2 \times 2 + 7$
3	$2 \times 3 + 7$
. . .	. . .
n	$2 \times n + 7$

In all cases, we know that the student's savings will be "the sum of twice a number and seven." The number is, of course, the number of weeks he works. If he works n weeks his savings are $2 \times n + 7$. Thus "$2n + 7$" is the mathematical description of the phrase "the student's savings if he works n weeks."

■ **Question 1** Would "$2x + 7$" be as good a phrase for the student's savings?

Let us now ask the question, "How many weeks will the student need to work (and save) if he wishes to purchase a used bicycle costing $23?"

We know that his savings will be

$2n + 7$

where n is the number of weeks he works. In order that he be able to purchase the bicycle as soon as possible, we know that his savings must be $23; hence, we want a number n such that $2n + 7$ is equal to 23. In this way we get the mathematical sentence

$2n + 7 = 23$

This kind of mathematical sentence is called an *equation*. Notice that it is composed of three parts:

A mathematical phrase	The equality symbol	A mathematical phrase
$2n + 7$	$=$	23

We can think of this equation as a "claim" that for *some* number replacement of n, the expressions or phrases "$2n + 7$" and "23" name the same number.

We have seen that the *equation*

$2n + 7 = 23$

can be used as a model for a problem about saving money for a bicycle. We remember that if we replace the variable n by a number, then the equation becomes a *statement* which is either *true* or *false, but not both.* Of course, we know from arithmetic that if n is replaced by 8 in the equation

$$2n + 7 = 23$$

then the resulting statement

$$2(8) + 7 = 23$$

is true (that is, $16 + 7 = 23$ is true).   Furthermore, we know that any other number replacement of n will produce a false statement.

■ **Question 2**   What numbers make the sentence $23 = 7 + 2x$ true?

SOLUTION SETS   We also remember that we call the set of numbers which result in *true* statements *the solution set* of the sentence, and each member of the solution set is called *a solution* of the sentence.   Thus

$\{8\}$ is the solution set of $2n + 7 = 23$

and

8 is a solution of $2n + 7 = 23$

There are some equations where it is not possible to find *any* numbers that will result in a true statement.   One such equation is

$$x + 3 = x$$

This is obvious since no number can be 3 greater than itself.

For example, if x is replaced by 2 in the equation $x + 3 = x$, then the resulting statement

$$2 + 3 = 2$$

is false (that is, $5 = 2$ is false).

There are also some equations where *every* number will result in a true statement.   Consider

$$x + {}^-x = 0$$

We saw in the last chapter that any number added to its opposite gives zero, so no matter what number we put in place of x, the resulting statement is true.

As you can see from the above discussion, it is possible for the solution set of a mathematical sentence to have no members, exactly one member, or more than one member.

In the rest of this chapter we will develop some organized procedures for finding solutions of certain kinds of sentences.

■ **Question 3**   The symbol $[x]$ stands for the greatest integer that is less than or equal to x.   Thus $[{}^+\frac{3}{2}] = 1$, $[2.05] = 2$, and $[{}^-\frac{3}{2}] = -2$. How many solutions does this equation have?

$$[x] = x$$

EQUIVALENT EQUATIONS   It is important to consider the set of numbers which can be used as replacements for the variable in an equation.

For the equation

$2x = 5$

if only the integers can be used as replacements for the variable, the equation has no solution. On the other hand, if rational numbers are allowed as replacements for the variable, then the equation $2x = 5$ has $\frac{5}{2}$ as a solution.

We call the set of numbers which can be used as replacements for the variable in a sentence *the replacement set* of the sentence. If you are asked to find the solution set of an equation and the replacement set is not specified, you may assume it to be the set of rational numbers.

Equations a and b below have the same solution, 6.

(a) $2x = 12$

(b) $4x = 24$

Since $4x$ equals $2(2x)$ and 24 equals 2(12), we say that Equation b may be obtained by multiplying both sides of Equation a by 2. Likewise Equation a may be obtained by multiplying both sides of Equation b by $\frac{1}{2}$.

Equations that have the same replacement set and the same solution set are said to be *equivalent* to each other, and we refer to them as *equivalent equations*.

**MULTIPLICATION
PROPERTY OF EQUALITY**

The idea we used above, multiplying *both sides* of an equation by the same nonzero number, to conclude that

$$2x = 12 \quad \text{and} \quad 4x = 24$$

are equivalent to each other, is based on our knowledge of multiplication. We will state it in mathematical language as follows:

If $a = b$ and $c \neq 0$, then $ac = bc$.

Also, if $ac = bc$ and $c \neq 0$, then $a = b$.

(Note: Since we are assuming the commutative property of multiplication for rational numbers, we could place the letter c to the left of a and b.)

■ **Question 4**   Can we say that if $a = b$ and $c \neq 0$, then $a/c = b/c$?

Since this idea will be used many times in the work that lies ahead, we will refer to it as *the multiplication property of equality*.

Now let us see how this property can be used to solve an equation like

$4x = 24$

The multiplication property of equality guarantees that

$4x = 24$

and

$c \times (4x) = c \times (24) \qquad c \neq 0$

are equivalent equations. What is the simplest equation we can write that is equivalent to $4x = 24$? Can we write one which has only $1 \times x$, or x, on one side? Suppose we multiply both sides of the equation

$$4x = 24$$

by $\frac{1}{4}$, the reciprocal of 4, to obtain the equation

$$\tfrac{1}{4}(4x) = \tfrac{1}{4}(24)$$

Now, of course, we can use the properties of the operations on rational numbers to find the simplest form of this equivalent equation. The details of the work are shown below:

$$4x = 24$$
$$\tfrac{1}{4}(4x) = \tfrac{1}{4}(24) \qquad \text{multiplication property of equality}$$
$$(\tfrac{1}{4} \times 4)x = \tfrac{1}{4}(24) \qquad \text{associative property of multiplication}$$
$$1 \times x = \tfrac{1}{4}(24) \qquad \text{multiplication property of reciprocals}$$
$$x = 6 \qquad \text{multiplication property of 1}$$

Thus, by the multiplication property of equality, if any number is a solution of $4x = 24$, it is a solution of $x = 6$; conversely, if any number is a solution of $x = 6$, we may multiply both sides of the equation by 4 to get

$$4x = 24$$

and see that if any number is a solution of $x = 6$, it is also a solution of $4x = 24$.

Since 6 is obviously the only solution of $x = 6$, we know that $\{6\}$ is the solution set of $4x = 24$.

■ **Question 5** Are these equations equivalent?

$$27x = 0 \qquad \text{and} \qquad \frac{x}{27} = 0$$

Let us use this method (in abbreviated form) to solve the equations in the following examples.

EXAMPLE 1:
Find the solution set of $\frac{3}{2}x = -15$.
The reciprocal of $\frac{3}{2}$ is $\frac{2}{3}$ since
$\frac{3}{2} \times \frac{2}{3} = 1$.

$$\frac{3}{2}x = -15$$
$$\tfrac{2}{3}(\tfrac{3}{2}x) = \tfrac{2}{3}(-15)$$
$$x = -10$$

Check:
$$\frac{3}{2}x = -15$$
$$\tfrac{3}{2}(-10) = -15$$
$$-15 = -15$$

The solution set is $\{-10\}$.

EXAMPLE 2:
Solve: $-x = 16.3$ [recall that $-x = (-1)x$].

$$(-1)x = 16.3$$
$$(-1)(-1)x = (-1)16.3$$
$$x = -16.3$$

Check:
$$-x = 16.3$$
$$-(-16.3) = 16.3$$
$$16.3 = 16.3$$

The solution set is $\{-16.3\}$.

In solving equations, the distributive property is often used to simplify expressions. For example, in solving the equation $3x - 8x = 45$, the distributive property is used to simplify $3x - 8x$ to $-5x$ as follows:

$$3x - 8x = 45$$
$$(3 - 8)x = 45$$
$$-5x = 45$$
$$-\tfrac{1}{5}(-5x) = -\tfrac{1}{5}(45)$$
$$x = -9$$

The solution set is $\{-9\}$.
Study the following example carefully.

EXAMPLE 3:
Solve $-2x/3 + x/2 = 5$ [recall that $-2x/3 = (-\tfrac{2}{3})x$ and $x/2 = (\tfrac{1}{2})x$].   Check:

$$\frac{-2x}{3} + \frac{x}{2} = 5 \qquad\qquad\qquad \frac{-2x}{3} + \frac{x}{2} = 5$$
$$(-\tfrac{2}{3} + \tfrac{1}{2})x = 5 \qquad\qquad \frac{-2(-30)}{3} + \frac{(-30)}{2} = 5$$
$$(-\tfrac{4}{6} + \tfrac{3}{6})x = 5$$
$$-\tfrac{1}{6}x = 5 \qquad\qquad\qquad 20 - 15 = 5$$
$$-6(-\tfrac{1}{6}x) = -6(5) \qquad\qquad\qquad 5 = 5$$
$$x = -30$$

The solution set is $\{-30\}$.

■   **Question 6**   Solve: $5 = -2x/3 + x/2$

ADDITION PROPERTY OF EQUALITY

The solution set for both Equation a and Equation b below is $\{4\}$. Check it.

(a) $3x + 5 = 17$

(b) $3x = 12$

Equation b may be obtained by adding -5 to both sides of Equation a. Likewise Equation a may be obtained by adding 5 to both sides of Equation b.

The idea of adding the same number to both sides of an equation to obtain an equivalent equation is based upon our knowledge of addition, and is called the *addition property of equality.* It is stated in mathematical language as follows.

If $a = b$ then $a + c = b + c$.

Also, if $a + c = b + c$ then $a = b$.

Note that since we are assuming the commutative property of addition for rational numbers, we could place the letter c to the left of a and b.

The use of the addition property of equality is illustrated in the following examples.

EXAMPLE 4:

Solve: $x - 11 = 23$

The additive inverse of -11 is 11 since

$$-11 + 11 = 0$$
$$x - 11 = 23$$
$$x - 11 + 11 = 23 + 11$$
$$x = 34$$

Check:
$$x - 11 = 23$$
$$34 - 11 = 23$$
$$23 = 23$$

The solution set is $\{34\}$.

EXAMPLE 5:

Solve: $\frac{1}{6} - x = -\frac{1}{2}$

The additive inverse of $\frac{1}{6}$ is $-\frac{1}{6}$.

$$\frac{1}{6} - x = -\frac{1}{2}$$
$$-\frac{1}{6} + \frac{1}{6} - x = -\frac{1}{6} - \frac{1}{2}$$
$$-x = -\frac{4}{6}$$
$$x = \frac{2}{3}$$

Check:
$$\frac{1}{6} - x = -\frac{1}{2}$$
$$\frac{1}{6} - \frac{2}{3} = -\frac{1}{2}$$
$$\frac{1}{6} - \frac{4}{6} = -\frac{1}{2}$$
$$-\frac{1}{2} = -\frac{1}{2}$$

The solution set is $\{\frac{2}{3}\}$.

To solve an equation such as

$$2n + 7 = 23$$

we use both the addition and multiplication properties of equality. Study the following chain of equivalent equations carefully.

$$2n + 7 = 23$$
$$2n = 23 + (-7) \qquad \text{add } -7$$
$$2n = 16$$
$$n = \frac{1}{2} \times 16 \qquad \text{multiply by } \frac{1}{2}$$
$$n = 8$$

The solution set is $\{8\}$.

■ **Question 7** Solve: $1 - 2x + 3 = 4 - \frac{1}{4}$

ANOTHER ANALYSIS

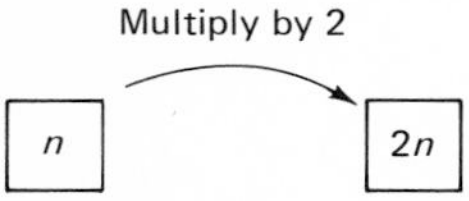

Before moving ahead, let us examine this equation from a new viewpoint. We want our method to appear logical rather than mysterious.

We know $2n + 7 = 23$ can be thought of as a "claim" that for *some* number replacement of n the expressions $2n + 7$ and 23 name the same number. To get the expression $2n + 7$, we see that we start with a number, n,

$$\boxed{n}$$

multiply it by 2 (see Figure 25.1), and then add seven (see Figure 25.2). Since we are assuming that for some number, n, $2n + 7$ and 23 name the same number, we have what is shown in Figure 25.3.

Multiply by 2

$$\boxed{n} \qquad \boxed{2n}$$

Figure 25.1

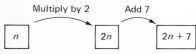

Figure 25.2

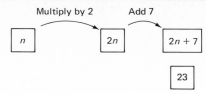

Figure 25.3

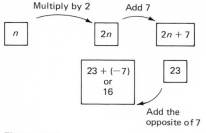

Figure 25.4

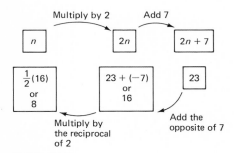

Figure 25.5

Reversing our steps we find the expression in the second box (Figure 25.4), and then the expression in the first box (Figure 25.5).

There are two things that stand out in this "box" analysis of $2n + 7 = 23$:

1 The addition of the same number, -7, to both sides of the equation to get an equivalent equation of the form $an = c$ (in our example, $2n = 16$).

2 The multiplication of both sides of the resulting equation, $2n = 16$, by the same number, $\frac{1}{2}$, to get an equivalent equation of the form $n =$ some number (in our example, $n = 8$).

■ **Question 8** Use boxes to solve: $\frac{1}{6} - x = -\frac{1}{2}$

Here is another problem.

Betty has $54 in savings and her brother Tom has $30. They would like to save more money for their vacation. If Betty can save $3 a week and Tom can save $5 a week, how long will it take until Tom has the same amount of savings as Betty?

Let us see if we can write a mathematical equation which can serve as a model for the problem. The chart below shows the amounts Betty and Tom will have in their savings accounts at the ends of various weeks.

Number of weeks worked	Betty's savings	Tom's savings
1	$3 \times 1 + 54$	$5 \times 1 + 30$
2	$3 \times 2 + 54$	$5 \times 2 + 30$
3	$3 \times 3 + 54$	$5 \times 3 + 30$
· · ·	· · ·	· · ·
x	$3x + 54$	$5x + 30$

We want to know for what number x the amount of Betty's savings and the amount of Tom's savings are the same. That is, we wish to solve the following equation.

$$3x + 54 = 5x + 30$$

This equation differs from the ones we have been solving in that x appears on both sides of the equation. We proceed by adding $-5x$, the opposite of $5x$, to both sides.

$$3x + 54 = 5x + 30$$

$-5x + 3x + 54 = 30$	add $-5x$
$-2x + 54 = 30$	distributive property
$-2x = -24$	add -54
$x = 12$	multiply by $-\frac{1}{2}$

It will take 12 weeks for Tom to have the same amount of savings as Betty.

The procedures we have illustrated can be used to solve many of the equations that arise from everyday problems. There are still, it is true, lots of equations which cannot be solved by these methods. Here is an example:

$$x(x - 2) = 3$$

Methods for solving such equations are not usually studied until high school, so we will not bother about them here.

What we have done so far in this chapter is to illustrate how negative numbers help us in solving the kind of equations we have been concentrating on.

■ **Question 9** What is the solution set of this equation?

$$\frac{x}{x} = 1$$

INEQUALITIES Suppose that Henry had a score of 68 on a test. He wonders how high a score he must have on a second test in order that his average[1] score for the two tests will be higher than 75.

Let x be the score he must make on the second test. The average of the two tests is $\frac{1}{2}(68 + x)$, which must be greater than 75.

Remember that the symbol for "is greater than" is ">." So the sentence which describes this situation is

$$\tfrac{1}{2}(68 + x) > 75$$

We call such a sentence an *inequality*.

If we substitute 90 for x in the inequality above, then $\frac{1}{2}(68 + 90)$ equals 79, which is greater than 75. Therefore 90 is a solution of the inequality. In fact, any number greater than 90 is also a solution.

If we substitute 82 for x in the above inequality, then $\frac{1}{2}(68 + 82)$ equals 75, which is *not* greater than 75. Therefore, 82 and any number less than 82 is not a solution of the inequality. However, any number greater than 82 is a solution.

Table 25.1

Inequality	English equivalent
$3x - 7 < 8$	"$3x$ minus 7 is less than 8"
$2x \leq 14$	"$2x$ is less than or equal to 14"
$x + 2 \neq 3$	"x plus 2 is not equal to 3"
$5x + 8 \geq 13$	"$5x$ plus 8 is greater than or equal to 13"

[1] You probably remember that the average of a set of numbers is obtained by adding them and then dividing by the number of numbers in the set. The notion of average is discussed at some length in Chapter 26.

In Table 25.1 are some other examples of inequalities along with the way they are read.

The inequality

$$2x \leq 14$$

has as a solution any number which, when substituted for x, makes either

$$2x = 14 \qquad \text{a true statement}$$

or

$$2x < 14 \qquad \text{a true statement}$$

Thus 7 is a solution because $2(7) = 14$ is true, and 4 is a solution because $2(4) < 14$ is true.

Note that 7 is a solution of $2x \leq 14$, but 7 is *not* a solution of $2x < 14$.

■ **Question 10** What are the solutions of $2x < 0$?

We know that "$-3 < 1$" is a true statement and is read

-3 is less than 1

Of course, this means that on the horizontal number line

-3 is to the left of 1

What happens when we add 2 to each number? How is $(-3) + 2$ related to $1 + 2$? In Figure 25.6 you can see that $(-3) + 2$ is to the left of $1 + 2$ on the number line.

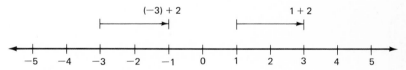

Figure 25.6

Let's see what happens when -2 is added to each number. How is $(-3) + (-2)$ related to $1 + (-2)$? In Figure 25.7 you can see that $(-3) + (-2)$ is to the left of $1 + (-2)$ on the number line.

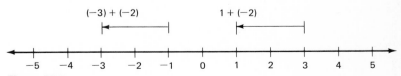

Figure 25.7

ADDITION PROPERTY OF ORDER

It has been helpful to picture *addition* and *order* on the number line. In the examples above we could have thought of two men walking on the number line carrying a ladder between them. At the start the man at -3 is to the left of the man at 1. If they walk 2 units in *either* direction, the fixed length of the ladder guarantees that the man to the left will still be to the left when both men arrive at their new locations.

The property illustrated here is named *the addition property of order,* and it can be stated as follows:

If $a < b$, then $a + c < b + c$.

Also, if $a + c < b + c$, then $a < b$.

In the statement of this property, the symbol "$<$" can be replaced by "$>$"; that is,

If $a > b$, then $a + c > b + c$.

Also, if $a + c > b + c$, then $a > b$.

Note: Since we are assuming the commutative property of addition for rational numbers, we could place the letter c to the left of a and b.

■ **Question 11** What can you say about $a - c$ and $b - c$ if you know that $a < b$?

MULTIPLICATION PROPERTY OF ORDER

Let us start again with the true statement

$-3 < 1$

How does $(-3) \times 5$ compare to 1×5? How does $(-3) \times (-4)$ compare to $1 \times (-4)$? You can see that although

$(-3) \times 5 < 1 \times 5$ is true

$(-3) \times (-4) < 1 \times (-4)$ is false

These illustrate another very important property. It is named *the multiplication property of order,* and it can be stated as follows:

If $a < b$ and $c > 0$ (c is a positive number), then $ac < bc$.

Also, if $ac < bc$ and $c > 0$, then $a < b$.

If $a < b$ and $c < 0$ (c is a negative number), then $ac > bc$.

Also, if $ac > bc$ and $c < 0$, then $a < b$.

Note: Since we are assuming the commutative property of multiplication for rational numbers, we could place the letter c to the left of a and b.

To illustrate how these properties can be used, consider this problem:

By 10:00 A.M. on a certain day, Mr. Hanning had already driven 85 miles. By 1:00 P.M. that afternoon he had driven a total distance of more than 215 miles. How far did he travel during the 3-hour interval from 10:00 A.M. to 1:00 P.M.?

If we let x represent the number of miles he drove during the 3-hour interval, then $x + 85$ represents the total number of miles he had driven by 1:00 P.M. Now we want to find all replacements for x such that the value of the phrase, $x + 85$, is greater than 215. Thus, the mathematical sentence which serves as a model for our problem is the inequality

$x + 85 > 215$

How do we go about solving this inequality? Would it help to add −85, the opposite of 85, to both sides? If we do this, we get

$x + 85 + (−85) > 215 + (−85)$

or $\qquad x > 130$

Do these inequalities have the same solution set as the original inequality? The answer is: yes. This follows from using the *addition property of order*. Two (or more) inequalities are said to be equivalent inequalities if they have the same replacement set and the same solution set.

■ **Question 12** If you know that $a < b$ but that $ac = bc$, what can you say about c?

Consider the two inequalities

$$x < −3$$
and $\qquad 4x < −12$

The number −5 is a solution of each inequality, since −5 < −3 is true, and 4 × (−5) < −12 is true. On the other hand, the number 2 is not a solution of either inequality, since 2 < −3 is false, and 4 × 2 < −12 is false. Perhaps by this time you have concluded that the two inequalities are equivalent (that is, they have the same replacement set and the same solution set). This is correct. This follows from the *multiplication property of order* (with c a positive number). You can obtain $4x < −12$ from $x < −3$ by multiplying both sides of $x < −3$ by the positive number 4.

The inequalities

$$x < −3$$
and $\qquad −5x > 15$

are also equivalent inequalities. This follows from the multiplication property of order (with c a negative number). You can obtain $−5x > 15$ from $x < −3$ by multiplying both sides of $x < −3$ by the negative number −5.

APPLICATIONS Earlier we constructed the inequality

$x + 85 > 215$

as a model for a problem. (The number x represents the number of miles Mr. Hanning drove during the 3-hour interval from 10:00 A.M. to 1:00 P.M.) We pointed out that

$x + 85 + (−85) > 215 + (−85)$

is equivalent to the original inequality, and that this second inequality can be rewritten as

$x > 130$

You will agree that this last inequality is simpler than the first one. In fact, we can see at once that its solution set is the set of all numbers greater than 130. We will use the following notation to describe this set:

$$\{x : x > 130\}$$

This is read "the set of all numbers x *such that* x is greater than 130." On the number line this set is shown in Figure 25.8. The unshaded circle at 130 indicates that 130 is not included in the set. The heavy shaded arrow means that all numbers greater than 130 are in the set.

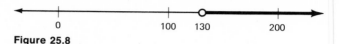

Figure 25.8

■ **Question 13** What set does Figure 25.9 represent?

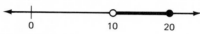

Figure 25.9

How do we interpret our solution set in the framework of the problem? Mr. Hanning drove more than 130 miles during the 3-hour interval from 10:00 A.M. to 1:00 P.M. There is a point, however, that should be made concerning the replacement set of the inequality (that is, the set of numbers that may be used as replacements for the variable in the inequality)

$$x + 85 > 215$$

The physical limitation of the car used by Mr. Hanning restricts the set of numbers we can use as replacements for x. Suppose that the maximum speed of the car is less than 100 miles per hour. In that case, the distance traveled during the 3-hour period is less than 300 miles. This means that x is less than 300. We write this as

$$x < 300$$

With this limitation on the replacement set, the solution set of the inequality

$$x + 85 > 215$$

is the set of all numbers between 130 and 300. The notation for this set is

$$\{x : x > 130 \text{ and } x < 300\}$$

This is read "the set of all numbers x such that x is greater than 130 and x is less than 300."

We want to emphasize that the key step in solving the inequality

$$x + 85 > 215$$

is the addition to both sides of -85, the opposite of 85, to get the simpler equivalent inequality

$x > 130$

Consider the inequality

$13x < 250$

How do we solve this inequality? We can multiply both sides by the same nonzero number. The best choice of multiplier is $\frac{1}{13}$, the reciprocal of 13, because $\frac{1}{13}$ times $13x$ is $1 \times x$, or x. Since $\frac{1}{13}$ is a positive number, the inequality

$13x < 250$

is equivalent to

$\frac{1}{13}(13x) < \frac{1}{13}(250)$

which can be rewritten as

$x < \frac{250}{13}$

Hence, the solution set of the inequality

$13x < 250$

is $\{x : x < \frac{250}{13}\}$.

■ **Question 14** What is the solution set of this inequality?

$1 - 2x < 3$

Here is a final example:

When Eddie and Doug were planning to buy a sailboat, they asked a salesman about the cost of a new type of boat that was being designed. The salesman replied, "It will cost less than $380." If Eddie and Doug had agreed that Eddie was to contribute $130 more than Doug when the boat was purchased, how much would Doug have to pay?

Let x be the number of dollars Doug pays. Then Eddie pays $x + 130$ dollars. Together the two pay $x + (x + 130)$ dollars. So we have

$$\text{Purchase price} = x + (x + 130)$$
$$= (x + x) + 130$$
$$= 2x + 130$$

According to the statement of the problem, the purchase price is less than $380. So we get the inequality

$2x + 130 < 380$

Of course, since x represents the number of dollars Doug pays, the replacement set of this inequality contains only positive numbers (that is, $x > 0$).

It is not a difficult matter to solve this inequality. Can we write a simpler inequality which is equivalent to the one above? We can begin by using the addition property of order. Let's add -130, the opposite of 130, to both sides.

$2x + 130 + (-130) < 380 + (-130)$

This can be rewritten as

$2x < 250$

This inequality is equivalent to the original one.

Next, we multiply by $\frac{1}{2}$, the reciprocal of 2, and we get

$\frac{1}{2}(2x) < \frac{1}{2}(250)$

which can be written simply as

$x < 125$

The solution set of this inequality, as well as that of the original inequality

$2x + 130 < 380$

is $\{x : x > 0 \text{ and } x < 125\}$. Its graph on the number line is shown in Figure 25.10. We know then that the amount Doug pays is between 0 and \$125.

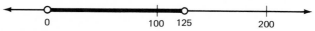

Figure 25.10

■ **Question 15** Draw the graph of what Eddie might pay.

In conclusion, we have seen that negative numbers help us to solve inequalities as well as equations.

PROBLEMS

1 Solve each of these:
   (a) $x + 17 = 32$ (b) $13 = x - 2$
   (c) $14 + x = 17$ (d) $16 = 28 + x$

2 Solve each of these:
   (a) $x + \frac{1}{3} = -\frac{1}{2}$ (b) $\frac{2}{3} + x = 2$
   (c) $\frac{13}{17}x + 1 = 0$ (d) $\frac{1}{6} = -\frac{1}{6}x$

3 Solve each of these:
   (a) $12 - 5x = -18$ (b) $-13 - 4x = 31$
   (c) $23 = -5x + 48$ (d) $-5 - x = -11$

4 Solve each of these:
   (a) $\frac{3}{8} - \frac{1}{4}x = \frac{1}{2}$ (b) $-\frac{2}{3} - \frac{3}{4}x = \frac{1}{4}$
   (c) $\frac{5}{8} = \frac{1}{4} - \frac{1}{2}x$ (d) $-\frac{3}{4} = \frac{1}{3} - 6x$

5 Solve each of these:
   (a) $3(x + 3) = 6$ (b) $5(9 + x) = 22$
   (c) $\frac{1}{2}(\frac{1}{2}x + 2) = 4$ (d) $4 + 3(4 - x) = 13$

6 Solve each of these:

(a) $6x + 3x = -27$ (b) $-3x - 9x = 48$

(c) $\frac{3}{5} = x/3 - 2x/5$ (d) $-3x + 3x/2 = 5$

7 Solve each of these:

(a) $-x + 17 = 1 - 4x$ (b) $x - 2 = \frac{1}{3}x + 2$

(c) $-x = 5x - 3$ (d) $-x + (4x + 5) = 5 + 3x$

8 Solve each of these:

(a) $7x - 13x = \frac{1}{2}$ (b) $18(2x + 3) = -36$

(c) $\frac{1}{2}x - \frac{1}{3} = \frac{1}{6}$ (d) $\frac{1}{3}(x - 6) - 8 = 7$

(e) $-11 - x = -8$ (f) $4(2x + 3) + 6 = 3(4x + \frac{8}{3})$

9 Below is a chain of equivalent equations. Starting with the second equation, Equation a, explain why we can say that each equation is equivalent to the one immediately above it.

$$5 - 4x = -9$$

(a) $5 + (-4x) = -9$ _____

(b) $(-5) + [5 + (-4x)] = (-5) + (-9)$ _____

(c) $[(-5) + 5] + (-4x) = -14$ _____

(d) $0 + (-4x) = -14$ _____

(e) $-4x = -14$ _____

(f) $(-\frac{1}{4}) \times (-4x) = (-\frac{1}{4}) \times (-14)$ _____

(g) $[(-\frac{1}{4}) \times (-4)]x = \frac{14}{4}$ _____

(h) $1 \times x = \frac{14}{4}$ _____

(i) $x = \frac{7}{2}$ _____

10 For each problem below, write an equation which can serve as a model of the situation. Then solve the equation and interpret your numerical solution in terms of the stated problem.

(a) If 1 is added to twice the number of years in a boy's age, the result is 19. What is the boy's age?

(b) Two cars start from the same point at the same time and travel in opposite directions at constant speeds of 44 and 66 miles per hour, respectively. In how many hours will they be 242 miles apart?

(c) A board 40 inches long is to be cut in two pieces so that one piece is 7 inches longer than twice the length of the other piece. Find the length of each piece.

(d) Arthur's allowance is $1.00 more than twice Betty's, but is $2.00 less than 3 times Betty's. What are their allowances?

(e) A square and an equilateral triangle have equal perimeters. A side of the triangle is 3.5 inches longer than a side of the square. What is the length of the side of the square?

(f) Tom bought a ticket for a football game. Altogether he paid $1.10 (or 110 cents), including the tax. If the cost of the ticket is $1.00 more than the amount of the tax, what is the amount of tax on the ticket?

11 Solve each of these:

(a) $x + 23 < 72$ (b) $x + 25 < 10$

(c) $x - 6 \geq 2$ (d) $2x - 5 > -6$

12 Solve each of these:

(a) $x - 3.5 \leq 6.8$ (b) $3\frac{3}{4} + x < 9\frac{1}{2}$

(c) $-\frac{7}{8} + x > \frac{3}{4}$ (d) $3x/4 - 5 < 6$

13 Solve each of these:

(a) $13 - x \geq 9$ (b) $3 - 2x < 8$

(c) $7 - 4x \leq -9$ (d) $13 - 2x > 5$

14 Solve each of these:

(a) $\frac{3}{4} - x/2 < -5\frac{1}{4}$ (b) $6 - \frac{2}{3}x \leq -18$

(c) $-x/4 + 7 \geq 2\frac{1}{2}$ (d) $11 - 3x/10 > -4$

15 Solve each of these:
(a) $2(3 - x) < 5$
(b) $3x + 10 < 2x + 14$
(c) $5(x + 3) > 0$
(d) $-x \leq 4x - 7$

16 Solve each of these:
(a) $17 - 3x \leq 1 - 2x$
(b) $3x + 4 > -2x + 14$
(c) $x + 3 < 2x + 7$
(d) $-x \leq 4x - 7$

17 Solve each of these:
(a) $-x + 2 \geq x + 2$
(b) $5x + 11 < 3(x + 3)$
(c) $99x - 10 > 100x - 14$
(d) $3x - x/3 < -4$
(e) $x + (x + 6) > 0$
(f) $x/5 + x/6 < 3$

18 For each part of Problem 17, show the solution set on a number line.

19 (a) Is there any value of x for which this inequality becomes a false statement? $x + 2 > x$

(b) Is there any value of x which makes this inequality a true statement? $2x + 1 < 2x$

20 For each of the following problems write a mathematical sentence which can serve as a model of the situation. Then find the solution set of each sentence and give the answer to the question.

(a) The sum of a number and 5 is less than twice the number. What is the number?

(b) On Thursday a boy read an additional 15 pages of a novel. At this point he found that he had read less than 51 pages. How many pages could he have read prior to Thursday of that week?

(c) A student has test grades of 82 and 91. What must he score on a third test to have an average of 90 or higher? The maximum grade on a test is 100.

(d) If a certain variety of bulbs is planted, less than $\frac{5}{8}$ of them will grow into plants. If, however, the bulbs are given proper care more than $\frac{3}{8}$ of them will grow. If a careful gardener has 15 plants, how many bulbs did he plant?

26.

STATISTICS

Our society continuously creates and spews out a vast amount of information. Much of this information is numerical. Every day, on radio or TV or in newspapers or magazines, we hear or read reports on such topics as the weather, unemployment, sports results, births and birth rates, wages and prices, automobile accidents, crime, etc.

Most of us are interested in certain pieces of this information. Thus, in order to vote intelligently, any citizen needs to be aware of some of this information such as, for example, the state of the nation's economy. But most of this information, in its original undigested form, is very hard to grasp. Therefore, we need ways of organizing information to make it easier to understand. Even then, there is so much information that each of us wants, or should be aware of, ways of summarizing large bodies of information.

ORGANIZING NUMERICAL
INFORMATION

In this chapter we discuss ways of organizing numerical information in the form of tables and graphs. Then we will discuss ways of summarizing masses of data by picking out certain important aspects and ignoring the rest.

You have already had experience with tables and graphs. For example, newspapers, magazines, and other publications often report numerical data to us in tables. The sports pages of a newspaper devote a large portion of their space to batting averages, box scores, bowling results, football "statistics," golf scores, league standings, etc. In many of your texts you have seen tables and graphs of many kinds. In Chapter 19 you used tables and graphs to record the results of experiments.

■ **Question 1** In which section of your paper do you find what the weather was like yesterday across the nation?

Let us review some ideas about tables and graphs. Of course, if we only have a few numbers, there is little need to do more than simply

Table 26.1

Score	Number of students (frequency)
90	1
80	2
70	3
60	1
50	1
Total	8

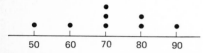

Figure 26.1

present them. Here is a list of scores on a test given to eight students:

70, 80, 50, 90, 70, 60, 70, 80

Even with this short list, it is somewhat neater to arrange them in some better order. One obvious way is to list them in order of size:

50, 60, 70, 70, 70, 80, 80, 90

Another idea is to report the data in the form of a table (see Table 26.1). The second column shows the number of times each score occurs. Or, we could say, the second column shows the frequency with which each score occurs. For this reason, such a table is called a *frequency* table. The sum of the entries in the frequency column is the total number of scores—the *total frequency*. Since the table shows how the scores are distributed, we may speak of a *frequency distribution*.

Representing numbers on the number line is also a way to display them. In Figure 26.1 we have shown each score as a dot above the proper point on the number line. If the number of scores is large, the *dot-frequency diagram* becomes awkward. It is then convenient to replace the dots with vertical line segments. The length of each segment shows the frequency of the corresponding score (see Figure 26.2).

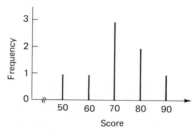

Figure 26.2

■ **Question 2** If 10 students take a test, what is the largest frequency that might occur?

Usually the line segments are thickened to make them easier to see. Hence, these graphs are often called bar graphs (see Figure 26.3). If we can consider only the upper endpoints of each vertical segment and connect them we have a *broken line graph* (Figure 26.4).

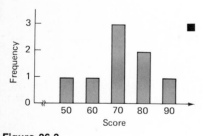

Figure 26.3

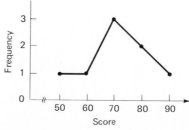

Figure 26.4

CUMULATIVE AND RELATIVE FREQUENCY

Table 26.2

Score	Frequency	Cumulative frequency
90	1	8
80	2	7
70	3	5
60	1	2
50	1	1

Let us return to our table of test scores. To find the total frequency we add the entries in the frequency column. If we add, starting at the bottom of the table, we count "1," "2," "5," "7," "8." The total frequency is 8.

We can put a new column in our table. This new column is the *cumulative frequency* column, since we have recorded the frequencies as we "accumulated" them by adding (see Table 26.2). We can read off from this table that, for example, five of the scores are less than or equal to 70, two are less than or equal to 60, etc. We can now construct a new graph—the graph of the cumulative frequencies (Figure 26.5).

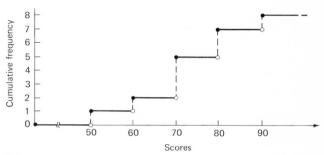

Figure 26.5

Notice that the heavy dots are placed so that we can see, for example, that the cumulative frequency stays "2" from 60 until we reach 70, then the cumulative frequency "jumps" or "steps" up to 5. The dashed vertical segments are not really part of the graph but are added for easier reading.

Of the eight scores two were 80. That is, $\frac{2}{8}$ equals $\frac{1}{4}$ of the *total frequency* is located at 80, and $\frac{1}{4}$ represents the proportion, or *relative frequency*, of a score of 80. Let us put a new column into our frequency table. We use $\frac{2}{8}$ instead of $\frac{1}{4}$ to emphasize that the total frequency is 8. Observe that the size of the "jump" for each score in the cumulative frequency graph (Figure 26.5) is proportional to the relative frequency of that score. See Table 26.3.

Table 26.3

Score	Frequency	Relative frequency	Cumulative frequency
90	1	$\frac{1}{8}$	8
80	2	$\frac{2}{8}$	7
70	3	$\frac{3}{8}$	5
60	1	$\frac{1}{8}$	2
50	1	$\frac{1}{8}$	1

■ **Question 3** Can the relative frequency ever be larger than 1?

CONDENSING INFORMATION

So far, we have considered only short lists of numbers. The techniques for constructing tables and graphs for somewhat larger lists are similar. However, for very long lists, the job becomes too difficult and we have to adopt some shortcuts.

Generally, we condense the information by "rounding off" or by grouping the numbers. For example, suppose we were interested in the populations of the various U.S. states. The figures obtained in the 1970 census are shown in Table 26.4.

Table 26.4

Populations of U.S. States

State	Population	State	Population
Alabama	3,444,165	Montana	694,409
Alaska	302,173	Nebraska	1,483,791
Arizona	1,772,482	Nevada	488,738
Arkansas	1,923,295	New Hampshire	737,681
California	19,953,134	New Jersey	7,168,164
Colorado	2,207,259	New Mexico	1,016,000
Connecticut	3,032,217	New York	18,241,266
Delaware	548,104	North Carolina	5,082,059
Florida	6,789,443	North Dakota	617,761
Georgia	4,589,575	Ohio	10,652,017
Hawaii	769,913	Oklahoma	2,559,253
Idaho	713,008	Oregon	2,091,385
Illinois	11,113,976	Pennsylvania	11,793,909
Indiana	5,193,669	Rhode Island	949,723
Iowa	2,825,041	South Carolina	2,590,516
Kansas	2,249,071	South Dakota	666,257
Kentucky	3,219,311	Tennessee	3,924,164
Louisiana	3,643,180	Texas	11,196,730
Maine	993,663	Utah	1,059,273
Maryland	3,922,399	Vermont	444,732
Massachusetts	5,689,170	Virginia	4,648,494
Michigan	8,875,083	Washington	3,409,169
Minnesota	3,805,069	West Virginia	1,744,237
Mississippi	2,216,912	Wisconsin	4,417,933
Missouri	4,677,399	Wyoming	332,416

If we "round off" these figures to the nearest million, the frequency table is shown in Table 26.5.

If we wanted to condense the information even more, we could use the bar graph in Figure 26.6. Note that the more we condense the data, the less accurate our report of the information becomes. Nevertheless, for many purposes the bar graph (Figure 26.6) may be sufficient, and it certainly is easier to read than is the original table.

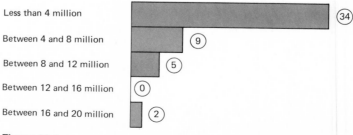

Less than 4 million   ③④

Between 4 and 8 million   ⑨

Between 8 and 12 million   ⑤

Between 12 and 16 million   ⓪

Between 16 and 20 million   ②

Figure 26.6

Table 26.5

Population (in millions)	Frequency	Cumulative frequency
0	4	4
1	12	16
2	7	23
3	7	30
4	5	35
5	5	40
6	1	41
7	2	43
8	0	43
9	1	44
10	0	44
11	3	47
12	1	48
13	0	48
14	0	48
15	0	48
16	0	48
17	0	48
18	1	49
19	0	49
20	1	50

ARITHMETIC MEAN Sometimes, however, instead of condensing the information, it is better to give a brief description of the frequency distribution. The following story illustrates one way of doing this.

Mr. A. Scout observed a college football game and was interested in the performance of a certain halfback, O. Jay. Mr. Scout kept a record of the yards gained and lost by O. Jay during the game. His list (recorded to the nearest yard) showed:

7, 3, 0, 12, −2, 14, 6, 3, −1, 45, 10, 4, 1, 8, −7, 39, 7, 20, 6, 25

■ **Question 4** What is the relative frequency of O. Jay's 7-yard gains?

Examination of the list tells us many things. (Incidentally, a frequency table or a graph is of little help here.) We can determine how often he carried the ball, how many times he lost yardage, the length of his longest run, and the total yards he gained—if we add the individual numbers. Actually, Mr. Scout sent a telegram to his employer.

O. JAY CARRIED THE BALL 20 TIMES. HIS AVERAGE WAS 10 YARDS PER CARRY.

This telegram reveals much important information in a very brief way. There are only two numbers, the number of carries and the average of the list of numbers. The two numbers summarize O. Jay's performance. It is easy to use these two numbers, for example, to find the total number of yards gained by O. Jay. This kind of brief description would be even more useful if we were dealing with his record of perhaps 250 or more carries in the course of a season.

Perhaps you have had so much experience in using "averages" that the football story was unnecessary. We wished to emphasize that a major use of an average is to summarize information from a list of numbers. In the example,

$$\text{Average number of yards gained} = \frac{\text{total yards gained}}{\text{number of carries}}$$

More generally, we obtain the *average* of a list of numbers by adding all the numbers and then dividing by the total frequency. Using the test results shown in the first table in this chapter, the average of 50, 60, 70, 70, 70, 80, 80, 90 is

$$\frac{50 + 60 + 70 + 70 + 70 + 80 + 80 + 90}{8} = 71.25$$

In practice we might round this off to 71. For the purposes of this chapter, we shall retain 71.25.

■ **Question 5** What is the average of these numbers?

0, 1, 0, 2, 0, 99

Unfortunately, the word "average" has many different dictionary meanings. To clarify, we shall refer to the average we have been discussing as the *arithmetic mean,* or, most often, as the *mean* of the numbers.

In computing the mean of 50, 60, 70, 70, 70, 80, 80, and 90 we had to add $70 + 70 + 70$. We could have instead noticed that there were three 70s and replaced $70 + 70 + 70$ by 70×3. What is the significance of "3"? It is the frequency of 70. This suggests we can compute the value of the mean by using a frequency table (see Table 26.6).

Table 26.6

Score	Frequency	Score times frequency
90	1	90
80	2	160
70	3	210
60	1	60
50	1	50
	Total	Total
	8	570

$$\text{Mean} = \frac{\text{sum of "score times frequency"}}{\text{total frequency}} = \frac{570}{8} = 71.25$$

For this particular example, the frequency table did not save much work. As we shall soon see, however, a table is useful if certain scores are repeated many times. But first, it is useful to use some symbols for the ideas we have been developing. We have used examples of lists of test scores, football yardages, heights, golf scores, etc. In order to discuss frequency distributions in general, we head the left column of our frequency table with the letter x (see Table 26.7).

Table 26.7

x	Frequency
90	1
80	2
70	3
60	1
50	1

We must be careful to notice that, although there are *five* different values of x, there are *eight* numbers in our list; eight is the total frequency. Let us use the letter n to stand for the total frequency of a distribution. In our example, $n = 8$. Finally, we use the symbol $\overline{x}$ ("x-bar") for the mean of the list. For our example we have seen two ways to compute $\overline{x}$:

1 Not using a frequency table:

$$\overline{x} = \frac{90 + 80 + 80 + 70 + 70 + 70 + 60 + 50}{8} = 71.25$$

2 Using a frequency table:

$$\overline{x} = \frac{90 \times 1 + 80 \times 2 + 70 \times 3 + 60 \times 1 + 50 \times 1}{8} = 71.25$$

Now we can rewrite Equation 1 in the form

$$\overline{x} = 90 \times \tfrac{1}{8} + 80 \times \tfrac{1}{8} + 80 \times \tfrac{1}{8} + 70 \times \tfrac{1}{8} + 70 \times \tfrac{1}{8} + 70 \times \tfrac{1}{8} + 60 \times \tfrac{1}{8} + 50 \times \tfrac{1}{8}$$

We see that each of the eight individual scores is multiplied by the same number, $\tfrac{1}{8}$. To compute the mean by Equation 1 we are in fact "weighting" each of the eight numbers the same. In general, each score is multiplied by $1/n$ and then the products are added.

The other method, Equation 2, can be rewritten as

$$\overline{x} = 90 \times \tfrac{1}{8} + 80 \times \tfrac{2}{8} + 70 \times \tfrac{3}{8} + 60 \times \tfrac{1}{8} + 50 \times \tfrac{1}{8}$$

From this we see that each number from the x column is multiplied by its *relative frequency*. Each value of x is "weighted" by its relative frequency.

■ **Question 6** What is the average of this set of numbers?

1, 10, 10, 10, 10, 10, 10, 10, 10, 19

Table 26.8

x	Frequency
120	10
80	15
40	15
15	20

Suppose we were asked to find the arithmetic mean of the numbers represented by this frequency table (see Table 26.8). The total frequency for this table is 60. We would scarcely want to add 60 numbers. Instead, it would be far easier to compute

$$\overline{x} = \frac{(120 \times 10) + (80 \times 15) + (40 \times 15) + (15 \times 20)}{60}$$

MEDIAN AND MODE

Earlier in this section, we mentioned that the word "average" has many different meanings. Three kinds of averages are used in statistics: the *mean,* the *median,* and the *mode.* Our attention has been focused (and we will continue to focus) on the mean. This is because, it turns out, the mean is more important in the problem of analyzing data. We remarked before that a major use of an average is to summarize information from a list of numbers, and we saw how this is accomplished by the mean. For completeness, we will define the *median* and the *mode,* and illustrate their uses.

Suppose x in Table 26.9 represents the monthly income for the 25 employees of a certain company. We can easily compute and find the mean to be $1,000. Clearly, $1,000 is not representative of the monthly salary of this company's employees when one considers that 23 of the 25 employees earn less than $1,000 per month. A better index is provided by the average known as the *median;* you have seen this word used in mathematics before, for example, referring to a median of a triangle. Perhaps you live in a part of the country where this word is commonly used as the name for the strip down the middle of a divided highway. The word has a similar meaning in statistics. The median of a list of numbers is the "middle" number when the numbers are arranged in order of size.

Table 26.9

x	Frequency
$5,000	1
4,000	1
900	1
800	9
700	1
600	12

■ **Question 7** What is the median of the set of composite numbers larger than 10 but less than 20?

In our example, where we have 25 numbers ($n = 25$), the "middle" number is the thirteenth on the list (namely, 700); there are 12 numbers above and 12 numbers below this. Whenever there is an odd number of numbers in the data, the middle number can be located without controversy.

If n is even, there would be the question as to which number is to be considered the "middle." For example, in 22, 37, 48, 54, 59, 71 where $n = 6$, we might be justified in choosing either 48 or 54 as the median. Some statisticians pick a number halfway between the numbers in question. We shall adopt this procedure for our purposes. In this case, we would say the median is $51 = (48 + 54)/2$.

We have seen an instance when extreme values can so distort the situation that the median presents a more "typical" picture than the mean. There are also occasions when it is impossible to compute the mean. Sometimes extreme values are not reported precisely. The following table is of this type. Here, x represents the number of words typed per minute by the typists in a certain office. The extreme values "over 100" and "under 50" do not give any definite numbers for use in computing the mean (see Table 26.10).

Table 26.10

x		Frequency
over	100	1
	90	1
	80	4
	70	2
	60	3
under	50	2

While we have made use of the median, we will introduce another average that might be more appropriate for this example. Consider again the above table on words per minute and suppose that all these typists are registered with an employment agency. Suppose also that you ask this agency to send you a typist. What is the most probable number of words per minute for a typist selected from this list? Since 80 has the largest relative frequency, 80 is the most probable value. This way of describing a "typical" number in a list identifies the *mode* of a list of numbers. The mode is that number which appears with greatest frequency. The use of the word, mode, is in the same sense as in "pie a la mode"; à la mode means in the fashionable or popular way. Sometimes there will be two or more entries with the "greatest" frequency. In that case, we will agree that the list has no mode. This is the case, for example, in the list: 12, 14, 14, 14, 15, 16, 16, 16, 18.

Occasionally, a summary of information might make use of more than one of the averages described above. This is a practice that we can be accustomed to, but it is not always clear which of the averages is intended. Unless we keep this in mind, we can be misled in interpretation of the data.

Hereafter, unless it is stated otherwise, by "average" we shall agree that the "mean" is intended.

■ **Question 8** What is the mode of the set of prime numbers less than 10?

CALCULATING THE MEAN

A previous section was devoted to defining the mean of n numbers and to performing certain calculations. Is there any more to be said? Fortunately (for the study of statistics) there is indeed a great deal more we can learn about the mean.

These two distributions are shown in Tables 26.11 and 26.12. (The mean for each has been computed for you to save you the trouble.)

Table 26.11

x	Frequency
90	1
80	2
70	3
60	1
50	1

$\overline{x} = 71.25$

Table 26.12

y	Frequency
20	1
10	2
0	3
-10	1
-20	1

$\overline{y} = 1.25$

How are these two distributions related? It takes only a glance to see that each value of y can be obtained by subtracting 70 from the value of x. We could write

$$y = x - 70$$

But notice also that

$$\overline{y} = \overline{x} - 70$$

This result is perfectly general. If x and y are used as they have been in this chapter and b is any fixed number (positive, negative, or zero), then we conclude:

If $y = x + b$, then $\overline{y} = \overline{x} + b$.

This can save us a lot of work. For example, suppose we needed to know the mean for the distribution shown in Table 26.13.

It would certainly be simpler to find the mean for the distribution shown in Table 26.14 and then to add 3,420 to the result.

Notice also that if we pick b to be $-\overline{x}$ and if

$$y = x - \overline{x}, \quad \text{then} \quad \overline{y} = 0$$

Table 26.13

x	Frequency
3,427	4
3,425	7
3,422	8
3,421	1

Table 26.14

z	Frequency
7	4
5	7
2	8
1	1

■ **Question 9** Find the mean of this set of numbers.

$-101, -100, -99$

Now let us compare these two distributions shown in Tables 26.15 and 26.16. We see immediately that each y is 10 times the corresponding x. If you compute $\overline{y}$ you will find that it is $10\overline{x}$.

Table 26.15	
y	Frequency
40	3
20	8
−10	7
−30	2

Table 26.16	
x	Frequency
4	3
2	8
−1	7
−3	2

This result is also perfectly general. For any fixed number a: if $y = ax$, then $\overline{y} = a\overline{x}$.

Of course, we can combine these results. If a and b are any two fixed numbers;

if $y = ax + b$, then $\overline{y} = a\overline{x} + b$

There is one common misunderstanding about averages that we should guard against. To illustrate, suppose two classes took the same test and that the distribution of scores came out as shown in Tables 26.17 and 26.18.

Table 26.17	
Class X	
Score	Frequency
90	1
80	2
70	3
60	1
50	1

Table 26.18	
Class Y	
Score	Frequency
90	0
80	0
70	9
60	3
50	12

If we use x and y to represent the two distributions, we find that

$$\overline{x} = \tfrac{570}{8} = 71.25 \qquad \text{and} \qquad \overline{y} = \tfrac{1410}{24} = 58.75$$

Now suppose that we want to know the average score for both classes combined. Will it be the average of $\overline{x}$ and $\overline{y}$? No!

When we combine the two classes, the new frequency table is illustrated by Table 26.19. Using z for this case, we find that $\overline{z} = \tfrac{1980}{32} = 61\tfrac{7}{8} \approx 61.9$. But the average of $\overline{x}$ and $\overline{y}$ is

$$\frac{71.25 + 58.75}{2} = \frac{130}{2} = 65$$

So we see that when we combine two sets of data, the mean of the combined set is not necessarily the average of the two means.

Table 26.19	
Combined results	
Score	Frequency
90	1
80	2
70	12
60	4
50	13

■ **Question 10** If $A = \{10, 10, 10, 10\}$ and $B = \{0\}$, find the mean of A, of B, and of the union of A and B.

SCATTER The mean, $\overline{x}$, and the total frequency, n, give us brief, useful information about a list of numbers. Quite often, however, we need more information, particularly when we wish to compare different collections of data.

Table 26.20

Distribution A	
Score	Frequency
0	1
5	6
10	1

Table 26.21

Distribution B	
Score	Frequency
0	2
2	1
5	2
8	1
10	2

Table 26.22

Distribution C	
Score	Frequency
2	1
5	6
8	1

For example, a class of 25 students takes two quizzes. On each quiz the mean is 70. On the first quiz the high score is 95, the low score 35. On the second quiz the high is 85, the low 65. Although the means of the quiz scores are the same, the *scatter* (spread, dispersion) of the scores is quite different.

We say that the *range* of the scores on the first quiz is 60 (95 − 35) and the range of the scores on the second quiz is 20 (85 − 65).

It may happen that two collections of data have the same (or nearly the same) means but quite different ranges.

However, experience has shown that there is another method of comparing the scatter of two distributions which is more useful than comparing their ranges.

Before we give a precise definition of "scatter," it will be helpful to look at some specific distributions. Three of them, specified first by their frequency tables (Tables 26.20, 26.21, and 26.22), and then displayed in the form of dot-frequency diagrams (Figure 26.7).

Figure 26.7

These three distributions are clearly quite different, each from the other. However, an easy calculation shows that the mean for each of these is 5, so merely knowing that the means are alike is not enough to ensure that the distributions look alike. Also, the ranges for distributions A and B are the same, 10. Knowing both that the means are the same and the ranges are the same is not enough to ensure that the distributions look alike.

■ **Question 11** What is the range of the data shown in Table 26.10?

Actually distributions A and C look more similar to each other than either one does to distribution B, in that most of the dots, in the dot-frequency diagrams, are near (actually at) the mean, while in the dot-frequency diagram for distribution B most of the dots are relatively far away from the mean.

We say that distribution B is more scattered than either distribution A or distribution C. What we need, however, is a numerical measure of the scatter, a number which will tell us how closely clumped around the mean the dots in the dot-frequency diagram are.

It is natural, then, to compute for each value of x the number $x - \overline{x}$. This number shows the amount by which x *deviates* from the mean. The totality of these deviations should tell us something about how scattered the distribution is.

Our first thought might be to compute the average of these deviations. Unfortunately, this idea does not work. For example, for distribution C, all the deviations are 0 except for one which is +3 and another which is −3. They add up to 0. For the other two distributions it is also easy to see that for each positive deviation there is a corresponding negative one, so they average out to 0. (We should really have expected this. We saw above that if $y = \bar{x} - x$, then $\bar{y} = 0$.)

What we need then is a way of replacing each deviation $x - \bar{x}$ by a related number which is always positive, so that the new numbers will not cancel out when we average them. Long experience has demonstrated that the best way to do this is to replace each deviation, $x - \bar{x}$, by its product with itself, $(x - \bar{x}) \times (x - \bar{x})$. The result will never be negative, since, as we saw in Chapter 24, the product of two negative, as well as the product of two positive, numbers is always positive.

■ **Question 12** What is the largest deviation for distribution B?

We will therefore turn aside for a brief time in order to look more carefully at this arithmetic operation.

SQUARES AND SQUARE ROOTS

The operation of multiplying a number by itself is so common in mathematics that it has its own name. It is called the *squaring* operation and the number $a \times a$ is called the square of a. It is easy to see where the name came from. We saw in Chapter 11 that the area of a rectangle is obtained by multiplying the base by the height. So a square, which is a special kind of rectangle, has area $a \times a$ if the length of its side is a.

There is also a special symbolism for the square of a number, a^2. So a^2 is just another name for $a \times a$.

There are a few observations to be made about the squaring operation. First, as we pointed out above, whether a is positive or negative, a^2 is positive (or 0 if a is 0).

For any number a, $a^2 \geq 0$.

Next, let us look at the results of applying the squaring operation to a variety of numbers. You should check each of these.

$$2^2 = 4 \qquad \left(\frac{1}{2}\right)^2 = \frac{1}{4}$$
$$10^2 = 100 \qquad (.3)^2 = .09$$
$$1.4^2 = 1.96 \qquad \left(\frac{2}{14}\right)^2 = \frac{4}{196}$$

Note that, on the left, the numbers that are squared are all larger than 1. On the right, the numbers that are squared are all less than 1. On the left, the result of the squaring operation in each case is larger than the original number. On the right, the result is always smaller.

This is always the case.

If $a > 1$, $a^2 > a$.
If $0 < a < 1$, $a^2 < a$.

■ **Question 13** If $-1 < a < 0$, what can you say about a^2?

Finally, note that the square of the product of two numbers is the product of their squares. For example:

$6 = 2 \times 3$ and $6^2 = 2^2 \times 3^2$

Because of the commutative and associative properties of multiplication, this will always be the case.

If $c = a \times b$, then

$$c^2 = (a \times b) \times (a \times b) = a \times a \times b \times b = a^2 \times b^2$$

Since we know how to multiply any two numbers, whether they are whole numbers, fractions, or decimals, it is easy to compute the square of any number which might be given us. However, the reverse process is a different matter. Suppose we are given a number, for example 7, and are asked to find the number whose square is 7. There is no easy way to do this.

First, however, some notation and terminology. If a is a number whose square is 7, ($a^2 = 7$), then we say that a is a *square root* of 7, and we write

$$a = \sqrt{7}$$

Returning to the problem of finding square roots, sometimes our memory can help us. For example, we have multiplied 3 by itself so often that when we see 9 we remember that it is 3×3. So we remember that $3 = \sqrt{9}$. Similarly, we can remember that $2 = \sqrt{4}$, $5 = \sqrt{25}$, $10 = \sqrt{100}$ etc. (Of course, we can also see that $(-3)^2 = 9$, $(-2)^2 = 4$, etc. Square roots come in pairs, one positive and one negative. We will reserve the square root symbol, $\sqrt{}$, for the positive one.)

But we cannot remember anything that tells us what $\sqrt{7}$ is. Now it turns out that we can locate this number $\sqrt{7}$ as closely as we please on the number line, but it also turns out that $\sqrt{7}$ is not a rational number. It is instead what is called an *irrational* number, which merely means that it is not the ratio of two whole numbers.

■ **Question 14**   What is $\sqrt{1}$?

CALCULATING SQUARE ROOTS

Let us consider how we can locate $\sqrt{7}$ on the number line. Suppose that we need only find $\sqrt{7}$ correct to one decimal place. First, we know that $1 < \sqrt{7} < 7$, from the discussion above. We start by choosing a number about halfway between 1 and 7. For example, we might choose 4. We attach the label A_1 to this number, and think of it as our first approximation to $\sqrt{7}$.

Next, we compute $7 \div A_1$, and carry out the division to two decimal places (one more place than finally needed).

$\frac{7}{4} = 1.75$

We label this number B_1. On the number line (Figure 26.8), we now have three numbers: 7, A_1, and B_1. Although we do not demonstrate this, it is true that $\sqrt{7}$ is somewhere between A_1 and B_1. Therefore, if we take the number halfway between A_1 and B_1 (the average of these two numbers) it will be closer to $\sqrt{7}$ than A_1 is. So we compute this number to two decimal places and call it A_2, the second approximation to $\sqrt{7}$.

$$A_2 = \tfrac{1}{2}(A_1 + B_1) = 2.88$$

Figure 26.8

Now we repeat the process. We find B_2 by dividing 7 by A_2, and we compute that

$$B_2 = 2.43$$

Then

$$A_3 = \tfrac{1}{2}(A_2 + B_2) = \tfrac{1}{2}(5.31) = 2.66$$

and

$$B_3 = 2.63$$

On the number line, we now have the situation shown in Figure 26.9.

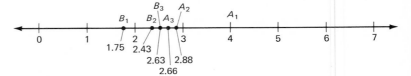

Figure 26.9

Now $\sqrt{7}$ is not only between A_1 and B_1, it is also between A_2 and B_2 and between A_3 and B_3. Now we are finished. A_3 and B_3 differ only after the first decimal place, so, to one decimal place,

$$\sqrt{7} = 2.6$$

We can check this by computing 2.5^2, 2.6^2, and 2.7^2. We find that of the three results, 2.6^2 is closest to 7.

■ **Question 15** Compute A_4 to three decimals.

The same method will work for any number N. We start by choosing a number A_1 about halfway between 1 and N. Then we compute B_1, A_2, B_2, A_3, ... as we did above. We know that $\sqrt{N}$ is between A_1 and B_1, between A_2 and B_2, between A_3 and B_3, etc. We keep going until the A and B approximations are as close together as we need them.

The same procedure will work if N is a fraction less than 1. In this case A_1 will be larger than N, but the same process can still be used.

Of course, no negative number has a square root, because the square of no number is ever negative.

VARIANCE AND STANDARD DEVIATION

Now we are ready to return to the discussion of the scatter of a distribution. We can measure it by the *variance,* the average of the squares of the deviations from the mean.

Table 26.23

x	$x - \overline{x}$	$(x - \overline{x})^2$
9	3	9
7	1	1
5	−1	1
3	−3	9

Let us compute the variance for two simple examples. Suppose we have only four numbers: 9, 7, 5, 3 (see Table 26.23).

$$\overline{x} = \frac{9 + 7 + 5 + 3}{4} = 6$$

Then the variance of this distribution is the average of the numbers in the right-hand column:

$$\frac{9 + 1 + 1 + 9}{4} = 5$$

We have used the symbol $\overline{x}$ to denote the average of the numbers in the x column of a table. So the average of the numbers in the $(x - \overline{x})^2$ column is denoted by

$$\overline{(x - \overline{x})^2}$$

or

$$\text{Variance} = \overline{(x - \overline{x})^2}$$

If the values of x are given by a frequency table (see Table 26.24), one must remember to multiply each value of $(x - \overline{x})^2$ by the frequency of the corresponding x. Notice that we have to begin by finding $\overline{x}$. For the following example, $\overline{x} = 10$.

Table 26.24

x	Frequency	$x - \overline{x}$	$(x - \overline{x})^2$	Frequency times $(x - \overline{x})^2$
15	1	5	25	25
12	2	2	4	8
11	3	1	1	3
10	6	0	0	0
9	3	−1	1	3
8	2	−2	4	8
5	1	−5	25	25
	$n = 18$			Total 72

The variance is $\frac{72}{18} = 4$.

■ **Question 16** What is the variance of the numbers in the set $\{0, 2, 4\}$?

Going back to the three distributions A, B, and C shown in Figure 26.24, we can compute their variances in the same way. For distribution A, we have

$$\tfrac{1}{8}[(0 - 5)^2 + 6 \times (5 - 5)^2) + (10 - 5)^2] = \tfrac{1}{8}(25 + 25) = \tfrac{1}{8}(50)$$

For distribution B, we have

$$\tfrac{1}{8}[2(0 - 5)^2 + (2 - 5)^2 + 2(5 - 5)^2 + (8 - 5)^2 + 2(10 - 5)^2] = \tfrac{1}{8}(118)$$

For distribution C, we have

$$\tfrac{1}{8}[(2-5)^2 + 6(5-5)^2 + (8-5)^2] = \tfrac{1}{8}(9+9) = \tfrac{1}{8}(18)$$

These results agree with our comment above that distributions A and C look more like each other than either does to distribution B. Although the ranges are different, the means of A and C are the same and the variances are both considerably smaller than the variance of distribution B.

The variance of a distribution is a perfectly good measure of scatter. One bothersome matter remains. Suppose, for some distribution, the values of x refer to weights in pounds. Then $x - \bar{x}$ represents pounds, but $(x - \bar{x})^2$ and the variance represent "square pounds." To say that the mean weight is, for example, 10 pounds and the variance is 4 "square pounds" sounds a bit silly. For this reason (there are other, more important mathematical reasons), we shall take the square root of the variance and use *this* as the usual measure of scatter. This measure is called the *standard deviation* of x:

Standard deviation of $x = \sqrt{\overline{(x - \bar{x})^2}}$

It may be helpful in remembering the formula for standard deviation to know that some British writers call the standard deviation the "root mean square deviation": the square root of the mean of the squared deviations. Remember also that the standard deviation is a *number* associated with a frequency distribution.

■ **Question 17** To one decimal place, what is the standard deviation of the data in Question 16?

Computing the variance of a distribution can lead to some dull and lengthy calculations if the given numbers are not as "nice" as the ones in the examples above. Here is our set of test scores again (see Table 26.25).

Table 26.25

x	Frequency	$x - 71.25$	$(x - 71.25)^2$	Frequency times $(x - 71.25)^2$
90	1	18.75	351.5625	351.5625
80	2	8.75	76.5625	153.1250
70	3	−1.25	1.5625	4.6875
60	1	−11.25	126.5625	126.5625
50	1	−21.25	451.5625	451.5625
	$n = 8$			Total 1087.5000

$$\text{Variance} = \frac{1087.5000}{8} = 135.9375$$

We have already found that $\bar{x} = 71.25$.

We see that the calculation of the variance is somewhat nasty.

This example, incidentally, illustrates why anyone who has to do much with statistics wishes to have access to a computer. To a computer, the arithmetic in the last example is no nastier than the arithmetic in the other examples.

APPLICATIONS OF STATISTICS

While the mean and the variance (or the standard deviation) are useful in describing a distribution, it turns out that their usefulness in dealing with real world problems is far greater than we have been able to suggest so far. To make clear how they can be used requires a good deal more mathematics than we have time for in this text, but a few examples, to which we now turn, may give a slight suggestion.

In the first example, a researcher is interested in the physical fitness of eighth-grade girls. He decided to use a "sit-up" test as one measure of fitness. He knows from long experience that the mean performance of eighth-grade girls on this sit-up test is 30.3 with standard deviation 3.2.

The researcher selects a sample of 64 girls at random from the set of all eighth-grade girls in the school system. He then puts these girls through a special training program. At the end of the program, he finds that the mean score on the sit-up test for these girls is 32.17. How can we decide if the special training program was really helpful?

In the second example, a manufacturer of light bulbs advertises that the mean life of his product is 980 hours with a standard deviation of 60 hours. A hospital buys 100 of these bulbs and finds that their mean life is only 970 hours. How can the hospital tell if this set of 100 bulbs was defective?

These two examples are typical of a situation which arises time after time in all sorts of contexts, in medical research and practice, in business, in politics, in education, in science, etc. In each such situation we select a random sample from a very large population. To begin with, let us suppose that the population mean and the population standard deviation are known. It is customary to denote the mean of the population by the Greek letter μ and the standard deviation of the population by the Greek letter σ. Suppose further that we have decided on a definite sample size, n.

Now our random sampling process will produce a single sample for which the sample mean $\bar{x}$ may be computed. *Before* we actually select the sample it is apparent there are very many random samples of size n that may be chosen. Each of these samples yields a particular $\bar{x}$. These possible values of $\bar{x}$ make up a new distribution—the distribution of the sample means of all samples of size n. This distribution itself has a mean and a standard deviation.

■ **Question 18** Compute the mean for each sample of size 2 from the population given in Question 16. Find the mean and variance of these means.

Clearly the composition of the possible samples and therefore the distribution of the sample means depends on the population. It would seem impossible that we would be able to make a statement about the distribution of sample means which would be true for all populations. However, the problems of statistical inference are so important that the nature of the distribution of sample means has been studied very carefully. The astonishing fact is that mathematicians have been able to demonstrate the following three propositions.

PROPOSITION 1:
The *mean* of the distribution of the means of all random samples of size n equals the population mean, μ.

PROPOSITION 2:
The standard deviation of the means of all random samples of size n equals $\sigma/\sqrt{n}$ where σ is the standard deviation of the population.

PROPOSITION 3:
If n is sufficiently large, then the distribution of the means of all random samples of size n from a given population is approximately *normal* (with mean μ and standard deviation $\sigma/\sqrt{n}$).

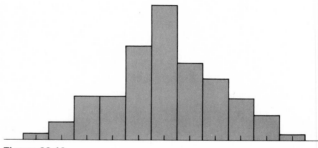

Figure 26.10

We will not give the technical definition of a "normal" distribution. The bar graph and broken line graph below show distributions that are close to normal (see Figures 26.10 and 11).

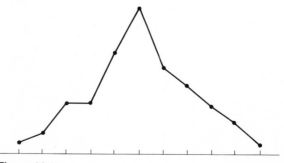

Figure 26.11

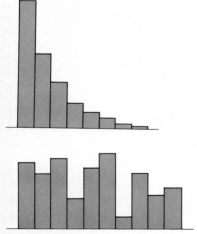

Figure 26.12

However, these two graphs represent distributions which are quite far from normal (see Figure 26.12).

For our purposes, it is enough to know only the following information about any normal distribution:

1 68 percent of the data lies within the interval which is centered at the mean of the distribution and which stretches out a distance of 1 standard deviation on each side.

2 95 percent of the data lies within the interval which is centered at the mean of the distribution and which stretches out a distance of 2 standard deviations on each side.

3 99 percent of the data lies within the interval which is centered at the mean of the distribution and which stretches out a distance of 3 standard deviations on each side.

■ **Question 19** In Figure 26.7, which of the three distributions seem to be close to "normal"?

Another way of putting this is that if we choose one sample at random from a normal population, then the probability is $\frac{68}{100}$ that its mean will be less than 1 standard deviation from the population mean, and so the probability that its mean will be more than 1 standard deviation away from the population mean is $1 - \frac{68}{100} = \frac{32}{100}$, while the probability that the mean of the sample will be more than 2 standard deviations away from the population mean is only $\frac{5}{100}$, and the probability that it will be more than 3 standard deviations away from the population mean is only $\frac{1}{100}$.

One point needs further discussion. Proposition 3 requires that n be "sufficiently large." How large is "sufficiently large"? There is really no easy answer to this question. The answer depends, among other things, on the actual distribution of the population. In practice, most cases are covered by a rule of thumb: use Proposition 3 if $n \geq 30$.

Now let us return to the case of the eighth-grade girls. With $\mu = 30.3$ and $\sigma = 3.2$, we know that 95 percent of the means of all samples of size 64 will fall within the interval from $30.3 - 2 \times 3.2/\sqrt{64}$ to $30.3 + 2 \times 3.2/\sqrt{64}$.

Since $\sqrt{64} = 8$, this is the interval from 29.5 to 31.1. *Our* sample, however, gave us a mean of 32.17. This falls outside the interval. We may conclude either

1 Our sample is really different from the general population of eighth-graders on the sit-up test, or

2 a rather rare event has occurred. Our sample just happens to be one of the 5 percent of all samples of size 64 that give a mean outside the interval from 29.5 to 31.1.

If we choose alternative 1 we might say that our sample is not representative of all eighth-graders. The performance of our sample is *significantly* different from that of the population. A common way to state this conclusion is to say our sample differs from the population at the *5 percent significance level.*

It is tempting to conclude that it was the special training that accounted for the difference between the sample and the population.

■ **Question 20** How large would the mean of the girls' scores have to be for us to conclude that this sample differs from the population at the 1 percent significance level?

Now take a look at the light bulb question. We have $\mu = 980$ and $\sigma = 60$. Then the means of 95 percent of all samples of size 100 will lie within the interval from $980 - 2 \times 60/\sqrt{100}$ to $980 + 2 \times 60/\sqrt{100}$ or from 968 to 992. Our sample, having a mean of 970, does lie within this interval, so it does not differ from the population at the 5 percent significance level and the hospital has no grounds for complaint.

These two examples can only hint at the variety of ways in which ideas about probability and statistics are used in real problems. Actually, getting a satisfactory grasp of the situation requires a great deal more mathematics than we have seen, but for our purposes this hint is enough.

PROBLEMS

1 Make a frequency table, a dot-frequency diagram, a bar graph and a broken line graph for this set of data:

10, 20, 30, 30, 30, 40, 40, 50

2 Do the same for this set of data:

0, 10, 20, 20, 20, 30, 30, 40

3 For this set of data, make a frequency table and extend it to show the cumulative and relative frequencies.

25, 30, 35, 35, 35, 40, 40, 45

4 Do the same for this set of data:

100, 120, 140, 140, 140, 160, 160, 180

5 (a) Make a frequency table, showing also cumulative and relative frequencies for this set of data: heights of members of a college basketball squad (to nearest inch):

77, 78, 73, 78, 80, 74, 74, 72, 78, 77, 72, 73, 78, 78, 76

(b) Make a bar graph for the above data.

Table P26.6

Interval	Midpoint	Frequency
80–84	82	
75– · · ·		
· · ·–· · ·		
· · ·–69		
· · ·–· · ·		
· · ·–· · ·		
· · ·–· · ·		
· · ·–· · ·		
· · ·–44		1

6 This is the list of scores on a mathematics test taken by 80 students.

84, 83, 82, 80, 78, 78, 78, 76, 75, 75, 74, 73, 73, 73, 70, 70, 70, 70, 70, 70, 69, 69, 68, 68, 67, 67, 67, 66, 66, 66, 65, 65, 64, 64, 64, 64, 64, 63, 63, 63, 63, 63, 62, 62, 62, 62, 62, 62, 62, 61, 60, 60, 60, 60, 60, 58, 58, 57, 57, 57, 56, 56, 56, 55, 55, 55, 55, 55, 55, 53, 52, 52, 52, 52, 51, 50, 50, 48, 45, 42

Complete Table P26.6.

7 Find the mean of the following 20 numbers:

26, 26, 26, 31, 31, 31, 31, 31, 35, 35, 35, 35, 42, 42, 42, 42, 42, 53, 53, 53

8 Find $\bar{x}$ for the following data given in Table P26.8.

9 Find the mode, mean, and median for the following.
(a) $-4, -2, -2, -2, 0, 1, 7, 10, 10$
(b) $10, 50, 30, 10, 40, 60, 10$
(c) $12, 13, 13, 13, 15, 16, 16, 17, 18$
(d) $4, 5, 5, 7, 7, 7, 9, 9, 10$

10 For each part of Problem 9, find the mean of the mode and the median.

11 (a) Trace through the flow chart for the following inputs:
3, 5, 5, 9
(b) What is the output in terms of the inputs?
(c) In Figure P26.11, how would the outputs differ if the box labeled 4 followed the box labeled 6?

12 Suppose you read the following account in a newspaper.
"The average American is a woman with 2.41 children in her family."
(a) Which method of "averaging" would give a result such as: "The average American is a woman"?
(b) Which method(s) of "averaging" would give a result such as ". . . 2.41 children"? Explain why this cannot be the mode. Explain why this is not likely to be the median.

13 If the mean weight of 7 football linemen is 220 pounds and the mean weight of the 4 backfield men is 187 pounds, what is the mean weight of the 11-man team?

14 At the Bar-X restaurant the average weekly wage for full-time employees is $130. There are also 10 part-time employees whose average weekly wage is $47. If the total weekly payroll of the restaurant is $5,800, how many full-time employees are there?

15 If the mean rainfall in a certain city was 35.2 inches for 12 years, what was the total rainfall for that time?

16 Compute the variance for the data sets given in Tables P26.16a and b.

Table P26.8

x	Frequency
1,111	3
1,110	4
1,109	8
1,108	6
1,107	5
1,106	4

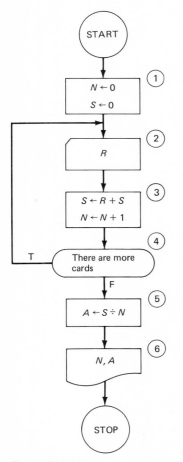

Figure P26.11

Table P26.16a

(a) Score	Frequency
23	1
20	2
19	3
18	6
17	3
16	2
13	1

Table P26.16b

(b) Score	Frequency
23	2
20	2
19	3
18	4
17	3
16	2
13	2

Draw dot-frequency diagrams.

For the following three problems, assume that a normal approximation is appropriate. Interpret "mean" as μ and "standard deviation" as σ.

17 For certain workers the mean wage is $4.25 per hour with standard deviation $.50. If a worker is chosen at random, what is the probability that his hourly wage (assuming the distribution of wages is normal)

(a) is between $3.75 and $4.75?

(b) is between $3.25 and $5.25?

(c) is between $2.75 and $5.75?

18 Refer back to the example about the eighth-grade girls. Another group of 100 girls were not trained, but were given special exercises. The mean score for this group was 30.6. Does this sample differ from the population at the 5 percent significance level?

19 At the end of Chapter 19 you did a 5-coin experiment (200 times) and a 10 coin experiment (100 times). You summarized the results in two tallies, which you should now recognize as frequency tables.

(a) Draw a bar graph for each of these distributions. Do they appear to be approximately normal?

(b) Compute the mean and standard deviation for each distribution.

20 At the end of Chapter 23, you measured the distance around your block five times. Compute the mean and variance of your five measurements.

27.

COORDINATE GEOMETRY

COORDINATES IN THE PLANE

Another place where negative numbers are useful is in geometry. When we first considered the number line, we have available only positive numbers so we could attach number labels only to points on one side (to the right) of 0 on the number line. We introduced negative numbers, in Chapter 24, and we observed that we can now attach number labels to points anywhere on the number line.

It turns out that we can do something similar for the plane. Let us start by drawing two lines, one horizontal and one vertical. We will name the horizontal line the X *axis* and the vertical line the Y *axis*. The point where the lines cross is called the *origin* and is labeled 0 (see Figure 27.1).

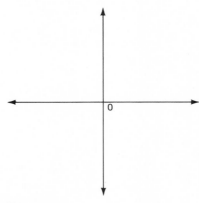

Figure 27.1

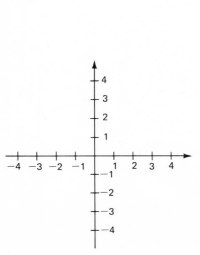

Figure 27.2

Now we choose a unit segment and use it to turn the X axis into a number line, with the positive numbers on the right. Using the *same* unit segment, we turn the Y axis into a number line, with the positive numbers above (see Figure 27.2).

Now we will see how we can locate a point P in the plane by specifying two numbers. First (Figure 27.3), we draw a vertical line through P. It will cross the X axis somewhere. The number attached to this point is called the *x coordinate* of P. In Figure 27.3, the vertical line crosses the X axis at the point 3. Next we draw a horizontal line through P. It will cross the Y axis somewhere. The number attached to this point is called the *y coordinate* of P. It is 4 in Figure 27.3.

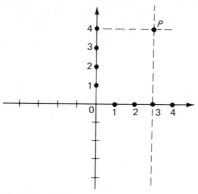

Figure 27.3

In this way, two numbers are attached to the point in the plane. We usually write them inside parentheses, with the x coordinate first. The point in the figure above has coordinates $(3, 4)$.

■ **Question 1** Was it necessary to have positive numbers to the right on the X axis?

Both the above coordinates are positive numbers. But this need not always be the case. Note that the x coordinate of Q, both coordinates of R, and the y coordinate of S are negative in Figure 27.4.

Naturally, the coordinates of a point need not be whole numbers. Fractions are more common.

Next we need to note that the process can be reversed. If we are given a pair of numbers, a and b, we can locate a point P whose coordinates are (a, b). All we need to do is to draw a vertical line through the point on the X axis labeled a and a horizontal line through the point on the Y axis labeled b. Where these two lines cross is the point P.

What we have seen so far is that there is a correspondence between points in the plane and pairs of numbers. Given the point, we can find the pair of numbers. Given the pair of numbers, we can find the point.

This is pretty easy and straightforward. It is also not very exciting. However, we are actually pretty close to something more important. This connection between geometry (points) and arithmetic (pairs of numbers) can be extended to more complicated geometric figures and more complicated arithmetic expressions.

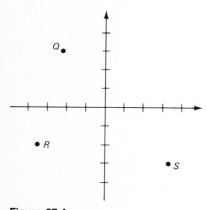

Figure 27.4

■ **Question 2** What are the coordinates of the origin?

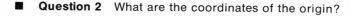

SLOPE OF A LINE We will illustrate this by showing how lines (geometric) can be connected with equations (arithmetic). We will start by investigating a particular line, the one which goes through the point $A(1, 0)$ and the point $B(5, 2)$. This is shown in Figure 27.5.

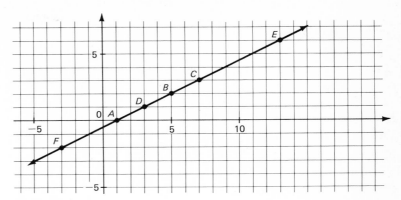

Figure 27.5

We have also marked some other points on this line. Their coordinates are:

$C(7, 3)$ $D(3, 1)$ $E(13, 6)$ $F(-3, -2)$

Naturally, different points on this line have different coordinates. However, it was discovered long ago that a certain combination of the coordinates of pairs of points on the line are always the same. To be specific, no matter which two points, P and Q, we may choose on the line, the ratio

$$\frac{(y \text{ coordinate of } P) - (y \text{ coordinate of } Q)}{(x \text{ coordinate of } P) - (x \text{ coordinate of } Q)}$$

will always be the same.

For example, if we choose P to be the point A and Q the point B, then the above ratio becomes

$$\frac{0 - 2}{1 - 5} = \frac{-2}{-4} = \frac{1}{2}$$

If we choose P to be C and Q to be F, then the ratio becomes

$$\frac{3 - (-2)}{7 - (-3)} = \frac{5}{10} = \frac{1}{2}$$

The reader should now proceed immediately to the problem set at the end of this chapter and work Problem 4.

For all pairs of points on this line, the ratio has the value $\frac{1}{2}$. We call this ratio

$$\frac{(y \text{ coordinate of } P) - (y \text{ coordinate of } Q)}{(x \text{ coordinate of } P) - (x \text{ coordinate of } Q)}$$

the *slope* of the line.

■ **Question 3** We saw that if we chose P to be A and Q to be B, then the slope ratio was $\frac{1}{2}$. What if we choose P to be B and Q to be A?

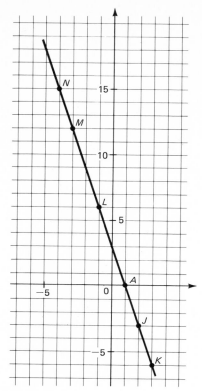

Figure 27.6

Now let us look at another line, the one going through point $A(1, 0)$ again and through the point $L(-1, 6)$. This is shown in Figure 27.6. Some other points on this line are:

$J(2, -3)$ $K(3, -6)$ $M(-3, 12)$ $N(-4, 15)$

If we take P to be A and Q to be J and compute the slope ratio, it becomes

$$\frac{0 - (-3)}{1 - 2} = \frac{3}{-1} = -3$$

If we take another pair of points on this line, say $P = L$ and $Q = N$, then the slope ratio is

$$\frac{6 - 15}{-1 - (-4)} = \frac{-9}{3} = -3$$

The reader should now go to the problem set at the end of the chapter and work Problem 5. The results will support the statement that the slope ratio is -3 for any pair of points chosen on this line, so the slope of the line is -3.

Finally, as is suggested by the sixth problem at the end of this chapter, if we pick a pair of points, one of which is on a particular line while the other is not, then the slope ratio for these two points will *not* be equal to the slope of the line.

HORIZONTAL AND VERTICAL LINES

There are two special cases we need to look at. Suppose we have a line which is horizontal. Then it is easy to see that any two points on it have the same y coordinate. This means that the numerator of the slope ratio will always be zero. Therefore the slope of any horizontal line is zero.

Next, consider a vertical line. In this case, the x coordinates of any two points on the line are the same. This means that the denominator of the slope ratio is always zero. But a fraction with zero denominator is meaningless, so a vertical line does not have a slope.

■ **Question 4** Choose P to be N and Q to be the origin in Figure 27.6. What is the slope ratio?

There are some other things suggested by the particular examples we have been examining. For the first line, displayed in Figure 27.5, the slope is positive and when we move along from left to right the line goes up. For the second line, displayed in Figure 27.6, the slope is negative and when we move along from left to right the line goes down.

This will always be the case. When we move from left to right lines with positive slopes will rise and lines with negative slopes will fall.

LINES AND EQUATIONS

Now that we have learned about the slope of a line we can return to the connection between lines and equations. We will illustrate the connection first with a particular line, the one displayed in Figure 27.5. We remember that the slope of this line is $\frac{1}{2}$. We have already located a number of points on this line. Let us choose one of them, say $B(5, 2)$. What can we write down that would assert that another point, P, is also on the line? Obviously, if P is on the line, then the slope ratio for P and B will have to be $\frac{1}{2}$. This slope ratio is

$$\frac{(y \text{ coordinate of } P) - 2}{(x \text{ coordinate of } P) - 5}$$

So, to assert that P is indeed on the line, we can write

$$\frac{(y \text{ coordinate of } P) - 2}{(x \text{ coordinate of } P) - 5} = \frac{1}{2}$$

We will abbreviate this to the equation

$$\frac{y - 2}{x - 5} = \frac{1}{2}$$

We call this the *equation of the line.* Whenever P is a point (other than B) on the line, then if we replace y in this equation by the y coordinate of P and x in the equation by the x coordinate of P, the resulting mathematical sentence is true. We also say, in this case, that the coordinates of P *satisfy* the equation.

■ **Question 5** Write an equation for the line shown in Figure 27.6, starting with the point K.

Notice also that if P is not on the line, then the coordinates of P do not satisfy the equation. This is because when P is not on the line then, as we saw, the slope ratio of P and B, which is what we get when we replace y and x with the coordinates of P in the left-hand side of the equation, will not be equal to the slope of the line, which is $\frac{1}{2}$.

An alternative, and slightly more convenient, form for the equation of this line is

$$y - 2 = \tfrac{1}{2}(x - 5)$$

Clearly, if the coordinates of a point, other than B itself, satisfy this equation, then all we have to do is divide both sides of this equation by the $(x - 5)$ term in order to see that the original equation is satisfied. However, the coordinates of B itself put a 0 in the denominator of the first form of the equation but they satisfy the second form, since

$$(2 - 2) = \tfrac{1}{2}(5 - 5)$$

This is why this second form is more convenient. From now on we will normally use this second form of the equation of a line.

Notice that the equation of a line is unlike the equations we have studied so far in Chapters 9 and 25 since it contains two variables, y and x, while the equations we studied before contained only one in each case. Nevertheless, some of the ideas about equations with one variable also apply to equations with two variables. Thus, a *solution* of an equation such as

$$y - 2 = \tfrac{1}{2}(x - 5)$$

is a pair of numbers which satisfy the equation. The *solution set* of the equation is the set of all such pairs of numbers.

■ **Question 6** Rewrite the equation of Question 5 and show that the coordinates of K satisfy it.

Two equations are *equivalent* if they have the same solution sets. The processes we used in Chapter 25 to transform equations into equivalent equations apply also to equations containing two variables. Thus, all of these equations are equivalent:

$$y - 2 = \tfrac{1}{2}(x - 5)$$
$$y - 2 = \tfrac{1}{2}x - \tfrac{5}{2}$$
$$y = \tfrac{1}{2}x - \tfrac{1}{2}$$
$$2y = x - 1$$

This helps to explain something that might otherwise be a puzzle. The line we have been dealing with has slope $\frac{1}{2}$ and goes through the point $(5, 2)$, so its equation is

$$y - 2 = \tfrac{1}{2}(x - 5)$$

But this line also goes through the point $(-3, -2)$, so another equation for this line is

$$y + 2 = \tfrac{1}{2}(x + 3)$$

These two equations, for the same line, are different. However, by subtracting 2 from each side of this last equation we get the equivalent equation

$$y = \tfrac{1}{2}x - \tfrac{1}{2}$$

This will always be the case. Any two equations for the same line, no matter how they are obtained, are equivalent.

EQUATIONS FOR HORIZONTAL AND VERTICAL LINES

Before we go any farther, there are two special cases we should look at. The first is that of a horizontal line. Suppose, for example, we consider the line which goes through the point (4, 7) and is horizontal, i.e., parallel to the X axis. Another point on this line is (0, 7), the point where this line crosses the Y axis.

We can use these two points to compute the slope of the line. It is

$$\frac{7 - 7}{4 - 0} = \frac{0}{4} = 0$$

An equation for this line, therefore, is

$$y - 7 = 0(x - 4)$$

or

$$y - 7 = 0$$

An equivalent equation is

$$y = 7$$

This kind of result is obtained for any horizontal line. An equation for the line will be

$$y = r$$

where r is the y coordinate of any point on the line.

■ **Question 7** Write an equation for the line shown in Figure 27.6, starting with the point N. Show that this is equivalent to the equation for Question 6.

Now let us consider a vertical line, say the one through the same point, (4, 7). Another point on this line is (4, 0), the point where the line crosses the X axis. Of course, as we remember, this line has no slope. Nevertheless, we can find its equation.

We remember that the x coordinate of any point is obtained by drawing the vertical line through the point and observing where it crosses the X axis. So, for any point on the vertical line through (4, 7), its x coordinate is 4, no matter what its y coordinate is. So the coordinates of any point on this vertical line satisfy the equation

$$x = 4$$

Therefore, this is the equation of this line.

In general, a vertical line has an equation

$$x = s$$

where s is the x coordinate of any point on the line.

LINEAR EQUATIONS Now let us note that for each line we have investigated in this chapter, any equation for it can be rewritten in the form:

$$Ax + By + C = 0$$

where A, B, and C are specific numbers. Thus, for example, the first line considered in this chapter had for an equation

$$y - 2 = \tfrac{1}{2}(x - 5)$$

Here is a chain of equivalent equations:

$$2y - 4 = x - 5$$
$$-x + 2y - 4 = -5$$
$$-x + 2y + 1 = 0$$

This last is also an equation for the same line, and it has the form stated above, with $A = -1$, $B = 2$, and $C = 1$.

■ **Question 8** Write equations for the X axis and the Y axis.

Similarly the equation for the horizontal line can be written as

$$y - 7 = 0$$

which is also in the stated form, with $A = 0$, $B = 1$, and $C = -7$.

Thus, to every line (geometric) there corresponds an equation (arithmetic) of the form $Ax + By + C = 0$ such that the coordinates of every point on the line satisfy the equation. But the reverse is also true. To any equation of the form $Ax + by + C = 0$ (except for the uninteresting case in which $A = B = 0$) there corresponds a line such that any pair of numbers which satisfy the equation are the coordinates of a point on the line.

We will illustrate this with some examples. Suppose we have the equation

$$3x + 4y = 5$$

Equivalent equations are

$$4y = -3x + 5$$
$$y = -\tfrac{3}{4}x + \tfrac{5}{4}$$
$$y - \tfrac{5}{4} = -\tfrac{3}{4}x$$

This last we recognize as an equation for the line through the point $\tfrac{5}{4}$ whose slope is $-\tfrac{3}{4}$.

Similarly, $3x = 5$ can be rewritten as $x = \tfrac{5}{3}$, which is an equation of the vertical line through the point $(\tfrac{5}{3}, 0)$. Also, $4y = 5$ can be rewritten as $y = \tfrac{5}{4}$, an equation for the horizontal line through the point $(0, \tfrac{5}{4})$.

Since any equation of the form $Ax + By + C = 0$ is an equation for a line, such an equation is usually called a *linear* equation.

■ **Question 9**   Is $2x + 3y + 4 = 5x + 6y + 7$ a linear equation?

APPLICATIONS   This connection between geometry and arithmetic is very useful. It makes it possible for each of these two branches of mathematics to help the other. Arithmetic can be used to solve geometric problems and vice versa. A few examples will illustrate how this can be done.

Here is a geometric problem. The points $P(-2, 4)$ and $Q(8, -3)$ determine a line. The points $R(-4, 1)$ and $S(4, 5)$ determine another line. Find the point where these two lines cross.

In order to work this problem arithmetically, we start by finding an equation for each line. The slope of the first line is $[4 - (-3)]/(-2 - 8) = -\frac{7}{10}$ so an equation for this line is

$$y - 4 = -\tfrac{7}{10}[x - (-2)] \qquad \text{or} \qquad y = -\tfrac{7}{10}x + \tfrac{26}{10}$$

The slope of the second line is $(1 - 5)/(-4 - 4) = \frac{1}{2}$. An equation for this line is

$$y - (1) = \tfrac{1}{2}[x - (-4)] \qquad \text{or} \qquad y = \tfrac{1}{2}x + 3$$

For the next step in this problem, we remember that the point we are looking for, the point where the two lines cross, is on both lines at once and so its coordinates satisfy both of these equations. The first equation says that the y coordinate of this point is obtained by multiplying its x coordinate by $-\frac{7}{10}$ and then adding $\frac{26}{10}$. The second equation says that the same y coordinate is obtained by multiplying the same x coordinate by $\frac{1}{2}$ and then adding 3.

What we need, then, is a value for the x coordinate for which

$$-\tfrac{7}{10}x + \tfrac{26}{10} = \tfrac{1}{2}x + 3$$

Now this is just the kind of equation that we studied in Chapter 25. We can solve it by means of the procedures we learned there and we find that the solution is $-\frac{1}{3}$. So the x coordinate of the crossing point is $-\frac{1}{3}$. The y coordinate (since the crossing point is on the second line) is $\frac{1}{2} \times (-\frac{1}{3}) + 3 = \frac{17}{6}$, and so we have calculated the coordinates of the crossing point exactly, $(-\frac{1}{3}, \frac{17}{6})$.

■ **Question 10**   Calculate the y coordinate of the crossing point by using the fact that it is on the first line.

Of course, another way to do this problem would be to plot the four points on graph paper, draw the two lines carefully with a straightedge and sharp pencil, and then to read off, from the graph paper, the coordinates of the crossing point. This method is easier but less accurate. Even if the drawing is done very carefully, answers such as $(-\frac{2}{7}, 2\frac{4}{5})$ or $(-\frac{1}{4}, 3)$ would not be unreasonable, and for some purposes would be quite satisfactory.

In general, if accuracy is required, use arithmetic. If an approximation is enough, use geometry.

We noticed that in solving the geometric problem of finding where two lines cross, we were led to an equation of the type we studied in Chapter 25. Other problems often lead to the same result. Suppose, for example, we are asked to find enough points on the line corresponding to the equation

$$2\tfrac{1}{4}x + 3\tfrac{3}{4}y + 5 = 0$$

so that the line can be drawn.

Of course we remember that it will be enough to find the coordinates of just two points on the line, and these can be any two we wish. One point, if we wish, could be the point where the line crosses the X axis. We know that the y coordinate of this particular point is 0. Also, we know that the two coordinates together satisfy the equation. This tells us that,

$$2\tfrac{1}{4}x + 3\tfrac{3}{4} \times 0 + 5 = 0 \qquad \text{or} \qquad 2\tfrac{1}{4}x + 5 = 0$$

We know how to solve this equation. The solution is

$$x = {}^{-}\tfrac{20}{9}$$

so one point on the line is

$$({}^{-}\tfrac{20}{9}, 0)$$

Another point on the line is the one where the line crosses the Y axis. Here the x coordinate is 0, so we have

$$2\tfrac{1}{4} \times 0 + 3\tfrac{3}{4}y + 5 = 0 \qquad \text{or} \qquad 3\tfrac{3}{4}y + 5 = 0$$

The solution is $y = {}^{-}\tfrac{4}{3}$, so the point $(0, {}^{-}\tfrac{4}{3})$ is also on the line.

■ **Question 11** Find a point on the line corresponding to the equation $2\tfrac{1}{4}x + 3\tfrac{3}{4}y + 5 = 0$ for which the x coordinate is 2.

INEQUALITIES

In Chapter 25 we learned how to find solutions not only of equations but also of inequalities when only one variable was involved. In this chapter we have studied equations involving two variables. Can we say anything about inequalities when two variables are involved?

Yes, we can. We will illustrate with an example: $2x + y > 2$. In order to find the solution set of this inequality, we first use the addition property of order to rewrite the inequality in this form.

$$y > -2x + 2$$

Now the solution set will consist of all those points in the plane whose coordinates satisfy this inequality. These are just the points for which the y coordinate is greater than the sum of -2 times the x coordinate and 2. In order to locate these points, we first locate the points where the y coordinate is *equal* to the sum of -2 times the x coordinate and 2. In other words, we locate the points for which

$$y = -2x + 2$$

But we know that these points are all on a line, and we have just learned how to find a pair of points on this line. In this case, $(1, 0)$ and $(0, 2)$ are both on the line, so the line is located as shown in Figure 27.7.

Now, if we consider any particular point on the line we know that its coordinates satisfy the equation

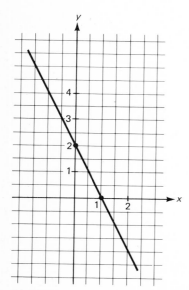

Figure 27.7

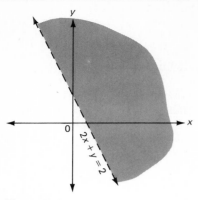

Figure 27.8

$y = -2x + 2$

Suppose we look at a point directly above some point on the line. It has exactly the same x coordinate, but its y coordinate is larger and so its two coordinates satisfy the inequality

$y > -2x + 2$

So the solution set of

$y > -2x + 2$

consists of all those points in the plane which are directly above points on the line whose equation is

$y = -2x + 2$

This solution set is indicated in Figure 27.8.

■ **Question 12** Find two points on the line whose equation is $y = 2x$.

Of course, if the inequality ran the other way,

$2x + y < 2$

then the graph would consist of all the points below the line given by

$2x + y = 2$

There is nothing special about this particular example. We remember from Chapter 3 that any line separates the plane into two half-planes. Whenever we have a *linear inequality*, of the form

$Ax + By + C > 0$

then the solution set is one of the two half-planes determined by the line whose equation is

$Ax + By + C = 0$

To find out which of the two half-planes, we could go through the kind of analysis we went through above. It is easier, however, to take some point, any point, in one of the half-planes and to find out if its coordinates satisfy the inequality. If so, we have the correct half-plane. If not, the other half-plane is the solution set.

In our example, we could have chosen the point $(0, 0)$ as the test point. But it is not true that

$2 \times 0 + 0 > 2$

So the solution set is the half-plane on the other side of the line from the origin.

As usual, there is one special case. When the linear inequality is such that the associated line is vertical, then the analysis we carried out before does not work. For any point on such a line, all the points above (or below) it are still on the line, rather than in one of the half-planes.

However, such inequalities can be solved just by inspection. Take

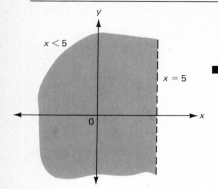

$x < 5$

$x = 5$

0

Figure 27.9

for example the inequality $x < 5$. It is easy to see that the solution set is the half-plane to the left of the line $x = 5$, as is indicated in Figure 27.9.

■ **Question 13** What is the solution set of $x < 0$?

Sometimes, instead of a strict inequality, such as $2x + y > 2$, we have an inequality with the $\geqslant$ sign, such as $2x + y \geqslant 2$. The solution set of this is easy to find. Any point on the line $2x + y = 2$ has coordinates for which $2x + y = 2$. Any point in the upper half-plane has coordinates for which $2x + y > 2$. In either case, the coordinates satisfy $2x + y \geqslant 2$. So the solution set is the union of the upper half-plane and the line.

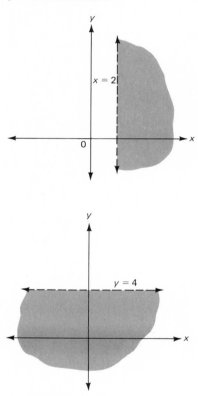

Figure 27.10

There is one final remark to be made about inequalities. Sometimes we are given two inequalities and we are asked to locate those points whose coordinates satisfy both of them. Other times we might be asked for those points whose coordinates satisfy at least one of them. In the first case, we form the intersection of the two solution sets. In the second case we form the union. Thus, if we have the inequalities $x > 2$ and $y < 4$ (Figure 27.10), the solution set of

$x > 2$ and $y < 4$

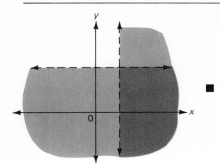

Figure 27.11

is the doubly shaded region in Figure 27.11. The solution set of

$$x > 2 \quad \text{or} \quad y < 4$$

is the union of the two half-planes.

■ **Question 14** What is the solution set of $x > 0$ and $x < 0$? What is the solution set of $x > 0$ or $x < 0$?

What we have done so far in this chapter is enough to demonstrate that there is a firm connection between arithmetic and geometry. This connection is far more extensive than what we have seen would indicate. In particular, there are many more links between arithmetic and geometry than the one between linear equations and lines. We will conclude our investigation of this part of mathematics by studying one other link.

THE PYTHAGOREAN THEOREM

Figure 27.12

For right triangles, there is an important property about how the lengths of the sides are related. Let $\triangle ABC$ be a right triangle with $\angle C$ the right angle (Figure 27.12). Construct three squares, one having $\overline{AB}$ as a side, one having $\overline{AC}$ as a side, and one having $\overline{BC}$ as a side (Figure 27.13). It has been known for some 4,000 years that the area of the square on $\overline{AB}$ (the *hypotenuse*) is equal to the sum of the areas of the squares on the other two sides ($\overline{AC}$ and $\overline{BC}$). When and by whom this was *proved* (as opposed to being *observed*) is not known. It is attributed to the Greek mathematician Pythagoras (who lived in the sixth century B.C.) in that it is called the Pythagorean property, although it may well have been established earlier.

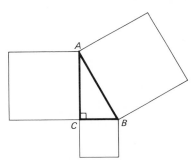

Figure 27.13

Since the area of a square is the square of the length of one of its sides, we can state this property most conveniently as:

The Pythagorean Theorem
In any right triangle the square of the length of the hypotenuse is equal to the sum of the squares of the lengths of the other two sides.

■ **Question 15** In Figure 27.12, the hypotenuse (the side opposite the right angle) is larger than either of the other sides. Draw some other right triangles and check to see if the same is true for them.

If we denote the length of the hypotenuse by c and the lengths of the other two sides by a and b (see Figure 27.14), then the theorem asserts that $a^2 + b^2 = c^2$.

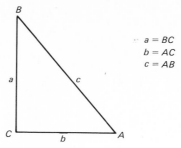

$a = BC$
$b = AC$
$c = AB$

Figure 27.14

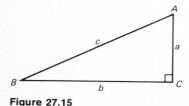

Figure 27.15

There are many proofs of the Pythagorean Theorem. We give the outline (without all the details of the argument) of a method which may appeal to the eye.

Let the measures of the sides of $\triangle ABC$ be a, b, and c as shown in Figure 27.15. Now draw a square with sides of length $a + b$ and subdivide it as shown in Figure 27.16. Notice that the left side of the square is subdivided so that the top portion is of length a and the bottom of length b. Instead, we could have subdivided this side so that the top portion is of length b and the bottom of length a. Likewise, we can reverse the subdivision on the lower side of the square, obtaining the figure shown in Figure 27.17.

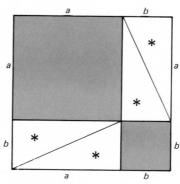

Figure 27.16

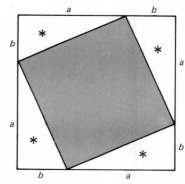

Figure 27.17

Each of the triangles in Figures 27.16 and 27.17 marked with an * is a right triangle whose legs have lengths a and b. Therefore, each of these triangles is congruent to $\triangle ABC$ (by SAS Congruence Property). Since congruent figures have the same area and there are four starred triangles in each of the large squares, the shaded region in Figure 27.16 and that in Figure 27.17 have the same area. In Figure 27.16, that area is $a^2 + b^2$.

In Figure 27.17, the shaded region is a square with sides of length c

and its area is c^2. Hence, by equating the areas of the shaded regions, $a^2 + b^2 = c^2$, and we have demonstration of the celebrated Pythagorean Theorem.

■ **Question 16** If two sides of a right triangle have lengths 3 and 4, how long is the hypotenuse?

COMPUTING DISTANCES

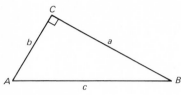

Figure 27.18

One consequence of this theorem is that we can compute the length of the hypotenuse of a right triangle if we know the lengths of the other two sides. For, since $c^2 = a^2 + b^2$, we know that $c = \sqrt{a^2 + b^2}$, and we saw in Chapter 26 how to find as close an approximation to $\sqrt{a^2 + b^2}$ as we wish (Figure 27.18).

The point of all this discussion is that we are now able to compute distances between pairs of points, and information about distances, as we shall see, can often be quite helpful.

Once again, we will use a particular case to illustrate the ideas. Suppose we are asked how far it is from $P(2, 3)$ to $Q(14, 8)$, or what is the same thing, for the length of the segment $\overline{PQ}$.

In Figure 27.19 we have drawn vertical lines from P and from Q down to the X axis and horizontal lines from P and from Q over to the Y axis. Also, we have drawn the horizontal line from P over to the vertical line from Q, meeting at the point R.

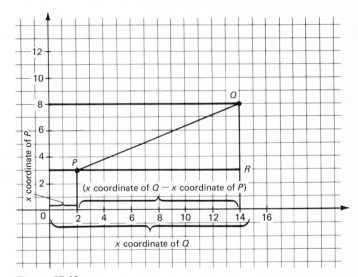

Figure 27.19

Now, PRQ is a right triangle, and the segment $\overline{PQ}$ is its hypotenuse. If we can find the lengths of the sides $\overline{PR}$ and $\overline{RQ}$, we can use the Pythagorean Theorem to find the length of $\overline{PQ}$.

If we look on the X axis, we see that the distance from the origin out to the foot of the perpendicular from P is, by definition, the x coordinate of P, which in this case is 2. The distance from the origin out to the foot of the perpendicular from Q is, by definition, the x coordinate of Q, which in this case is 14. The distance between the feet of the two perpendiculars, which is the same as the length of the segment $\overline{PR}$, is the difference between the distances out to the two feet, namely

(x coordinate of Q) − (x coordinate of P)

A similar look on the Y axis reveals that the length of the segment $\overline{RQ}$ is

(y coordinate of Q) − (y coordinate of P)

Now the Pythagorean Theorem tells us that

$$\overline{PQ}^2 = (x \text{ coordinate of } Q - x \text{ coordinate of } P)^2$$
$$+ (y \text{ coordinate of } Q - y \text{ coordinate of } P)^2$$
$$= (14 - 2)^2 + (8 - 3)^2$$
$$= 12^2 + 5^2 = 169$$

Some easy calculations show that $13^2 = 169$, so the distance from P to Q is 13.

■ **Question 17** How far is it from the origin to the point whose coordinates are (5, 12)?

In the problems at the end of this chapter are shown a number of ways of using information about distances. Probably the most important use is in providing another connection between equations and geometric objects, in this case circles. We remember that a circle is defined as the set of all points at a particular distance, the radius, from a particular point, the center.

EQUATION OF A CIRCLE Suppose we consider a particular circle, for example the one whose center is at the point (2, 1) and whose radius is 3. Can we write an equation in two variables which is satisfied by the coordinates of just those points that are on the circle? Here is such an equation

$$\sqrt{(x - 2)^2 + (y - 1)^2} = 3$$

If the coordinates of a point, when substituted in this equation, do satisfy it, then this equation says that the distance between the point and the point (2, 1) is exactly 3. In this case, the point is indeed on the circle. If the coordinates of a point do not satisfy this equation then we know that its distance from (2, 1) is different from 3, so the point is not on the circle.

Another way of writing the equation, obtained by squaring each side of the equation, is

$$(x - 2)^2 + (y - 1)^2 = 9$$

If we take another example, this time the circle with its center at the origin and its radius 5, we have the equation

$$(x - 0)^2 + (y - 0)^2 = 25$$

or

$$x^2 + y^2 = 25$$

Question 18 Are there any points whose coordinates satisfy the equation $x^2 + y^2 = 0$?

Thus we see that in addition to lines another kind of geometric object, circles, can be connected with arithmetic equations. This suggests that perhaps still other kinds of geometric objects can be connected with equations. This is indeed the case. However, we will go no further into this part of mathematics. What we have done in this chapter is to continue a process started in Chapter 10, that of showing the possibility of, and in fact the first steps in, building a firm connection between arithmetic and geometry, the two most important parts of mathematics that are studied in the elementary school.

PROBLEMS

1 (a) Locate each of these points in the plane. $D(-3, 3)$, $E(-1, -3)$, $F(5, -1)$, $G(3, 5)$
   (b) Draw the segments $\overline{DE}$, $\overline{EF}$, $\overline{FG}$, and $\overline{GD}$.
   (c) Which of the following correctly describe the figure $DEFG$?
     (1) Simple closed curve   (2) Polygon
     (3) Quadrilateral   (4) Square region
     (5) Square   (6) Isosceles triangle
     (7) Rectangle   (8) Union of four angles

2 Write the coordinates of each of the labeled points in Figure P27.2.

3 The X and Y axes separate the plane into four *quadrants* which are identified by roman numerals as shown in Figure P27.3. Points on the axes are not in any quadrant.

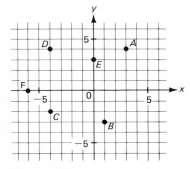

Figure P27.2

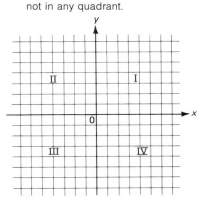

Figure P27.3

   (a) Tell in which quadrant, if any, each point lies: $A(-7, 2)$, $B(7, -2)$, $C(-7, -2)$, $D(7, 2)$, $E(5, 0)$, $F(0, -2)$, $G(0, 0)$
   (b) In which quadrent does a point lie if
     (1) both coordinates are positive?
     (2) both coordinates are negative?
     (3) the x coordinate is positive, the y coordinate is negative?
     (4) the x coordinate is negative, the y coordinate is positive?

4 For the points A, B, C, D, E, and F in Figure 27.5, compute the ratio

$$\frac{(y \text{ coordinate } P) - (y \text{ coordinate } Q)}{(x \text{ coordinate } P) - (x \text{ coordinate } Q)}$$

for each choice of P and Q in the Table P27.4.

Table P27.4

P	Q	Value of ratio
D	C	
C	A	
C	E	
F	B	
B	F	

Table P27.5

P	Q	Value of ratio
N	J	
M	A	
J	L	

5 For the points A, J, K, L, M, and N in Figure 27.6, compute the slope ratio for each choice of P and Q in Table P27.5.

6 (a) The point $B(5, 2)$ is on the line displayed in Figure 27.5. The point $J(2, -3)$ is not on this line. What is the slope ratio determined by these two points? Is it the same as the slope of the line?

    (b) The point $J(2, -3)$ is on the line displayed in Figure 27.6 but the point $F(-3, -2)$ is not. What is the slope ratio for these two points? Is it the same as the slope of the line?

7 Find a number k so that the points $(k, 10)$, $(4, 4)$, and $(2, 8)$ all lie on the same line. (Hint: If they are all on the same line, what about the various slope ratios they determine?)

8 Figure P27.8 is a reproduction of Figure 27.5 with two triangles FPB and CQE drawn in.

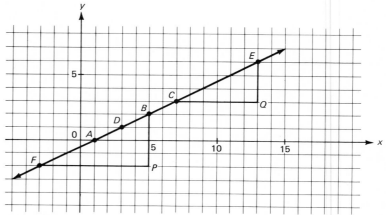

Figure P27.8

    (a) Are $\overleftrightarrow{FP}$ and $\overleftrightarrow{CQ}$ parallel? Why?

    (b) Are angles $\angle PFB$ and $\angle QCE$ congruent? Why?

    (c) Are angles $\angle FPB$ and $\angle CQE$ congruent? Why?

    (d) Are triangles $\triangle FPB$ and $\triangle CQE$ similar? Why?

    (e) Is $\dfrac{PB}{FP} = \dfrac{QE}{CQ}$? Why?

9 On a sheet of graph paper, draw one set of coordinate axes. Draw and label the graphs of the following equations:

    (a) $y = \frac{1}{4}x$ (b) $y = \frac{1}{2}x$

    (c) $y = \frac{2}{3}x$ (d) $y = x$

    (e) $y = 2x$ (f) $y = -\frac{1}{3}x$

    (g) $y = -\frac{1}{2}x$ (h) $y = -\frac{3}{4}x$

    (i) $y = -x$ (j) $y = -2x$

10 Answer the following questions about the lines and their equations.

    (a) What point is on all the lines?

    (b) Which equations have graphs that rise from left to right?

    (c) For the equations having graphs that rise from left to right, is the coefficient of x (the number by which x is multiplied) a positive number or a negative number?

    (d) For the equations in which the coefficient of x is positive, what relation do you observe between the coefficient and the steepness of the line?

    (e) What facts do you observe about the graphs of equations in which the coefficient of x is negative?

(f)   Note that each equation in Problem 9 is in the form $y = mx$ where m is a real number. Write the equation of the X axis in this form.

11   The common black cricket chirps at a rate which depends on the temperature. Observation shows that if we add 39 to the number of chirps in 15 seconds, the result is equal to the temperature (measured in degrees Fahrenheit).

(a)   Write an equation which fits this observation.

(b)   How many chirps would you expect in 15 seconds if the temperature is 83°F?

(c)   What would you expect the temperature to be if you hear 150 chirps in 1 minute?

12   Sketch the solution sets of the following sentences.

(a) $3x + y = 7$ (b) $3x + y > 7$ (c) $3x + y \leq 7$

13   Find the solution set of each of the following and make a sketch of it.

(a) $x < -1$ and $y > -2$ (b) $x \geq 0$ or $y < 3$

(c) $x > -1$ and $y > 2$

14   Write a double inequality for which the shaded area in Figure P27.14 is the solution set.

15   Sketch the solution set of $x \geq 0$ and $y \geq 0$ and $x + 3y \leq 11$ and $3x + y \leq 9$.

16   Find the distance between each pair of points. If it is not an integer, write the result in the form $\sqrt{n}$. Calculate $\sqrt{n}$ to one decimal place.

(a) $(1, 5)$, $(13, 0)$ (b) $(2, 5)$, $(-2, 2)$ (c) $(9, 8)$, $(6, 7)$

17   Modify Figure 27.19 by changing P to the point with coordinates $(-2, -5)$. Is the length of the base of the triangle still given by this formula?

(x coordinate of Q) $-$ (x coordinate of P)

18   In each part of this problem, the three vertices of a triangle are given. Which triangles are isosceles?

(a) $(2, 3)$, $(3, 10)$, $(10, 9)$ (b) $(-3, 1)$, $(2, 4)$, $(2, -2)$

(c) $(2, 3)$, $(5, -2)$, $(8, -7)$

19   A circle has its center at the point $(-2, 4)$ and it passes through the point $(5, -1)$. What is its radius? Write an equation for this circle.

20   If a, b, and c are positive integers such that $a^2 + b^2 = c^2$, then the triple of numbers (a, b, c) is called a Pythagorean triple. $(3, 4, 5)$ is a Pythagorean triple, for example, because $3^2 = 9$, $4^2 = 16$, $5^2 = 25$ and $9 + 16 = 25$. Verify that each triple of numbers below is a Pythagorean triple.

(a)   13, 5, 12   (b)   15, 12, 9

(c)   17, 8, 15   (d)   58, 42, 40

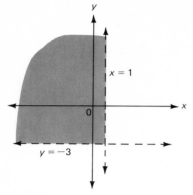

Figure P27.14

■ ANSWERS

An asterisk attached to the number of a problem indicates that there is no unique correct answer to the problem. Any answer that is given in such a case is merely illustrative.

CHAPTER 1

1 b and c **3** (a) Wednesday; (b) March, April, May, June, July; (c) { }. There are no whole numbers between 7 and 8; (d) Tokyo, London; (e) red, orange, yellow, green, blue, purple; (f) 6, 7, 8, 9, 10; (g) Alabama, Alaska, Arkansas; Arizona **5** (c) **7** See Figure A1a and b; **9** No. There are not enough cities with more than 1 million inhabitants. **11** a, b, c, e, f, g **13** Ask each boy to hold hands with a girl. **15** See Figure A2. **17** (a) C is less than B, which is less than A, which is less than D; (b) D is less than A, which is less than B, which matches C. **19** C

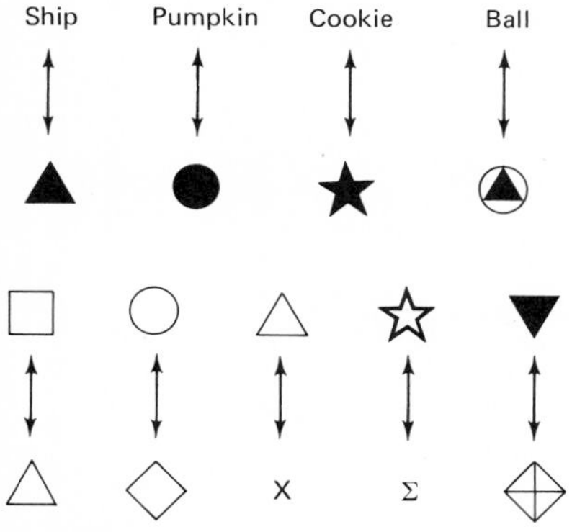

Ship   Pumpkin   Cookie   Ball

Figure A1a

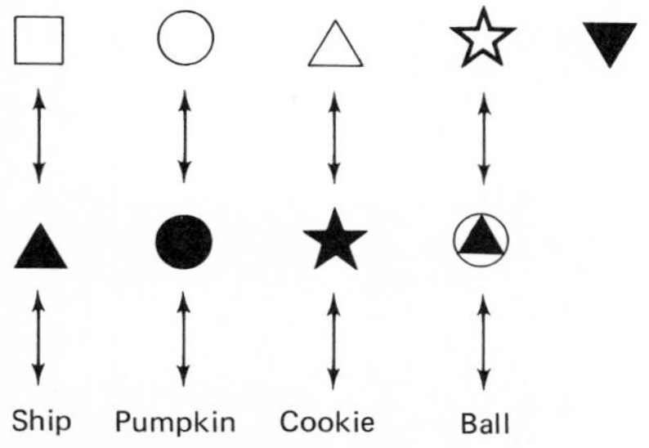

Figure A1b

Ship   Pumpkin   Cookie   Ball

r   e   d

r   d   e

d   e   r

d   r   e

e   r   d

e   d   r

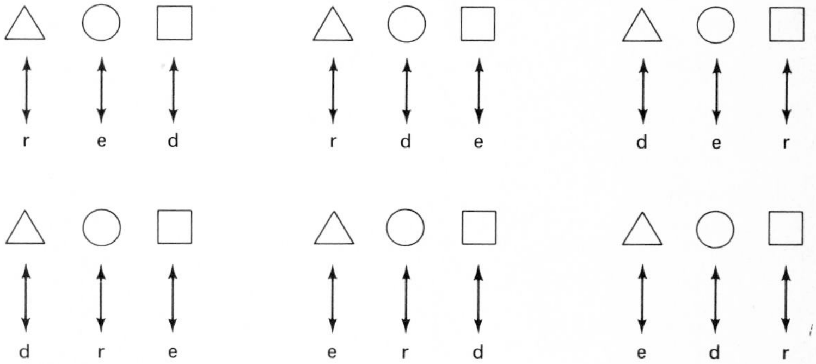

Figure A2

CHAPTER 2　**1** (a) 4; (b) 1; (c) 1; (d) 2; (e) 3; (f) 1　**3** $N(C) < N(B) < N(A)$　**5** 0
7 (a) 749; (b) 8,306; (c) 2,751; (d) 46,083; (e) 5,024; (f) 70,609; (g) 8,403;
(h) 95,060　**9** a, c, e, g　**11** (a) $>$; (b) $<$　**13** (a) 40; (b) 400 (c) 4
15 Six hundred quintillion, five hundred seventy quadrillion, one hundred
seven trillion, four billion, two hundred thirty million, five hundred thousand, five
17 (a) 1011_{three}; (b) 1212_{three}; (c) $10,000_{three}$　**19** (a) 21_{three}; (b) 22_{three};
(c) 100_{three}; (d) 210_{three}; (e) 20021_{three}

CHAPTER 3　**1*** 5 vertices; see Figure A3.　**3** None of them　**5** (a) $\{1, 3, 5, \ldots, 19\}$;
(b) $\{0, 5, 10, 15\}$; (c) $\{5, 10, 15\}$　**7** (a) B; (b) $\overline{BC}$; see Figure A4.　**9*** See
Figure A5.　**11*** See Figure A6.　**13** (a) 6; (b) 6　**15** (a) Each set of
three of the points (and there are four such sets) lies in exactly one plane;
(b) Each set of two of the points (and there are six such sets) lies in exactly one
line.　**17*** Note that there are many possible answers. We give only one of
the possibilities: (a) A,B,C　G,H,D; (b) A,B,H　C,B,H; (c) A,B,H
C,B,H　G,H,D; (d) B,H,D　G,H,D　F,H,D; (e) A,B　G,H,D; (f) A,B　G,H;
(g) A,B　C,D; (h) A,B　H,B　C,B　**19** (a) No, no; (b) four, not counting
the planes

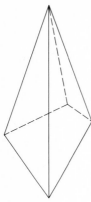

Figure A3

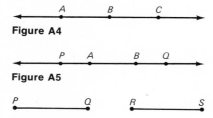

Figure A4

Figure A5

Figure A6

CHAPTER 4　**1** a, c, e, f, i, j, ℓ　**3** D, O are simple closed curves. C, I, L, M, N, S, U, V, W,
Z are simple but not closed. A, B, D, O, P, Q, R separate the plane.
5 (a) $\{W, X\}$; (b) $\triangle ABC$,　$\triangle AWX$,　$\triangle BWY$,　$\triangle XCY$; (c) none;
(d) $ABC: Y$, $AWX: B$, C, Y, $BWX: A$, $XCY: A$, W, B; (e) B, A
7* See Figure A7.　**9** (b)　**11** Yes　**13** $\overline{AB}$ and $\overline{GH}$　**15** Yes, yes
17 (a) $\overline{AC}, \overline{AD}, \overline{AB}, \overline{CD}, \overline{CB}, \overline{DB}$, six; (b) $\overline{DB}, \overline{AB}$　**19** The first is to the
right of the second.

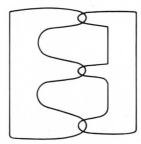

Figure A7

CHAPTER 5　**1** (a) $A \cup B = \{a, b, c, d, e, f\}$, $A \cap B = \{c, e\}$; (b) $A \cup B = \{a, b, c, d, e, f\}$,
$A \cap B = \{c, e\}$; (c) $A \cup B = \{a, b, c\}$; $A \cap B = \{a, b, c\}$; (d) $A \cup B =$
$\{a, b, c, d, e\}$; $A \cap B = \{\ \ \}$　**3** A　**5** (a) no; (b) no; (c) no; (d) no;
(e) yes

7 $2 + 3 = 5$ $5 + 4 = 9$ $3 + 4 = 7$ $2 + 7 = 9$
$3 + 5 = 8$ $8 + 2 = 10$ $5 + 2 = 7$ $3 + 7 = 10$
$8 + 1 = 9$ $9 + 7 = 16$ $1 + 7 = 8$ $8 + 8 = 16$

Because $4 + 7 = 11$ and $3 + 11$ is not given by the addition table.
9 (a) Commutativity of addition; (b) identity element for addition; (c) associative property of addition; (d) commutative property of addition; (e) identity element for addition; (f) commutative property of addition   **11** (a) 1; (b) 1, 4, and 5; (c) 1, 3; (d) 1   **13** $A \sim B = \{0, \square, \triangle, \boxtimes, \bigcirc, \frown \}$ **15** A is the empty set.   **17** (a) $\{2, 6\}$;   (b) $\{1, 3, 4, 5\}$;   (c) $\{3, 4, 5, 6, 7, 8\}$;   (d) B;   (e) $\{7\}$; (f) $\{1, 2, 8\}$; (g) B **19** See Figure A8.

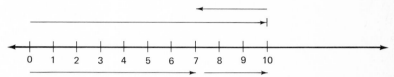

Figure A8

CHAPTER 6

1 See Figures A9a, b, and c.

3 (1) 5 tens + 8 ones
            1 ten + 7 ones
            6 tens + 15 ones
            7 tens + 5 ones = 75

(2) 50 + 8
     10 + 7
     60 + 15
     70 + 5 = 75

(3) 58
     17
     15
     60
     75

(4) 58
     17
     75

5 (1)   5 thousands + 6 hundreds + 7 tens + 8 ones
             9 thousands + 7 hundreds + 5 tens + 3 ones
            14 thousands + 13 hundreds + 12 tens + 11 ones
            14 thousands + 13 hundreds + 13 tens + 1 one
            14 thousands + 14 hundreds + 3 tens + 1 one
            15 thousands + 4 hundreds + 3 tens + 1 one
            1 ten thousand + 5 thousands + 4 hundreds + 3 tens + 1 one = 15,431

(2)   5,000 +   600 + 70 + 8
       9,000 +   700 + 50 + 3
      14,000 + 1,300 + 120 + 11
      14,000 + 1,300 + 130 + 1
      14,000 + 1,400 + 30 + 1
      15,000 +   400 + 30 + 1
      10,000 + 5,000 + 400 + 30 + 1 = 15,431

(3) 5,678
     9,753
        11
        12
       1 3
        14
    15,431

(4) 5,678
     9,753
    15,431

7 (1) 4 hundreds + 4 tens + 6 ones   3 hundreds + 13 tens + 16 ones
            1 hundred + 6 tens + 8 ones   1 hundred + 6 tens + 8 ones
                                            +2 hundreds + 7 tens + 8 ones = 278

$$(2) \quad \begin{array}{r} 400 + 40 + 6 \\ 100 + 60 + 8 \end{array} \qquad \begin{array}{r} 300 + 130 + 16 \\ 100 + 60 + 8 \\ \hline 200 + 70 + 8 = 278 \end{array} \qquad (3) \quad \begin{array}{r} 446 \\ -168 \\ \hline 278 \end{array}$$

9 (1)

1 ten thousand $+$	0 tens $+$ 0 ones
	1 ten $+$ 1 one

9 thousands $+$ 9 hundreds $+$ 9 tens $+$ 10 ones

1 ten $+$ 1 one

9 thousands $+$ 9 hundreds $+$ 8 tens $+$ 9 ones

$= 9{,}989$

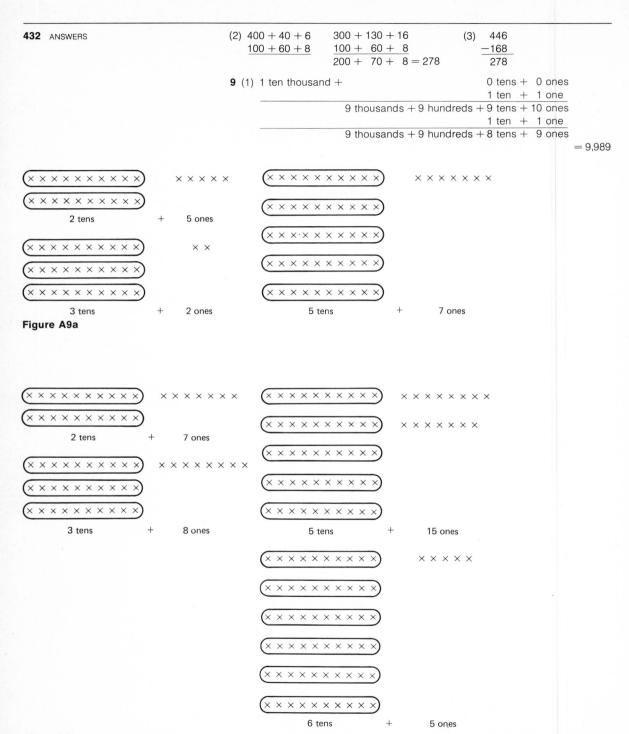

Figure A9a

Figure A9b

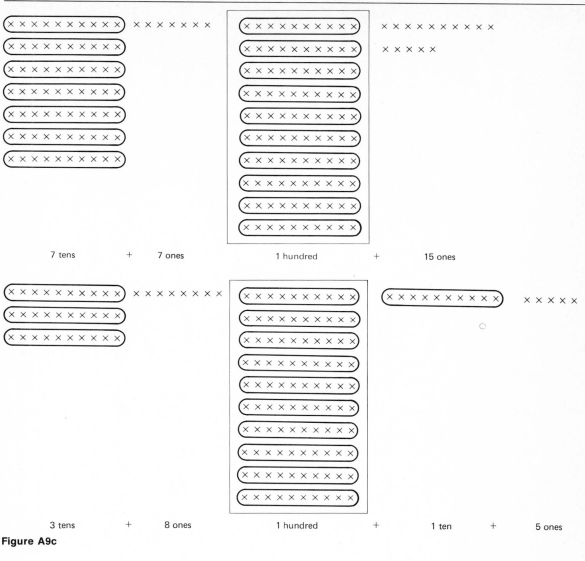

7 tens   +   7 ones   1 hundred   +   15 ones

3 tens   +   8 ones   1 hundred   +   1 ten   +   5 ones

Figure A9c

(2) 10,000

$$\begin{array}{r} 10 + 1 \\ \hline 9{,}000 + 900 + 90 + 10 \\ 10 + 1 \\ \hline 9{,}000 + 900 + 80 + 9 = 9{,}989 \end{array}$$

(3)
$$\begin{array}{r} 10{,}000 \\ 11 \\ \hline 9{,}989 \end{array}$$

11
$$\begin{array}{r} 300 + 40 + 2 \\ 100 + 70 + 4 \\ 400 + 10 + 8 \\ \hline 800 + 120 + 14 \\ 900 + 30 + 4 = 934 \end{array}$$

13 (a) 57   (b) 59   (c)   543
      -33 -15 -357
      27   49   243
      -3 -5 -57
      24   44   193
                         -7
                       186

15 298. He forgets that he "borrowed."   **17** See Table A6.17.   **18** See Table A6.19.

Table A6.17

+	0	1	2	10
0	0	1	2	10
1	1	2	10	11
2	2	10	11	12
10	10	11	12	20

Table A6.19

−	0	1	2
0	0	×	×
1	1	0	×
2	2	1	0

CHAPTER 7   **1** (a) E; (b) D; (c) B; (d) C **3** $20 \times (28 + 22) = 20 \times 50 = 1,000$ **5** At least one of them has to be 0.   **7** $52 = 7 \times 5 + 17$. $7 \times 5 = 35$ cards are to be dealt. 17 are left over. Each player may be dealt two more.   **9** (a) 40; (b) 31; (c) 98; (d) 68   **11** (a) $20 \div 5 = n$; $n = 4$; (b) $28 \div 4 = p$; $p = 7$; (c) $6 \div 1 = n$; $n = 6$ **13** Look along the b row, from left to right, until you find a. The number at the top of that column is $a \div b$.   **15** 30 and 45 are possible, since $30 \div 3$ and $30 \div 5$ are whole numbers, as are $45 \div 3$ and $45 \div 5$. 24, 32, and 16 are not possible.   **17** (a) 856; (b) 0; (c) 43; (d) 0; (e) 0; (f) 2   **19** $6 = 3 \times 2$ so $6 \times 9 = 54$
                                     $(3 \times 2) \times 9 = 54$
                                       $3 \times (2 \times 9) = 54$

CHAPTER 8   **1** It stands for 86,600 ($=200 \times 433$).   **3** Because $a + 0 = a$ for any number a.

5   6,439   7
       7   6,439
     63   63
     21   21
   2 8   2 8
   42   42
  45,073   45,073

7 (a)   207   (b)   642
          24   155
       828   3,210
      414   3,210
    4,968   642
                 99,510

       207   642
          24   155
      414   642
     828   3,210
    4,968   3,210
                 99,510

9 9. The billions place.   **11** It is similar to division by numbers greater than 10 but much easier.   **13** (a) $512 = 64 \times 8 + 0$; (b) $779 = 43 \times 18 + 5$; (c) $23 = 1 \times 14 + 9$; (d) $50 = 0 \times 100 + 50$; (e) $6,535 = 139 \times 47 + 2$ **15** See Figure A10.   **17** (a) $23 = 4 \times 5 + 3$; (b) $3 = 0 \times 8 + 3$; (c) $37 = 5 \times 7 + 2$

37

0 × 7   1 × 7   2 × 7   3 × 7   4 × 7   5 × 7   6 × 7

Figure A10

```
19  21        202
     2         12
     2         11
    11         11
   112          2
                2
             10,201
```

CHAPTER 9   **1** False: (a) The sum of 8 and 7 is 19.
True: (c) $4 + 5 = 10 - 1$ (d) $3 < 2 + 6$
Open: (b) $n + 8 = 16$ (e) $5 > 2 + \square$
3 (a) Addition; (b) subtraction; (c) subtraction; (d) subtraction; (e) subtraction; (f) subtraction;   **5** (a) $n + 4 = 6$; (b) $n = 4 + 3$ **7** (a) $49 - 35 = n$ or $35 + n = 49$; (b) $n = 14$;   **9** (a) $360 + 419 + 284 = X$; (b) $X = 1{,}063$
11 (a) $15 \div 3 = g$ (b) $g = 5$ **13** (a) $5 \times 82 = p$ (b) $p = 410$
15 $29 - 12 = a$ or $a + 12 = 29$; $a = 17$ **17** $5(12 + 12) = m$; $m = 120$
19 $(300 - 50) + (200 - 50) = \sigma$; $\sigma = 450$

CHAPTER 10   **1** (a) 1, 1, 1; (b) 45, 4; (c) The distance around is greater than the sum of the lengths of the sides in feet; (d) Because we measured to the nearest foot, which allowed three errors of up to half a foot.   **3** Half the length of the piece of chalk.   **5** One   **7** No   **9** (a) $\overline{BE}$; (b) B; (c) $\overrightarrow{CA}$; (d) $\overline{CD}$; (e) $\overrightarrow{AB}$
11 $\angle PQR$, $\angle PQS$, $\angle PQT$, $\angle RQS$, $\angle RQT$, $\angle SQT$.   **13** (a) A segment cannot be congruent to a ray; (b) A line cannot be a ray; (c) An angle cannot be compared with a segment.   **15** $\angle ABC > \angle XYZ$ **17** b, d, and g
19 The point A is on the ray $\overrightarrow{AB}$ but is not in the interior of $\angle MAL$.

CHAPTER 11   **1** A triangle all of whose angles are congruent.   **3** (a) They have the same area; (b) they have the same area.   **5** (a) 50; (b) 88; (c) 50, 88; (d) 38; (e) $9\frac{1}{2}$ **7** Distance around is 14 meters. Area is 12 square meters.   **9** The area of a two-foot square is four square feet.   **11** 2,079 square inches.
13 The areas are 24, 96, 6, and 24. If a and b are each doubled, the new rectangle is four times as large. If both are halved, the new one is only one-fourth as large. If one is doubled and the other halved, the new area is the same as the old.   **15** 930   **17** 30   **19** 20 square inches

CHAPTER 12   **1** (4)   **3** (a) $\frac{4}{4}$; (b) $\frac{2}{8}$; (c) $\frac{4}{8}$; (d) $\frac{4}{6}$; (e) $\frac{6}{10}$; (f) $\frac{3}{2}$; (g) none; (h) none; (i) none; (j) $\frac{6}{6}$ **5*** See Figure A11.   **7** See Figure A12.   **9** See Figure A13.
11 (a) 30; (b) 21; (c) 24; (d) No whole number can make $\frac{2}{15} = \frac{\square}{20}$ **13** $\frac{11}{4}$, $\frac{7}{12}$, $\frac{2}{3}$ **15** (a) $<$; (b) $=$; (c) $<$; (d) $>$; (e) $=$; (f) $>$ **17** (a) $1\frac{3}{4}$; (b) $1\frac{7}{8}$; (c) $2\frac{3}{9}$; (d) $2\frac{4}{15}$; (e) $4\frac{8}{12}$ **19** More than can be counted.

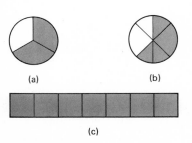

(a)   (b)

(c)

Figure A11

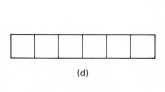

(d)

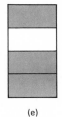

(e)

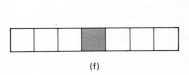

(f)

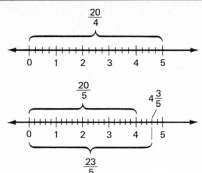

Figure A13

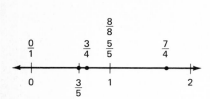

Figure A12

CHAPTER 13 **1** There are more red marbles than white ones in the bowl. **3** No. Your probability of winning is only $\frac{1}{3}$. **5** $\frac{2}{3}$ **7** (a) $\frac{3}{12}$; (b) $\frac{4}{12}$; (c) $\frac{8}{12}$ **9** Two black, one green, two red. **11** (a) No. The probability of exactly 2 heads and 2 tails is $\frac{3}{8}$. The probability of all the other outcomes together is $\frac{5}{8}$; (b) $\frac{1}{32}$ **13** (a) $\frac{5}{36}$; (b) $\frac{15}{36}$; (c) $\frac{4}{36}$; (d) $\frac{6}{36}$; (e) $\frac{27}{36}$ **15** (a) $\frac{5}{10}$; (b) $\frac{3}{10}$; (c) $\frac{2}{10}$ **17** (1) (a) $\frac{5}{16}$; (b) $\frac{4}{16}$; (c) $\frac{4}{16}$ **19** (1) (a) $\frac{1}{6}$; (b) $\frac{4}{6}$

CHAPTER 14 **1** (a)

A	B	C
9	11	13
11	11	13
11	13	13

(b)

A	B	C
9	11	13
9	13	13
13	13	13

3

N	1	2	3	4	5	6	7	8	9	10
T	0	1	3	6	10	15	21	28	36	45

5 $6! = 720$; $7! = 5,040$; see Figure A14 **7** Paul's solution numbers the output starting with 0 rather than 1; Pete's solution would not number the outputs consecutively; Lar's program would result in only one card being input; Bob's program would produce only one output. **9** (a) No output; (b) 5, 25 **11** See Table A14.11. **13** See Table A14.13. **15** See Table A14.15. **17** 6 **19** See Table A14.19.

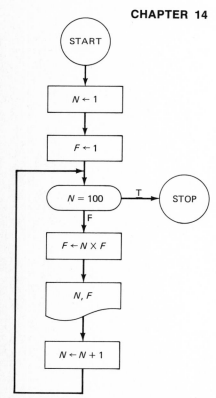

Figure A14

Table A14.11

N	D	R	Q	
39	13	~~39~~	~~0~~	
		~~26~~	1	
		~~13~~	2	
		0	3	$39 = 3 \times 13 + 0$

Table A14.13

N	D	R	Q	
125	125	~~125~~	~~0~~	
		0	1	$125 = 1 \times 125 + 0$

Table A14.15

N	D	R	Q	
14	1	~~14~~	0	
		~~13~~	1	
		~~12~~	~~2~~	
		. . .	. . .	
		~~2~~	~~12~~	
		~~1~~	~~13~~	
		0	14	$14 = 14 \times 1 + 0$

Table A14.19

N	D	R	Q	P	
856	7	856	0	1	
				10	
				100	
		156			
			100		
				10	
		86			
			110		
		16			
			120		
				1	
		9			
			121		
		2			
			122		$856 = 122 \times 7 + 2$

CHAPTER 15   **1** $\frac{2}{3} = \frac{6}{9}$, $\frac{12}{27} = \frac{4}{9}$, $\frac{14}{63} = \frac{2}{9}$, $\frac{14}{3} = \frac{42}{9}$, $\frac{4}{1} = \frac{36}{9}$ **3** (a) $n = 2$; (b) $n = 3$; (c) no; (d) no;   **5*** See Figure A15.   **7** See Figure A16.   **9** (a) $45\frac{3}{14}$; (b) $20\frac{50}{51}$ **11** $\frac{7}{36}$ **13** $7\frac{4}{5}$ **15** $\frac{21}{48}$ miles (or $\frac{7}{16}$ miles)   **17** $\frac{15}{16} = s \times \frac{5}{32}$ **19** $24 - (7 + 5) = f \times 24$

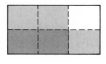

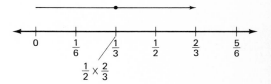

$$\frac{1}{2} \times \frac{2}{3}$$

(a)

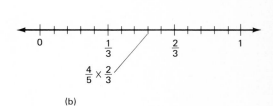

$$\frac{4}{5} \times \frac{2}{3}$$

(b)

Figure A15

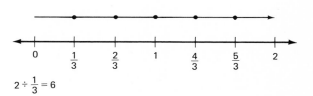

$$2 \div \frac{1}{3} = 6$$

(a)

Figure A16

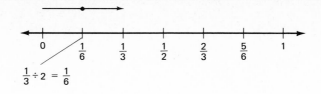

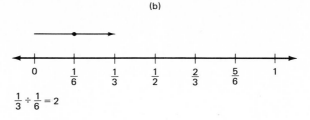

$$\frac{1}{3} \div 2 = \frac{1}{6}$$

(b)

$$\frac{1}{3} \div \frac{1}{6} = 2$$

(c)

Figure A16 *Continued*

CHAPTER 16

1 (a) 1, 2, 3, 4, 6, 12; (b) 1, 5, 7, 35; (c) 1, 2, 3, 4, 6, 9, 12, 18, 36; (d) 1, 2, 37, 74; (e) 1, 2, 73, 146; (f) 1, 2, 3, 4, 6, 7, 9, 12, 14, 18, 21, 28, 36, 42, 63, 84, 126, 252 **3** 233 and 269 are prime. $323 = 17 \times 19$ **5** (a) $100 = 2 \times 2 \times 5 \times 5$; (b) $162 = 2 \times 3 \times 3 \times 3 \times 3$; (c) $198 = 2 \times 3 \times 3 \times 11$; (d) $276 = 2 \times 2 \times 3 \times 23$ **7** (a) 4; (b) 12; (c) 12; (d) 8 **9** (a) 8; (b)4 **11** (a) $\frac{5}{7}$; (b) $\frac{11}{20}$; (c) $\frac{5}{8}$; (d) $\frac{4}{45}$ **13** (a) Insert the following between boxes 5 and 6:

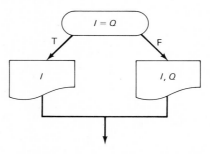

(c) 6 is not printed out because its factors were divided out first. **15** (a) 3; (b) 2, 3; (c) 3; (d) 2, 3; (e) 2, 3; (f) —; (g) 2; (h) 2, 3; (i) 2; (j) —; (k) 2, 3; (l) 2, 3 **17** Suppose N is any whole number. Multiply by 6 and write down the next five numbers after that:

$$6N \quad 6N + 1 \quad 6N + 2 \quad 6N + 3 \quad 6N + 4 \quad 6N + 5$$

The first of these is divisible by 2 (and 3). $6N + 2$ is equal to $2(3N + 1)$ and hence is divisible by 2. $6N + 3 = 3(2N + 1)$ and is divisible by 3. $6N + 4 = 2(3N + 2)$ and is divisible by 2. So if a number is prime and is one of these six numbers, it must be $6N + 1$ or $6N + 5$, and the latter is one less than $6(N + 1)$.

19 407, 671, 4,257, 5,291 are divisible by 11.

CHAPTER 17

1 (a) 9,876.543; (b) 702.907 **3*** (a) 400.061, .461; (b) 2,300.004, .2340; (c) 100.000016, .000116 **5** (a) (1) >, (2) >, (3) =, (4) >; (b) .02 < .25 < 1 < 1.02 < 2.002 < 2.2 < 2.25

7 (a) $8\frac{9}{10} + 3 + 5\frac{375}{1000} = 8 + 3 + 5 + \frac{9}{10} + \frac{375}{1000}$

$= 16 + \frac{900}{1000} + \frac{375}{1000}$

$= 16 + \frac{1275}{1000}$

$= 17\frac{275}{1000}$

$$\begin{array}{r} 8.900 \\ 3.000 \\ \underline{5.375} \\ 17.275 \end{array}$$

(b) $40 - 12\frac{57}{100} = 40 - (12 + \frac{57}{100}) = (39 - 12) + (1 - \frac{57}{100})$

$= 27 + \frac{43}{100} = 27\frac{43}{100}$

$$\begin{array}{r} 40.00 \\ \underline{-12.57} \\ 27.43 \end{array}$$

9 (a) $\frac{7360}{25}$; (b) $\frac{9.7}{326}$; (c) $\frac{6850}{82}$; (d) $\frac{350000}{7}$; (e) $\frac{64.9}{36}$ **11** (a) 23.132; (b) .0024; (c) .1925; (d)19,250; (e) 2.443; (f) 99.995; (g) .000016   **13** (a) \$3.33; (b) 6

15 (a) $\dfrac{2}{3 \times 3} + \dfrac{1}{3 \times 3 \times 3}$

(b) $2 \times 3 + 2 + \dfrac{1}{3 \times 3}$

(c) $1 \times 3 \times 3 + 1 + \dfrac{1}{3 \times 3} + \dfrac{1}{3 \times 3 \times 3}$

(d) $2 + \dfrac{1}{3} + \dfrac{1}{3 \times 3} + \dfrac{2}{3 \times 3 \times 3 \times 3}$

17 (a) (1) 0, 6, 6, 6; (2) 0, 2, 0, 0,; (3) 0, 0, 2, 0; (4) 2, 5, 0, 0; (5) 25, 0, 0, 0; (6) 0, 4, 2, 8 (b) (1) .666; (2) .200; (3) .020; (4) 2.500; (5) 25.000; (6) .428 (c) The number to the left of the decimal place of the answer in part b. (Note that the program prints O even though we usually do not bother to write this in practice.) (d) Add this box:

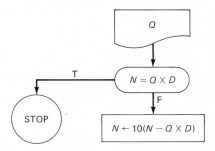

19 .45 pounds

CHAPTER 18   **1*** 10:20 (or 2:4)   **3** 8 for 36 cents   **5** $17\frac{31}{32}$ feet   **7** $12\frac{1}{2}$ blue grass, $7\frac{1}{2}$ clover   **9** Who knows? Pounds to shoes is not a sensible ratio.   **11** 25 **13** (a) 288.36; (b) 42; (c) \$88; (d) 4.2; (e) 400; (f) 1; (g) 1   **15** See Figure A17.   **17** No   **19** (a) $400 + 120 = \$520$; (b) 1,040 gallons; (c) 31,200 miles; (d) About 125 miles per day; (e) Few people would walk over 60 miles to work and back each day.

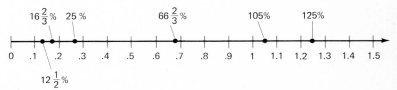

Figure A17

CHAPTER 19   **1** $\frac{1}{3} \times \frac{1}{6} = \frac{1}{18}$ **3** (a) $\frac{16}{81}$; (b) $\frac{1}{6}$ **5** $\frac{19}{36}$ **7** (a) $\frac{60}{100}$; (b) $\frac{15}{100}$; (c) $\frac{10}{100}$; (d) $A \cup B$; (e) $\frac{65}{100}$

9 $P(\text{RG}) = \frac{3}{16}$
$P(\text{GR}) = \frac{3}{16}$
$P(\text{GG}) = \frac{9}{16}$

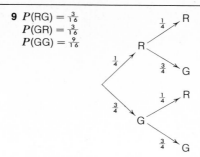

11 (a) $P(A) = \frac{2}{5}$, $P(B) = \frac{2}{5}$, $P(A \cap B) = \frac{1}{10}$, $P(A \cap B) \neq P(A) \times P(B)$, A and B are not independent; (b) $P(A) = \frac{2}{5}$, $P(C) = \frac{1}{5}$, $P(A \cap C) = \frac{1}{10}$, $P(A \cap C) \neq P(A) \times P(C)$; A and C are not independent. **13** (No answer) **15** 325 to 675 (not quite 1 to 2); 675 to 325 (a little more than 2 to 1) **17** (a) $\frac{380}{500}$; (b) 3,040 **19** (b) $\frac{15}{32}$

CHAPTER 20 **1*** (a) $\overline{OA}$, $\overline{OD}$, $\overline{OB}$, $\overline{OC}$; (b) $\overline{BC}$; (c) interior; (d) no; (e) yes; (f) $\widehat{AC}$, $\widehat{AD}$, $\widehat{ADB}$ **3** Interior **5** Four **7** (a) $\widehat{CKL}$ and $\widehat{HDL}$; (b) no **9*** See Figure A18. (b) A and C, B and D **11** (a) More than can be counted; (b) more than can be counted; (c) no. **13** The circle of latitude passes through P, Q, and R. The angle which measures the latitude of P is $\angle DOP$. The angle which measures the longitude of P is $\angle OED$. **15** (0°, 52° north) **17*** (1) All but Figure P20.17e; (2) In P20.17b, any face can be considered to be the base. For P20.17a and d, the apex is at the top. For P20.17c it is to the right. The base for each of these is the face which does not touch the apex. **19** (a) Pyramid; (b) prism; (c) pyramid; (d) prism; (e) sphere; (f) prism; (g) prism; (h) cylinder; (i) cone

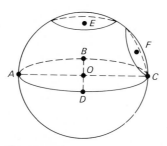

Figure A18

CHAPTER 21 **1** (a) $\triangle ABC \cong \triangle RST$, by SAS; (b) $\triangle QPL \cong \triangle EDF$ by ASA; (c) They are not necessarily congruent; (d) $\triangle MCP \cong \triangle MBP$ by SSS;

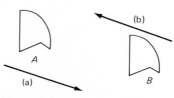

Figure A19

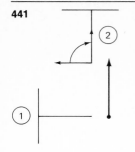

1. First the slide, then the turn

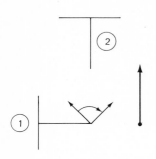

2. First the turn, then the slide

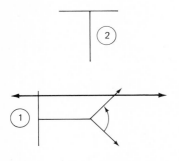

3. First the turn, then flip around the line

Figure A23a

(e) $\triangle JKM \cong \triangle MHJ$ by SAS.   **3*** (a) $\angle DAB \cong \angle DAC$, $\triangle DAB \cong \triangle DAC$, SAS; (b) $\overline{OK} \cong \overline{OP}$, $\triangle LOK \cong \triangle LOP$, SAS; (c) $\angle HJM \cong \angle KMJ$, $\triangle HJM \cong \triangle KMJ$, ASA   **5** Not for triangles. It is true for squares or circular regions.   **7** Yes.   Since $\overline{RA} \cong \overline{CS}$, $\overline{RC} \cong \overline{SA}$. So SSS applies.   **9** $\angle ACB \cong \angle DCE$ since they are vertical angles. $\triangle ABC \cong \triangle DEC$ by SAS.   **11** See Figure A19.   **13*** See Figure A20.   **15*** See Figure A21.   **17** See Figure A22.   **19*** (a) See Figure A23a; (1) First the slide, then the turn; (2) First the turn, then the slide; (3) First the turn, then flip around the line; (b) See Figure A23b; (1) First the turn, then the slide; (2) First the turn, then the slide; (3) First the slide, then the turn.

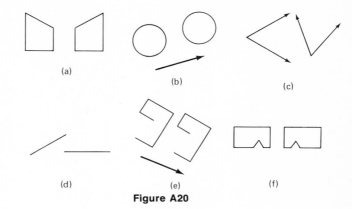

Figure A20

Figure A21

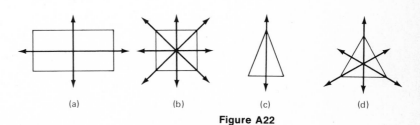

Figure A22

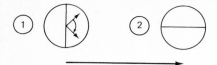

1. First the turn, then the slide.

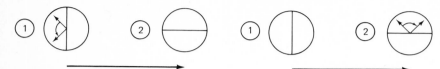

2. First the turn, then the slide.

3. First the slide, then the turn.

Figure A23b

CHAPTER 22   **1** Yes   **3** $\triangle ABC \sim \triangle XYZ$ **5** Yes, by SAS.   **7** (a) $\angle BEA \cong \angle DEC$ (vertical angles), SAS; (b) $\angle BAC \cong \angle DAE$, SAS; (c) SSS

9 (a) $\dfrac{PQ}{PR} = \dfrac{DE}{DF}$, $\dfrac{4}{2} = \dfrac{DE}{3}$, $DE = 6$ $\dfrac{QR}{PR} = \dfrac{EF}{DF}$, $\dfrac{3}{2} = \dfrac{EF}{3}$, $EF = \dfrac{9}{2}$

(b) $\dfrac{AC}{AB} = \dfrac{DC}{DE}$, $\dfrac{AC}{4} = \dfrac{2}{3}$, $AC = \dfrac{8}{3}$ $AD = AC - DC = \dfrac{2}{3}$

$\dfrac{BC}{BA} = \dfrac{EC}{ED}$, $\dfrac{BC}{4} = \dfrac{2\frac{1}{2}}{3}$, $BC = \dfrac{10}{3}$ $BE = BC - EC = \dfrac{5}{6}$

11 (a) $RT = \frac{15}{2}$;   (b) $\triangle ABC \sim \triangle RST$;   (c) $RT = \frac{10}{3}$, $\triangle ABC \sim \triangle RTS$
13 (a) 85 miles; (b) 442.5 miles   **15** (a) 18; (b) 114; (c) 35; (d) 150; (e) 116; (f) 95   **17** Yes, 27   **19** (a) They must all be equal; (b) $k \times AC$; (c) $k \times AC$; (d) $k \times BC$; (e) $k \times BC$; (f) SSS

CHAPTER 23   **1** See Table A23.1.   **3** (a) 65; (b) 16   **5** 22 centimeters wide, 28 centimeters long.   **7** (a) .035 ounces in 1 gram; (b) 907 kilograms in 1 ton.
9 (a) 49.6 cubic inches; (b) 45.26 cubic centimeters; (c) 120 cubic inches
11 $\frac{22}{7}$: 3.142857; $\frac{223}{71}$:3.140845; $\frac{355}{113}$:3.141593; $\frac{355}{113}$ differs from π by one unit in the sixth decimal place.   **13** 7,216 miles   **15** (a) 9.42 square centimeters; (b) It is one-third as large   **17** (a) 113.04 cubic centimeters; (b) 904.32 cubic centimeters; (c) 33.49 cubic yards; (d) 267.94 cubic yards; (e) 8; (f) 8
19 (a) Multiply by 60; (b) multiply by 60; (c) multiply by $\frac{1}{24}$; (d) multiply by $\frac{1}{60}$; (e) multiply by $\frac{1}{3600}$

Table A23.1

Measurement	Precise to the nearest	Actual length lies between		
. . .	. . .	. . .	and	. . .
. . .	$\frac{1}{8}$ yards	$1\frac{1}{8}$ yards	and	$1\frac{3}{8}$ yards
. . .	$\frac{1}{16}$ inches	$3\frac{1}{16}$ inches	and	$3\frac{3}{16}$ inches
. . .	.05 millimeters	52.25 millimeters	and	52.35 millimeters
. . .	.05 kilometers	141.95 kilometers	and	142.05 kilometers
. . .	500,000 miles	92,500,000 miles	and	93,500,000 miles
. . .	.00005 cm	.00035 centimeters	and	.00045 centimeters

1 (a) 4; (b) 3; (c) $1\frac{1}{4}$; (d) 4; (e) $^-2\frac{7}{15}$; (f) $^-12$; (g) $^-3\frac{5}{8}$; (h) 1; (i) $^-4$; (j) $^-2$
3 See Table A24.3.　　　**5** (a) $^-24$; (b) $^-5$; (c) 16; (d) $^-9$; (e) 4; (f) $^-4$

Table A24.3

+	$^-4$	$^-3$	$^-2$	$^-1$	0	1	2	3	4
$^-4$	$^-8$	$^-7$	$^-6$	$^-5$	$^-4$	$^-3$	$^-2$	$^-1$	0
$^-3$	$^-7$	$^-6$	$^-5$	$^-4$	$^-3$	$^-2$	$^-1$	0	1
$^-2$	$^-6$	$^-5$	$^-4$	$^-3$	$^-2$	$^-1$	0	1	2
$^-1$	$^-5$	$^-4$	$^-3$	$^-2$	$^-1$	0	1	2	3
0	$^-4$	$^-3$	$^-2$	$^-1$	0	1	2	3	4
1	$^-3$	$^-2$	$^-1$	0	1	2	3	4	5
2	$^-2$	$^-1$	0	1	2	3	4	5	6
3	$^-1$	0	1	2	3	4	5	6	7
4	0	1	2	3	4	5	6	7	8

7 (a) $-$; (b) $+$; (c) $-$; (d) $+$; (e) $+$; (f) $-$; (g) $-$; (h) $-$　　**9** (a) $^-4$; (b) $^-3$;
(c) 3; (d) $^-\frac{1}{2}$; (e) $^-2$; (f) $^-\frac{2}{49}$　　**11** See Table A24.9.　　**13** (a) $<$; (b) $>$;
(c) $<$; (d) $>$; (e) $>$; (f) $>$; (g) $<$; (h) $=$; (i) $<$; (j) $>$　　**15** (a) $^-50$; (b) $^-5$;
(c) $^-41$; (d) $^-7$; (e) $^-9$　　**17** 204°F　　**19** (a) The output is: (1) 3; (2) 5; (3) 1;
(4) $^-1$; (5) $^-1$; (6) 1; (7) $^-3$; (8) $^-5$;　　(b) The output is the sum of the two
inputs.

Table A24.9

×	$^-4$	$^-3$	$^-2$	$^-1$	0	1	2	3	4
$^-4$	16	12	8	4	0	$^-4$	$^-8$	$^-12$	$^-16$
$^-3$	12	9	6	3	0	$^-3$	$^-6$	$^-9$	$^-12$
$^-2$	8	6	4	2	0	2	$^-4$	$^-6$	$^-8$
$^-1$	4	3	2	1	0	$^-1$	$^-2$	$^-3$	$^-4$
0	0	0	0	0	0	0	0	0	0
1	$^-4$	$^-3$	$^-2$	$^-1$	0	1	2	3	4
2	$^-8$	$^-6$	$^-4$	$^-2$	0	2	4	6	8
3	$^-12$	$^-9$	$^-6$	$^-3$	0	3	6	9	12
4	$^-16$	$^-12$	$^-8$	$^-4$	0	4	8	12	16

CHAPTER 25　　**1** (a) 15; (b) 15; (c) 3; (d) $^-12$　　**3** (a) 6; (b) $^-11$; (c) 5; (d) 6　　**5** (a) $^-1$;
(b) $^-\frac{23}{5}$; (c) 12; (d) 1　　**7** (a) $^-\frac{16}{3}$; (b) 6; (c) $\frac{1}{2}$; (d) any number
9 (a) $a - b = a + (^-b)$; (b) addition property of equality; (c) associative
property of addition; (d) $a + {^-a} = 0$; (e) addition property of 0; (f) multiplication property of equality; (g) associative property of multiplication; (h) multiplication property of reciprocals (i) multiplication property of 1
11 (a) $x < 49$;　(b) $x < -15$;　(c) $x \geq 8$;　(d) $x > -\frac{1}{2}$;　　**13** (a) $x \leq 4$;
(b) $x > -\frac{5}{2}$; (c) $x \geq 4$; (d) $x < 4$　　**15** (a) $x > \frac{1}{2}$; (b) $x < 4$; (c) $x > -3$;
(d) $x \geq \frac{7}{5}$　　**17** (a) $x \leq 0$; (b) $x < -1$; (c) $x < 4$; (d) $x < -\frac{3}{2}$; (e) $x > -3$;
(f) $x < \frac{90}{11}$　　**19** (a) Yes. Any number makes the statement false; (b) no.

CHAPTER 26

1 See Figure A24 and Table A26.1.   **3** See Table A26.3.   **5** (a) See Table A26.5; (b) See Figure A25.   **7** 37.1

Table A26.1

x	Frequency
50	1
40	2
30	3
20	1
10	1

Table A26.3

x	Frequency	Cumulative frequency	Rating frequency
45	1	8	$\frac{1}{8}$
40	2	7	$\frac{2}{8}$
35	3	5	$\frac{3}{8}$
30	1	2	$\frac{1}{8}$
25	1	1	$\frac{1}{8}$

Table A26.5

x	Frequency	Cumulative frequency	Relative frequency
80	1	15	$\frac{1}{15}$
79	0	14	$\frac{0}{15}$
78	5	14	$\frac{5}{15}$
77	2	9	$\frac{2}{15}$
76	1	7	$\frac{1}{15}$
75	0	6	$\frac{0}{15}$
74	2	6	$\frac{2}{15}$
73	2	4	$\frac{2}{15}$
72	2	2	$\frac{2}{15}$

9

	Mode	Mean	Median
(a)	-2	2	0
(b)	10	30	30
(c)	13	14.78	15
(d)	7	7	7

11 (a) The output is $5\frac{1}{2}$; (b) The output is the mean of the inputs; (c) The output would be the means of the first number, the first two numbers, the first three numbers, etc.   **13** 208   **15** 422.4 inches   **17** (a) .67; (b) .95; (c) .99

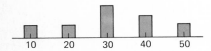

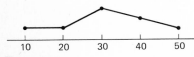

Figure A24

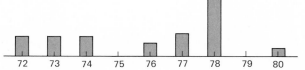

Figure A25

CHAPTER 27

Table A27.5

P	Q	Value of ratio
N	J	$\dfrac{15 - {}^-3}{-4 - 2} = \dfrac{18}{-6} = -3$
M	A	$\dfrac{12 - 0}{-3 - 1} = \dfrac{12}{-4} = -3$
J	L	$\dfrac{-3 - 6}{2 - -1} = \dfrac{-9}{3} = -3$

1 For answers to Problem 1a and b see Figure A26; (c) (1), (2), (3), (5), (7)
3 (a) A II, B IV, C III, D I, E none, F none, G none; (b) (1) I, (2) III, (3) IV, (4) II
5 See Table A27.5. **7** $k = 1$ **9** See Figure A27. **11** (a) $T = N + 39$;
(b) 44; (c) $76\frac{1}{2}°F$ **13** See Figure A28. **15** See Figure A29. **17** Yes
19 Radius $= \sqrt{74}$, $(x + 2)^2 + (y - 4)^2 = 74$

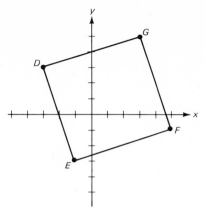

Figure A26

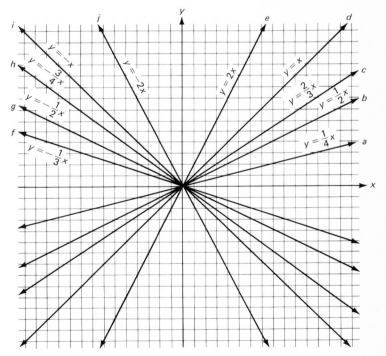

Figure A27

445

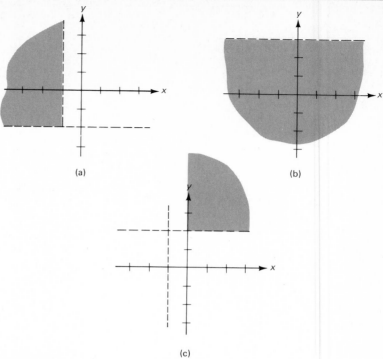

(a)

(b)

(c)

Figure A28

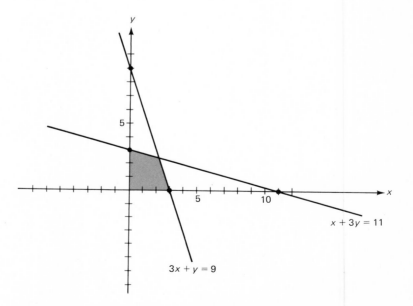

$x + 3y = 11$

$3x + y = 9$

INDEX

INDEX